Dynamics of Large
Mammal Populations

Dynamics of Large Mammal Populations

Charles W. Fowler
Tim D. Smith

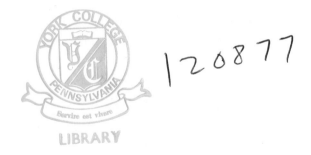

A WILEY-INTERSCIENCE PUBLICATION

JOHN WILEY & SONS

New York • Chichester • Brisbane • Toronto • Singapore

Library of Congress Cataloging in Publication Data:

Main entry under title:

Dynamics of large mammal populations.

   "A Wiley-Interscience publication."
   Based on papers presented at a conference at Utah
State University, Logan, on May 25-27, 1978.
   Includes index.
   1. Mammal populations—Congresses. I. Fowler,
Charles W. (Charles Winsor), 1941-   . II. Smith,
Tim D. (Tim Dennis) III. Title: Large mammal
populations.

QL700.D96   599.05'248   81-115
ISBN 0-471-05160-8   AACR2

Printed in the United States of America

10 9 8 7 6 5 4 3 2 1

This book is dedicated to Dr. Richard M. Laws in recognition of his outstanding work on the biology (and especially the population biology) of large mammals. Dr. Laws' productive and pioneering career has included work on a wide range of the largest of the large mammals, in both marine and terrestrial environments. His interests and efforts have varied and include all elements of the spectrum from basic biological work to comprehensive ecosystem studies.

To date, Dr. Laws's research covers a period of over 30 years. In 1948 he took charge of the Falkland Island Dependencies Survey Station in the South Orkney Islands where he studied the southern elephant seal. As a Whaling Inspector/Biologist on a 25,000 ton whaling factory ship he began work on whales in 1953. This work, which was conducted mostly within the Antarctic, involved the sperm, humpback, blue, fin and sei whales. In 1961 Dr. Laws became the first director of the Nuffield Unit of Tropical Animal Ecology in Uganda. This switch to more terrestrial work was made gradual by his work on hippopotamus. Elephants came under Dr. Laws' study in the early 1960's. His work on this species spread to involve populations in Uganda, Kenya and Tanzania. In 1967 Dr. Laws became director of the Tsavo Research Project funded by the Ford Foundation. Elephants occupied most of his time until 1969 when Dr. Laws reentered the study of marine mammals. With what is now the British Antarctic Survey, Dr. Laws has been involved in studies of antarctic seals, specifically the crabeater, leopard, and weddell seals. Now, as director of the British Antarctic Survey in Cambridge, Dr. Laws is very active in the management and conservation of the Southern Ocean Ecosystem.

The many publications produced under Dr. Laws' authorship (including a book concerning the elephants of east Africa) are a tribute to his successful endeavors in the study of large mammals. His paper in this volume (modified from his banquet address at the Logan conference) contains a sampling of his work. The breadth, depth, context, and original nature of Dr. Laws' work stands as an example which we wish to emphasize through the dedication of this book.

K. Radway Allen
192 Ewos Parade
Cronulla, Australia

Daniel B. Botkin
Environmental Studies Program
Department of Biological Sciences
University of California
Santa Barbara, California

F. L. Bunnell
Faculty of Forestry
University of British Columbia
Vancouver, B.C., Canada

George E. Burgoyne, Jr.
Department of Natural Resources
Surveys and Statistics
Lansing, Michigan

Graeme Caughley
C.S.I.R.O. Division of Wildlife
   Research
Lyneham, Australia

Douglas G. Chapman
Dean, College of Fisheries
University of Washington
Seattle, Washington

Harvey Croze
Global Environmental Monitoring
   System
United Nations Environment
   Programme
Nairobi, Kenya

Douglas P. DeMaster
National Marine Fisheries Service
Southwest Fisheries Center
LaJolla, California

L. L. Eberhardt
Ecosystems Department
Battelle Pacific Northwest
   Laboratories
Richland, Washington

Charles W. Fowler
National Marine Mammal
   Laboratory, Northwest and
   Alaska Fisheries Center
Seattle, Washington

P. R. Furniss
Department of Applied Math
University of Witwatersrand
Johannesburg, South Africa

Daniel Goodman
Scripps Institute of Oceanography
La Jolla, California

D. F. Gray
Fisheries System and Data
   Processing Group
Marine Fish Division
Bedford Institute of
   Oceanography
Dartmouth, Nova Scotia

Gordon C. Haber
Department of Mathematics and
   Natural Sciences
Anchorage Community College
Anchorage, Alaska

John Hanks
Institute of Natural Resources
University of Natal
Pietermaritzburg, South Africa

Larry D. Harris
School of Forest Resources and
Conservation
University of Florida
Gainesville, Florida

John Harwood
Natural Environment Research
Council
Sea Mammal Research Unit
Cambridge, England

Allison K. K. Hillman
I.U.C.N. African Rhino Group
Nairobi, Kenya

J. W. Horwood
Ministry of Agriculture, Fisheries
and Food
Fisheries Laboratory
Lowestoft, Suffolk, England

Irvin H. Kochel
School of Forest Resources and
Conservation
University of Florida
Gainesville, Florida

Ernst M. Lang
Zoologischer Garten
Basel, Switzerland

Richard M. Laws
Director, British Antarctic Survey
Natural Environment Research
Council
Cambridge, England

Patrick F. Lett
Marine Resource Analysts Ltd.
Dartmouth, Nova Scotia

Dale R. McCullough
Department of Forestry and
Resource Management
University of California, Berkeley
Berkeley, California

Jerry M. Mellilo
Ecosystems Center
Marine Biological Laboratory
Woods Hole, Massachusetts

R. K. Mohn
Fisheries System and Data
Processing Group
Marine Fish Division
Bedford Institute of
Oceanography
Dartmouth, Nova Scotia

Tom Polacheck
Department of Biology
University of Oregon
Eugene, Oregon

Thomas M. Pojar
Colorado Division of Wildlife
Ft. Collins, Colorado

Tim D. Smith
National Marine Fisheries Service,
Southwest Fisheries Center
La Jolla, California

G. L. Smuts
National Parks Board of South
Africa
Skukuza, South Africa

A. M. Starfield
Department of Civil and Mineral
Engineering
University of Minnesota
Minneapolis, Minnesota

Max Stocker
  Pacific Biological Station
  Nanaimo, B.C., Canada

D. E. N. Tait
  Faculty of Forestry
  University of British Columbia
  Vancouver, B.C., Canada

Carl J. Walters
  Institute of Animal Resource
    Ecology
  University of British Columbia
  Vancouver, B.C., Canada
Lilian S.-Y. Wu
  Mathematical Sciences Department
  IBM T. J. Watson Research Center
  Yorktown Heights, New York

As the title suggests, this book is about the ways in which the numbers of animals in populations of mammals with large bodies change over time in response to a variety of factors. A presentation of current work involving a wide variety of species as reported by scientists from around the world, the contributions to this volume deal with terrestrial species such as deer, elephants, bears, lions, and wolves, as well as marine mammals such as the northern fur seal, harp seals, and various other pinnipeds and whales. Throughout this book we hope to stimulate comparative work, wherein scientists involved in various specialized studies involving large mammal populations may learn more about the work and progress that has been realized in other quarters of the same field. The chapters in this book, along with the literature on which they are based, facilitate comparison of the population dynamics of various groups including herbivores, carnivores, ungulates, cetaceans, and pinnipeds.

As discussed in detail in the introductory chapter, the collection of chapters herein provides a perspective for the subject of the population biology of large mammals and specifically for that of their population dynamics. It helps provide an identity to this field, one that has been a long time in developing owing to the difficulty in studying species that live so long and exhibit so little change over the spans of time to which scientists are often restricted in their studies, and to the fact that there are relatively few species of large mammals available for study. Instead of borrowing concepts from work on other groups (such as fish and insects), we are enjoying the development of a field that has its own integrity. This integrity is based on both theoretical and empirical study, as exemplified by work presented in the following chapters. There is a particularly strong contribution among these chapters to the implementation of formal analytical tools in the quest for better understanding of large mammal populations. There are also examples of studies undertaken with a view toward casting our understanding in terms of the interaction of mammals with their ecosystems. Others examine the importance of factors that are internal to populations. There is a mixture of management-oriented chapters with those of more theoretical importance. Noting that a large portion of the work in this field is not published in readily available form at this date, we hope that this book is a positive step toward contributing to progress in the study, management, and protection of large mammals.

This collection of contributions culminates in the final product of desires that began to form in our minds during the early 1970s. During the spring of 1977 we spent several days at the University of Hawaii drawing up concrete plans to hold a conference at Utah State University in Logan, Utah. Approximately one year later, May 25-27, 1978, twenty-five people gathered on that campus to present papers. All but three of those papers are included in this book. Editorial review and time have resulted in some changes, but the basic messages remain the same.

The time and effort that went into the production of the individual chapters is not made apparent in the following pages. For their extensive efforts (often under pressure from us), we extend our heartfelt gratitude to the contributors to this volume. Each showed a great deal of dedication in working toward this final product. Dr. Sidney Holt, Dr. Walt Conley, and Dr. John Beddington are also to be thanked for the papers which were presented at the conference but which are not included in this volume.

Owing to space considerations, no acknowledgments appear at the conclusion of each chapter. We are grateful to the many individuals, agencies, and organizations that are thus left unidentified.

In addition to the authors who presented papers, the following acted as session chairmen at the Logan Conference: Dr. R. H. Hofman, Dr. L. M. Talbot, Dr. J. E. Powers, R. L. Phillips, Dr. Fred Wagner, and Dr. A. R. Tipton. Our thanks are extended to these individuals for their aid in helping make the conference a medium for exchange of ideas and cross-fertilization of concepts between groups of diverse backgrounds.

The review of the various chapters in this book was conducted largely by the group of authors and session chairmen as listed above and elsewhere in this book. Additional help in this work was provided by James Gessamen and James MacMahon. We are grateful to these individuals and to those who provided review at the individual author's request.

Monetary support and indirect aid in the production of the conference and this book were provided by the Animal Welfare Institute, the Center for Environmental Education (Whale Protection Fund), the International Union for the Conservation of Nature and Natural Resources, the National Audubon Society, the National Marine Fisheries Service, the National Wildlife Federation, the U.S. Marine Mammal Commission, the Utah State University Agricultural Experiment Station, the Utah State University Department of Wildlife Science, Utah State University Extension and Continuing Education, the Wildlife Management Institute, the U.S. Fish and Wildlife Service, and John Wiley and Sons, Inc. Neither the conference nor this book would have been possible without the help of the various individuals within these agencies and organizations who shared our desire to promote a communication of ideas and concepts on the population dynamics of large mammals.

A great deal of the organizational and editorial work was conducted while CWF was at Utah State University and while TDS was at the University of Hawaii prior to our joining the National Marine Fisheries Service. Help in organizing and conducting the conference was provided by various students and members of the faculty and staff of the Department of Wildlife Science at Utah State University. Stephanie Payne, Dede Olson, and Mary Conway provided invaluable editorial aid. Our thanks are extended to all of these people. We are especially grateful to all who helped provide moral support, particularly our wives, Jean and Margene.

Seattle, Washington   CHARLES W. FOWLER
La Jolla, California   TIM D. SMITH
July 1981

CONTENTS

ONE

An Overview of the Study of the
Population Dynamics of Large Mammals

TIM D. SMITH

CHARLES W. FOWLER

HISTORICAL BACKGROUND

Verhulst's interest in human populations and the publication of his application of the logistic equation in the 1830s may be viewed as some of the original work in the formal study of the dynamics of populations of large mammals (see also Chapter 14). Most of the progress in this field, however, has been realized within the last few years. Over time, the character of studies involving large mammal populations has changed as have studies on animal populations in general. Early investigations focused primarily on the capacity of populations to increase. The study of factors involved in preventing unlimited growth acquired more importance later. These same trends characterize the study of population dynamics at other taxonomic levels with an apparent tendency for studies of large mammal populations to lag behind studies of other groups. An excellent discussion of the general history of the study of populations can be found in Hutchinson (1978).

From a taxonomic perspective, the study of population dynamics has shown a tendency to focus on humans, microorganisms, insects, fish, birds, and, to some extent, on small mammals. Comparatively little of the effort expended in the field of population dynamics has involved large mammals other than humans. There are several factors that, in combination, have contributed to this discrepancy. Although often of importance to man, undomesticated large mammals in general are neither as influential as pests nor as important as food sources as are many smaller organisms. The study of the natural history and general biology of large mammals is easier than, and has often precluded or taken precedence over, the study of their population dynamics. The long-lived nature of large mammals (see Chapters 2, 22) prevents the rapid accumulation of data that is possible with populations of small organisms, especially those that can be reared in the laboratory. Obviously, there are fewer species of large mammals than of the smaller-bodied species. Together these factors seem to be the basic cause of a rather short and relatively unproductive history of the study of population dynamics of large mammals.

Most studies involving population dynamics within all taxonomic groups have involved either species of economic interest or man's own population. This has encouraged a narrow perspective oriented toward solving specific problems or

1

achieving specific goals. By comparison, a philosophically more general or academic view of population dynamics has evolved through studies of birds, microorganisms reared in laboratories, and other groups of species that are relatively less important to man.

This trend is especially obvious among studies involving large mammals. As we will discuss them, most studies of the dynamics of large mammals involve three groups: (1) humans; (2) large terrestrial mammals (as predators, tourist attractions, or game animals); and (3) marine mammals (being larger and of economic value). In all cases, most studies involve species of direct importance to man. Few studies of the population dynamics of large mammals are conducted out of pure academic interest. Some of the studies on the George Reserve deer herd (McCullough, 1979), for example, provide a notable and very valuable exception.

The study of the dynamics of human populations has an interesting history. Having started with a conceptually holistic approach embodied in the logistic equation, the work on humans has become deeply engrossed in the effects of structure by age, sex, and reproductive potential, migration, socioeconomics, and other dimensions of importance to short-term projections. The concept of an upper limit to the population (or carrying capacity) has been of much less importance than would have been predicted on the basis of Verhulst's first consideration. Today the bulk of the theory involving dynamics in age-structured populations is to be found under the authorship of people who have worked with human populations.

Increasing impact, brought about by man's growing population and growing interest and concern by the public for the protection of certain species, has created new demands for a better understanding of the ways in which large mammal populations change over time. In terrestrial environments man has long had an interest in protecting, harvesting, or controlling the populations of specific species of large mammals. As in the case of studies involving humans, the study of these species has often progressed to a level of resolution that has, to a degree, prevented the development of holistic views. Studies in these situations have often emphasized minute details of the relationships between the animals and their environments, especially their resources. These studies have generally been less quantitatively oriented than have population studies of humans, insects, and fish. Most studies of terrestrial large mammals have thus resulted in a fairly detailed understanding of the natural history of the species involved. As a result we often know more about the husbandry of these species than we do about either how the populations are regulated or patterns in their dynamics. Specific needs for management and for short-term predictions have worked against the development of a holistic understanding of the population dynamics of large mammals.

Nonetheless, the general need for a holistic view of population regulation is growing in importance as increasing numbers of species of terrestrial large mammals are being managed within the context of their ecosystem. There is a growing need for a general perspective within which such management and future research may be conducted if management is to be realistically based on ecosystem principles.

The mathematical modeling of terrestrial populations of wild large mammals has been attempted only in recent years. Much of the delay in such activities has been a product of insufficient training (see Chapter 21). Most people who have worked with populations of large terrestrial mammals have not been given the mathematical background that is more characteristic of those studying fisheries, human populations, or marine mammals. Those who have attempted to undertake quantitative studies have tried to use life tables without realizing that such an approach is less of a dynamic model than a static description. This weakness is being avoided more in recent work as age-structured matrix models and other more elaborate but less analytically tractable models are being used with the help of computers. Many chapters in this volume exemplify the progress currently being made.

Many large mammals in marine environments are of economic value. The history of the harvest of these species is long and, in some cases, relatively well documented. Technological advances in harvesting techniques have resulted in the extinction, or near extinction, of a number of species. Compared with our view of the population dynamics of terrestrial species, we have a more holistic view of the dynamics of marine mammal populations. This situation has been forced on us by the nature of their environment; detailed examinations of the relationships between marine animals and their environment is prohibitively difficult. Most of the information useful for studying the population dynamics of marine mammals has come from harvest records. As in the case of fish, these data are collected in terms of numbers, ages, pregnancy rates, and weights. The dynamics of each of these attributes of harvested populations has been scrutinized in relationship to the other attributes without the benefit of a detailed understanding of their behavior, physiology, and interaction with other species.

It was not until the 1960s that quantitative syntheses of the data for marine mammals began to take the shape of population models (e.g., Chapman, et al., 1964). The studies that were of greatest importance involved the effects of harvest on exploited populations. Unfortunately, many of these studies were based on previous studies of fish and, to some extent, insect populations. The study of fish was also characterized by data involving numbers, weight, and age, as indicated by the catch. This similarity in data led to a natural tendency to borrow models from older work on fisheries to be applied to marine mammals. As is the case for terrestrial populations, much of the current work on marine mammals is beginning to take on its own identity, overcoming the sometimes inappropriate approaches inherited from early studies in the field of fisheries.

GENERAL APPROACH

Most approaches taken in the study of populations take one of three basic forms. The first is the approach taken by natural historians. This is least formal and is basically descriptive. Little if any attention is paid to the utility of any paradigm developed. The intuitively appealing, intriguing, and visually apparent aspects of a given species' population biology are emphasized. The second approach involves conceptual models formalized in written narrative. This makes a conscious recognition of the existence of the animals in the context of a population. The third approach uses formal mathematical models to describe these populations. Although parts of a continuous and complementary spectrum, these overlapping categories of approaches are useful for discussing the field of population dynamics. It is the more formal approach that we attempt to emphasize through the chapters that make up this volume. Our bias is toward this type of study and, when overwise left unspecified, we will be referring to this approach with the term "population dynamics."

Studies of terrestrial large mammals have often been primarily a process of developing a description of their natural history. Partly because of the length of time required for sufficient data to accumulate, only recently have more holistic, formal treatments of the population dynamics of terrestrial large mammals been developed. An examination of the chapters in this volume, the literature cited herein, and a bibliography of population dynamics in large mammals (Fowler et al., 1980a) illustrates that relatively few organized or specialized efforts have been made along such formal avenues. In many cases, however, there exist relatively unutilized data sufficient for studies involving the production of a formal synthesis. Many populations in Europe, Asia, and North America have been monitored for years. An important need is to identify such sets of data and to utilize this information in formal syntheses to improve our understanding of the dynamics of terrestrial large mammal populations.

Compared with that of terrestrial large mammals, our knowledge of the natural history of marine mammals is limited. Basic biological data often have come solely from animals captured in the harvest, rather than from observations of animals in their natural environment. Motivated by international as well as national concerns, various studies focusing primarily on the dynamics of marine mammal populations were first seriously approached during the 1960s when several formal mathematical models were first applied. The pioneering work of D. G. Chapman on the northern fur seal (Chapman, 1961, 1964, 1973) and with others on the great whales (Chapman et al., 1964; and subsequent volumes of the Reports of the International Whaling Commission) was important in this regard. As mentioned earlier, these studies drew heavily on techniques and concepts developed in earlier studies of fish populations. As will be amplified later, the governments sponsoring such research needed relatively precise evaluations of the size and rate of production of the populations in question in order to provide specific advice for management.

For both terrestrial and marine large mammals, there is a growing need to refine the underlying conceptual models. In most instances, this means that the basic biology or natural history of the species involved should be better utilized in the construction of these models. For terrestrial species this involves using the existing information to extract general principles, while for marine populations this involves the need for more such information. By virtue of this contrast we encounter one of the basic differences in studies of large mammals in these two environments — the difference in resolution — an attribute discussed in the next section.

The study of the dynamics of human populations has developed very differently from the study of the population dynamics of marine and terrestrial large mammals. Studies of human populations involve very formal mathematical models, the third approach discussed above. These models are firmly grounded with precise and detailed descriptions of vital rates, almost always on an age-specific basis. But the underlying conceptual models do not appear to be self-contained. For example, the fact that vital rates (birth, mortality, and so forth) do change is accounted for, but as external input rather than as changes that can be accounted for within the models. (See Chapters 15 and 19 for work of this type on animals other than humans.)

A great deal of attention in the models used in the study of human populations is given to structure, distribution, and details of importance in short-term predictions. Precision over short time spans is more important than are long-term changes, trends or patterns. Very little emphasis is placed on factors of ultimate importance in establishing an upper limit to the human population. Feedback by way of density dependence is rarely if ever included. In comparison to the classic models of marine mammal populations, the models of human populations are quite resolute. In contrast to most models of terrestrial large mammals, and many for marine mammals, models of human populations involve a great deal of internal dynamics relating to age and sex structure.

Because of technological support behind the growth of human populations, relatively little evidence has been found for the existence of density dependence (see Fowler, 1981, Chapter 23). This has led to the abandonment of the logistic that was one of the earliest models to be applied to populations of large mammals (Pollard, 1973). By contrast, considerable evidence for density dependence in large mammals (Fowler et al., 1980b, Chapter 23) exists. Most of the formal models of large mammal population dynamics for nonhuman species involve density dependence as seen in many of the chapters in this book.

As a result of research involving human populations and the supporting theory, we have a very detailed understanding of the process of population growth as it relates to such factors as age structure, reproductive schedules (reproductive value by age), age at first birth, and life span. This is an example of study that has taken a very resolute formal quantitative approach to specific aspects of population dynamics. It is to be contrasted to attempts to incorporate a great deal of resolution into our representations of the interactions between populations and their environments, a matter of scope as discussed in the next section. Such approaches include a larger system, but often with less detail or resolution involving the population itself.

SCOPE AND RESOLUTION

Among the most important aspects of the design of studies of large mammal populations are those of scope and resolution. The breadth of a study and the degree of detail involved are determined by the questions to be addressed. Several dimensions of scope and resolution as involved in studies of large mammal populations are addressed here.

It is important to point out that separate studies of the same population, with different scope and resolution, can be mutually complementary to our understanding of the dynamics of that population. Studies of marine mammals, of terrestrial mammals, and of humans, however, have each tended historically to retain separate sets of goals that are different among, and consistent within, groups. Too frequently this has led to marine mammals being studied nearly consistently at one level of resolution, and humans, for example, at an entirely different level. In view of the fact that studies at different levels of scope and resolution are complementary, it would seem useful to design two studies of an elephant population, for example — one to answer questions generally asked about whales, and a second to answer questions generally asked about humans. Such complementary studies, applied on a large scale, would enhance our overall understanding of the dynamics of large mammal populations.

In all cases the design of studies of populations of large mammals must deal with the overriding problem of resolution, or the degree to which detail is involved. The proper resolution of any particular study is determined by the questions being addressed. The realizable resolution is determined by the existing data, logistical constraints, time, and money. Practical considerations or management-oriented questions often demand precision and predictability at the expense of generality. General principles that often emerge from questions that are more academic in nature, and that demand more generality in approach, are often sacrificed.

Of similar importance is the scope of factors thought to be influential in determining the nature of the dynamics of populations. Can a population be represented adequately by a model involving only the population's numbers and its essential internal characteristics alone? Or do we need to include other elements, such as those of the ecosystem, to capture the essential features of the dynamics of interest. As with the resolution of a model, the scope of a model must be determined largely by the questions being addressed. However, a growing body of theoretical and empirical background would indicate that the proper scope of such studies depends in part on the species involved. Small-bodied "r-selected" species seem to be more at the mercy of their environment than are large "K-selected" species, and hence may require more scope in the models developed to represent them.

Even so, greater scope may be required in meeting the objective of much of the current work on large mammals. In managing the harvest of the great whales and the Alaska fur seals, for example, it has become increasingly clear that the degree of interspecific interactions, both among harvested species and between the harvested populations and their food sources, is considerable (see Chapters 2, 14, 17). It has been estimated that southern hemisphere sei whales may have increased by more than 60% in population size owing to changes in their reproductive rates prior to their being exploited (Smith, 1977; but see also Mizroch, 1980). Similarly, a major area of research is the possible effect of changes in fish populations in the Bering Sea, due to increased fishing effort, on fur seal population size. In both cases the traditional models of limited scope, in providing advice for management, cannot be realistic in accounting for these interactions; but these models can and do provide advice that is precise (discussed below).

Such considerations do not negate, however, the utility of single species models for many purposes. A production model (such as the generalized production model discussed in Chapman, 1960; Richards, 1959; Pella and Tomlinson, 1969) with stochastic properties may capture the essential elements of the dynamics of a broad spectrum of species. Greater degrees of stochasticity (more variance) may be needed for those species that are more subject to their physical and biological environment. The interaction of the species with its biological environment (the ecosystem aspect) may, in large part, be captured by the parameters of such models (e.g., see Fowler, 1980a, 1981).

Obviously, permanent or unusual changes in the ecosystem are not going to be reflected in simple models (see Chapter 17). In addition, it may facilitate our understanding of the ways in which simple models should be parameterized to conduct studies in which a greater scope is included. Thus the more complex ecosystem models are of obvious value, and work along these lines should be encouraged. Progress along these lines will, by necessity, be slow, however, because the quantity of information required often proves prohibitive. (See Fowler et al., 1980b; Chapter 14.)

PRECISION, REALITY, AND GENERALITY

Levins (1966) identified three features of models constructed to represent natural systems. These were generality, precision, and realism. He points out the difficulty (if not impossibility) of obtaining strength in all three of these attributes in one model. One or more must be sacrificed to obtain a strong representation of the others. Different tradeoffs have strength and utility in particular applications. In the study of population dynamics, these principles can be used in evaluating and comparing the various approaches being taken.

Some of the chapters in this book argue that more reality should be included in models of the population dynamics of large mammals, through an increase in scope, especially in terms of the interaction of populations with their environment (Chapters 2, 15-19, 23). It is argued that simple models, such as the logistic, although they implicitly reflect interactions with environments, are often not sufficiently realistic. Some generality and realism are sacrificed despite the fact that such models capture several basic concepts of importance. Botkin et al. (Chapter 19) present examples of how specific types of information concerning factors external to populations can potentially affect our views of the ways they behave. Both Caughley (Chapter 18) and Fowler (Chapter 23) argue that interactions between trophic levels may produce overriding effects on the types of dynamics to be expected. These latter arguments attempt to achieve some degree of generality and realism by sacrificing precision. Horwood's chapter (Chapter 17) deals with other interactions between species, sacrificing precision and some generality in an attempt to be realistic. Generality is of less importance than either precision or reality in McCullough's work on grizzly bears (Chapter 9). Somewhat of a balance seems to have been struck by the model behind the chapter by Pojar (Chapter 12), in which only a limited proportion of all three features is achieved.

As argued earlier, one way of dealing with the impossibility of producing a complete representation of reality in one model is to produce several models of the same system. As Walters et al. (Chapter 16) have described, several models of different levels of resolution (and hence precision, reality, or generality) may be produced for examining questions from different points of view. Frequently not well recognized is the possibility of constructing several different models of the same system, instead of developing one in great detail. Although difficult to implement, such an approach emphasizes the problems created by the fact that no model is perfect. Relying on one imperfect model regardless of its resolution can create serious problems.

An interesting example of the balance among realism, precision, and generality appears in the population studies developed to provide advice for the management of several groups of marine mammals. The constraint of providing specific defensible advice to regulatory bodies such as the International Whaling Commission, the United States National Marine Fisheries Service, and the North Pacific Fur Seal Commission has prompted models that emphasize realism and precision — especially the latter — at the expense of generality.

Studies by Chapman and others referenced above for the International Whaling Commission were designed to provide advice to the political decision-making body. The results of these studies had increasing impact on the actual decisions about quota levels that were made through the 1960s and into the 1970s. For different points of view on this, see McHugh (1974), McVay (1974), and Smith (1976). In order to be useful in a sometimes adversarial context, it was necessary that the advice given be quite specific, requiring very high precision in the models being used. It was generally more important that estimates of the effect of various harvest levels be precise than that they be obtained using realistic models. Certainly both of these aspects far outweigh the need for generality in this context.

Very likely, in the interest of providing advice in a usable fashion, the precision of the results of studies on the population dynamics of the great whales has, from time to time, been oversold (Smith, 1976). Certainly decisions have been made primarily on the basis of point estimates, rather than on statistically more meaningful interval estimates. Indeed, in some instances simple sensitivity analyses of the models being used have only recently been undertaken.

Another result of this emphasis on precision is that the scope of the models developed within the context of management has remained quite narrow. They are almost always single-species descriptions, and are heavily dependent on the concepts of density dependence, carrying capacity, and maximum sustainable yield. In fact, the emphasis on precision has, at times, resulted in these concepts being built into the legal framework (national and international) that provides the context in which management decisions are made. Thus the stated management objective for the northern fur seal is maximum sustainable yield. Similarly, the much more general concept of optimum sustainable population level (OSPL), as used in the U.S. Marine Mammal Protection Act (MMPA, Public Law 92-522, 1972), has been interpreted for purposes of managing the incidental kill of porpoises by tuna fishermen as a population size between the level giving maximum sustained yield and the carrying capacity (Smith, 1979).

The problems of management in the context of not being able to provide relatively precise advice are very great. When it becomes apparent or likely that the models being used are rather more precise than they are realistic, or accurate, what is to be done? In the case of the great whales the information available is insufficient to allow creation of new models of greater realism without greatly sacrificing precision. Yet without the level of precision that management decisions have come to depend on, these decisions are more likely to be based on political and economic pressures, too often to the detriment of the long-term productivity and viability of the populations.

Studies of the populations of porpoise in the eastern tropical Pacific illustrate this point in directly acknowledging the problems of realism. Animals of these populations are known to associate closely with schools of yellowfin tuna. Fishermen using purse seines utilize the surface-swimming porpoise as indicators of the presence of tuna, resulting in the incidental killing of some porpoise. Management decisions about allowable incidental kill levels are based, in part, on fairly precise, but rather simple, population models (Smith, 1979; Smith and Polalcheck, 1978). Because of the apparent symbiotic relationship between the tuna and some populations of porpoise, the co-occurrence of several species of porpoise and of tuna, and the reduction in abundance of both the porpoise and the tuna populations, it is reasonable to expect some interspecific relationships to be important. Current management actions are based on a model that is precise, but more general than realistic.

This single-species model was selected as the best for generating management advice when it was realized how difficult it would be to deal with the interspecific interactions. It was *not* decided that a less precise but more realistic multispecies or ecosystem model would be appropriate for management advice. The loss in precision to gain this realism would have been so great as to preclude any utility of the management advice that would be produced. Interestingly, in court challenges to the management decisions that have been made on this advice, the adequacy of the scientific advice is being attacked not in terms of the realism of the models, but rather in terms of the precision of specific parameter estimates used. Thus the decision that precision could not be sacrificed to gain realism is probably reasonable from the perspective of management. Frequently this emphasis on precision over realism has been shown to be a necessity if rational management decisions in adversarial context are to prevail. Indeed, without reliance on such concepts as carrying capacity and maximum sustained yield, the management of marine mammal populations would be governed far more by short-term economic interest than by interests concerning long-term renewability.

COMPARATIVE APPROACHES

Another way of gaining insights into the nature of any specific population and its dynamics is to view it as an example of particular types of populations, and on the basis of the properties of each of the larger groups to deduce the properties of the particular case. This approach presupposes that groupings of populations can be described on the basis of samples of populations drawn from that group. Comparison of the dynamics of populations of various categories should lead, in theory, to general principles of use in deducing properties of specific populations.

It is the apparent lack of attempts to make comparisons and to formulate generalizations concerning the nature of the population dynamics of large mammals that provided the general stimulus for convening the conference at which the contributions to this volume were originally presented. In this volume the chapters by Bunnell and Tait (Chapter 4), Chapman (Chapter 14), Goodman (Chapter 22), and Fowler (Chapter 23) (as well as the unpublished papers by Holt and Beddington presented at the conference) deal with this issue directly. Others treat the issue less directly. As more papers of similar nature are produced, the potential for such approaches will become more apparent. The work of Harestad and Bunnell (1979) and Smith (1974) falls into this category along with work recently completed by Fowler (1980a, 1981). The principles having their origin in such studies can help provide a perspective within which each special case may be considered. Such studies must rely, however, on a general background of information contained in papers similar to those in this volume and in the general literature. It is hoped that this book will be of help in stimulating further work along such lines.

In attempting to generate comparisons, schemes must be formulated that result in classifications across which comparisons can be made. Examples can be produced on the basis of the chapters in this book. One of our first objectives was to provide a sampling of marine and terrestrial work that could be used for initial contrasts and comparison in the format of the conference and this book. It was hoped that consideration of what is known in terrestrial settings could be of material help in gaining a better understanding of the dynamics of the marine species, and vice versa. As noted earlier, one difference seen in comparing these two groups of papers is that the terrestrial species are represented by studies with greater resolution than are the marine species. A further difference to be kept in mind is that the major problem facing many species of marine mammals is the danger of their being overharvested. They are less likely to be subject to habitat destruction, a problem common to many terrestrial species.

It is difficult to determine which comparisons will result in the greatest progress in our understanding. There are, however, a number of dimensions across which theoretical and empirical comparisons could be made to help develop our view of the population dynamics of large mammals in particular and of animals in general. In addition to longevity, initial survival, age at first reproduction, birth, and mortality, as discussed by Fowler in Chapter 23, there are such factors as social organization, territoriality, and taxonomic categories. The 12 characteristics mentioned by Eberhardt and Siniff (1977) as potentially useful for evaluating a population may serve as bases for comparison as well. Some of the attributes mentioned above are part of the "life-history strategy" that must serve as a basis for comparison as discussed by Goodman (Chapter 21) and Fowler (1981). Various categories of specialization and behavioral types may serve as fruitful categories for comparison. Growth rates and specific productivity are further dimensions for potential comparisons.

There are numerous studies on species of terrestrial herbivores (such as the ungulates in general). There is relatively less work that deals with terrestrial carnivores, but progress is being made (see, e.g., Chapters 6 and 16). As emphasized in the chapters by Fowler (Chapter 23) and Caughley (Chapter 18), the dimension of trophic level may be an important dimension across which to make comparisons. Specific arguments for expecting differences between trophic levels were developed by May (1973), Fowler (1980a, 1980b), and May et al. (1979).

Numerous types of comparisons can be made across the dimensions identified above. As discussed in Fowler (1980a; Chapter 23), the shape of productivity curves can be compared. These are of special interest in management and from the point of view of the roles played by the populations in their respective ecosystems. Comparison may be made between such attributes, since they may be correlated. There may be a tendency for populations of large-bodied species to exhibit regulation through change in birthrate, as opposed to change in juvenile survival, which may be of more importance in the regulation of smaller-bodied species. Such correlations, if they exist, can only be recognized through comparison. As shown by Harestad and Bunnell (1979), there are relationships between the size of territories (range) and body size, trophic status, and resource productivity.

There are a number of specific questions and hypotheses that need to be addressed as they relate to matters of practical importance. In particular, the scientific committee of the International Whaling Commission is in desperate need of insights useful in developing more realistic approaches to setting quotas or deciding which stocks should receive protection. As carnivorous large-bodied and socially organized large mammals, do cetaceans exhibit dynamics that can be described on the basis of what we know about large-bodied species combined with what we know about the socially organized species and carnivores in general? Do social groups of cetaceans exhibit territoriality in ways in which our growing knowledge of the effects of territoriality in terrestrial systems may be applied? Are there genetically determined differences between the various taxa, or are population dynamics more a product of the nature of a species as independent of its taxonomic relationship to other groups?

Before leaving the topic of comparative population dynamics, it should be noted that we do not wish to underemphasize the need to make comparisons across a much wider spectrum of species types than is encompassed by large mammals alone. The study of large mammals in particular and animals (and possibly plants) in general should profit from broad comparisons. As pointed out by Fowler (1980a, 1981), patterns concerning population dynamics seem to be emerging as a result of comparisons made across the spectrum of body size, r-K strategies, longevity, and productivity. These comparisons need both theoretical and empirical underpinnings to provide the progress we need toward a better understanding of the population we wish to manage. Through such studies we may better appreciate the errors made in adapting models developed in the study of fish and insects to the study of large mammals.

THE SCOPE OF THIS BOOK

The chapters included in this book are examples of studies of large mammals that provide a basis for comparisons of two types. First, each chapter contains information concerning individual species such that species may be compared as outlined in the previous section and as discussed in Chapter 23. Second, the studies themselves can also be compared, as outlined in earlier sections. Such comparisons are a necessary part of coming to grips with a truly comparative study of population dynamics. The basis for comparisons in each case is not always well defined; indeed, it is the principal problem of comparative science to explore the possible bases, searching for those from which new insights can be obtained.

Some sample bases for comparison of both types are shown in Table 1. In this table the columns are grouped into six categories. The first two categories represent material for comparisons over the characteristics of the species involved. The remainder represent comparisons that may be drawn between and among the studies themselves. In all cases Table 1 contains an indication of how each chapter has been categorized, to show the scope included in this book. In addition to the bases for comparison outlined above we have indicated the continent of origin of the senior author of each chapter.

The absence of chapters dealing with studies of human populations (see columns labeled *Group of Mammals*, Table 1) is regrettable. Although two contributions were planned, conflicts in schedules resulted in cancelations. This lack is unfortunate, as much of the theory of population growth has been developed within the context of research involving humans (see, e.g., Pollard, 1973; Keyfitz, 1968). The work on humans, however, has taken on a very different character from studies of other large mammals as discussed above. Inter-

Table 1 Classification of Chapters in This Volume Across Several Dimensions

First Author	Chapter	Group of Mammals			Trophic Level			Approach			Scope			Levins's Categories			Continent					
		Terrestrial	Marine	Human	Herbivores	Carnivores	Omnivores	Descriptive	Conceptual	Formal	Single Species	Multispecies	Ecosystem	Realism	Precision	Generality	Africa	Europe	Australia	N. America	S. America	Asia
Allen	13		×			×			×	×	×		×	×	×	×			×			
Botkin	19	×			×		×	×	×	×		×		×		×				×		
Bunnell	4	×							×		×			×		×				×		
Burgoyne	21	×	×		×			×	×	×		×		×		×			×	×		
Caughley	18	×			×				×		×				×	×				×		
Chapman	14	×			×				×	×		×		×		×	×					
Croze	15					×			×		×			×		×				×		
DeMaster	20	×	×		×	×			×	×	×			×		×				×		
Eberhardt	10	×	×		×	×	×		×	×	×			×		×				×		
Fowler	23	×	×		×	×	×		×	×		×		×						×		
Goodman	22	×	×		×	×		×	×			×	×	×		×	×					
Hanks	3					×			×	×		×		×		×		×				
Harris	11	×			×				×	×		×		×	×			×				
Harwood	8		×			×			×	×				×		×		×				
Horwood	17		×			×		×	×	×	×			×		×				×		
Laws	2	×	×		×				×	×	×					×				×		
Lett	7		×			×			×	×	×			×	×					×		
McCullough	9	×			×		×	×	×			×			×	×				×		
Pojar	12	×							×			×			×					×		
Smith	5		×		×				×	×	×			×	×	×			×			
Starfield	6	×				×			×	×	×			×	×					×		
Walters	16	×			×	×			×	×		×		×		×	×			×		

action between students of human populations and students of other large mammal populations is needed, as humans are large mammals.

The collection of chapters in this book may be distributed along a spectrum ranging from single species to ecosystems. As mentioned earlier, the single-species approach makes the assumption that the essential dynamics of a population can be characterized as abstracted from its environment. The ecosystem approach carries the philosophy that the ecosystem may provide the overriding driving forces that determine the dynamics of any particular species. As shown in the columns labeled *Scope* in Table 1, there is preponderance of chapters that take the single-species approach. This partly expresses our bias toward working along these lines as well as the fact that most work is being conducted at this level. The latter should be viewed as a gap in the approaches being taken, and the recommendations being given by Botkin et al. (Chapter 19), Croze et al. (Chapter 15), Caughley (Chapter 18), and others in this volume should be taken seriously. Ecosystem models and results of research conducted at this level should be examined for a better understanding of the cause-and-effect relationships contributing to particular types of dynamics. Simple, general, single-species models should be examined for the differences and similarities that exist between various groups or types of populations as categorized on the basis of their interactions with their ecosystems.

There is a relatively large number of species of terrestrial herbivores in comparison to carnivores. In spite of this differential, this book contains a number of chapters that deal with carnivores (see columns labeled *Tropic Level*, Table 1). McCullough's chapter on bears (potentially classified as omnivores, Chapter 9), the chapter by Starfield et al. on lions (Chapter 6), and the treatment by Walters et al. on wolves (Chapter 16) exemplify advances being made in our understanding of the population dynamics of these higher trophic levels. It is apparent that interactions *within* the populations of such species is of great importance. It is important to know to what degree these interactions override those with other elements of their ecosystem.

Three levels of approaches to the study of populations were identified above: descriptive, conceptual, and formal. The chapters in this book are categorized in these terms in Table 1 under the column labeled *Approach*. Most include aspects in two of these three categories, and all include some level of conceptual modeling. All the chapters, of course, rely on descriptive studies at some point, even if this is not specifically discussed.

Levins's (1966) three attributes of realism, precision, and generality discussed above provide another basis for comparison of the chapters in this volume. Our categorization of the studies presented in this volume as they achieved a balance among these aspects are indicated in Table 1 in the column labeled *Levins's Categories*. Although classifying chapters into these categories is a matter of judgment, it appears that most of the contributions favor realism and generality at the expense of precision. Somewhat fewer favor realism and precision, while fewest favor precision and generality. These relationships are different for studies on terrestrial and marine mammals.

In trying to obtain a diverse sampling of workers and populations we attempted to attract to the conference a number of individuals from countries other than the United States. As shown in the columns labeled *Continent* in Table 1, we succeeded in obtaining three contributions from Africa, two from Australia, and three from Europe (plus two not published). Financial constraints and a large population of workers in the United States resulted in an imbalance, wherein almost one-half the participants represent North America.

It should be noted that a great deal of work has been conducted by the Japanese. In particular, there has been a considerable effort spent in dealing with marine mammals, especially fur seals as well as both large and small cetaceans. The Soviets have conducted some work along these lines, and considerable data exist among the western Asian and European countries concerning several terrestrial species. There seems to be an opportunity for progress in these areas.

DIRECTIONS FOR THE FUTURE

We have tried to present a general description of our view of the history and perspective of studies involving the population dynamics of large mammals and to place this book within that perspective. In presenting the history we have tried to point out the important differences between approaches taken by various disciplines within the general field. The strengths of each approach may be used to aid progress in the others. We have provided a general perspective with the view that each type of model has its own merits. Specific questions require specific avenues of research. We have discussed general comparisons and emphasized this aspect of the study of the population dynamics of large mammals because, in the past, little attention has been paid to academic issues, general principles, and overall perspective.

To promote the progress that seems possible, several things are required. The first is a need to standardize the definitions and concepts behind terms being used within the field of population dynamics, especially as they are used by biologists studying large mammals. The term "carrying capacity" is a good example. Most theoretical population biologists and population dynamicists working in the fields of fisheries and entomology use the term to refer to that mean population level that would be observed under conditions unaltered by man. It is often referred to as the constant K in the deterministic logistic equation. Range specialists, however, as investigators who deal with many species of terrestrial ungulates, define carrying capacity in terms of the conditions of the forage. Such differences must be resolved before any meaningful communication between disciplines can take place.

Second, the specialists working in the field of population dynamics of large mammals need to recognize the existence of other work both within their specific field as well as in the general field of population dynamics. As mentioned earlier, there is a need to identify existing sets of historical data and to subject them to formal analysis to help overcome the problems created by the long-lived nature of large mammals. A specialized course in this field at several major universities would prove fruitful and could draw on the resources now available in the published literature. Symposia and conferences such as that which resulted in this book are to be encouraged. One or two specialized centers for research in this field would help provide coherency. The publication of a journal specializing in the population dynamics of large mammals would be of considerable help. The field has an identity and integrity of its own, and recognition of the existence of the collections of published papers, data, and workers doing this type of work will help promote the progress needed.

Hand in hand with recognizing and being familiar with the breadth of work in other aspects of the field of population dynamics in its application to large mammals is the need to undertake complementary studies, as mentioned earlier. Work on any single population or species can be made much more productive by taking separate approaches with different scope, resolution, levels of formalism, realism, precision, and generality. Since most of these attributes are determined by practical considerations, there is a need to emphasize academic studies — studies presenting more general and less restricted questions and hypotheses. Broader interest in conducting such studies should be encouraged.

There is clearly a great need for catalyzing and supporting further work on the population dynamics of large mammals. As man's population continues to rise, we encounter increasing numbers of problems that involve the populations of other large mammals. Many species such as the elephant and bison are restricted to ranges much smaller than previously occupied. In many cases these ranges continue to decrease because of habitat destruction. Other populations are either depleted or are overharvested. There are conflicts involving other resources, such as exemplified in the cases of tuna, porpoise, and gray seals (see Chapters 8 and 17). Predator control and its direct and indirect effects continues to be of concern. Growing public concern over sound management or complete protection places us in a particularly good position for requiring more understanding of the effects of various alternatives. More thought is needed regarding future options in place of current demand. One of the purposes of this chapter and of the book as a whole is to provide new insights of value in making this progress, to underline the need for continued effort, and to help provide some alternatives for fruitful study.

LITERATURE CITED

Chapman, D. G. 1960. Statistical problems in dynamics of exploited fisheries populations. Proceedings of the Fourth Berkeley Symposium on Mathematical Statistics and Probability.

Chapman, D. G. 1961. Population dynamics of the Alaska fur seal herd. Trans. N. Am. Wildl. Nat. Resour. Conf. 26:356-369.

Chapman, D. G. 1964. A critical study of Pribilof fur seal population estimates. U.S. Fish. Wildl. Serv. Fish. Bull. 63:657-669.

Chapman, D. G. 1973. Management of international whaling and north Pacific fur seals: Implications for fisheries management. J. Fish. Res. Board Can. 30:2419-2426.

Chapman, D. G., K. R. Allen, and S. S. Holt. 1964. Report of the Committee of three scientists on the special scientific investigation of the Antarctic whale stocks. Rep. Int. Whaling Comm. 14:32-106.

Eberhardt, L. L., and D. B. Siniff. 1977. Population dynamics and marine mammal management policies. J. Fish. Res. Board Can. 34:183-190.

Fowler, C. W. 1980a. Non-linearity in population dynamics with special reference to large mammals. Appendix C *in*: Fowler et al. 1980b.

Fowler, C. W. 1980b. Exploited populations of predator and prey: Implications of a model. Appendix F *in*: Fowler et al. 1980b.

Fowler, C. W. 1981. Density dependence as related to life history strategy. Ecology (in press).

Fowler, C. W., W. T. Bunderson, and M. B. Cherry. 1980a. Selected bibliography on population dynamics of large mammals. Appendix A *in*: Fowler et al. 1980b.

Fowler, C. W., W. T. Bunderson, M. B. Cherry, R. J. Ryel, and B. B. Steele. 1980b. Comparative dynamics of large mammals: A search for management criteria. U.S. Mar. Mam. Comm. Rep. No. MM7AC013. NTIS No. PB80-178627. National Technical Information Service, Springfield, Va.

Harestad, A. S., and F. L. Bunnell. 1979. Home range and body weight—A reevaluation. Ecology 60:389-402.

Hutchinson, G. E. 1978. An Introduction to Population Ecology. Yale University Press, New Haven, Conn.

Keyfitz, N. 1968. Introduction to the Mathematics of Populations. Addison-Wesley, Mass.

Levins, R. 1966. The strategy of model building in population biology. Am. Sci. 54:421-431.

May, R. M. 1973. Time delay versus stability in population models with two and three trophic levels. Ecology 54:315-325.

May, R. M., J. R. Beddington, C. W. Clark, S. J. Holt, and R. M. Laws. 1979. Management of multispecies fisheries. Science 205:267-277.

McCullough, D. R. 1979. The George Reserve Deer Herd: Population Ecology of K-Selected Species. University of Michigan Press, Ann Arbor, Mich.

McHugh, J. L. 1974. The role and history of the International Whaling Commission. Chapter 13 *in*: W. E. Schevill (ed.). The Whale Problem: A Status Report. Harvard University Press, Cambridge, Mass.

McVay, S. 1974. Reflections on the management of whaling. Chapter 17 *in*: W. E. Schevill (ed.). The Whale Problem: A status Report. Harvard University Press, Cambridge, Mass.

Mizroch, S. A. 1980. Some notes on Southern hemisphere baleen whale pregnancy rate trends. Rep. Int. Whaling Comm. 30:561-574.

Pella, J. J., and P. K. Tomlinson. 1969. A generalized stock production model. Inter-Am. Trop. Tuna Comm. Bull. 13:420-456.

Pollard, J. H. 1973. Mathematical Models for the Growth of Human Populations. Cambridge University Press, Cambridge, Mass.

Richards, F. J. 1959. A flexible growth function for empirical use. J. Exp. Bot. 10:290-300.

Smith, T. D. 1974. Researchers in comparative population dynamics. Southwest Fish. Ctr. Admin. Rep. LJ-74-52. Nat. Mar. Fish. Serv., La Jolla, Calif.

Smith, T. D. 1976. The adequacy of the scientific basis for the management of sperm whales. Working paper ACMRR/MM/121 of the scientific consultation on the conservation and management of marine mammals, FAO, Rome.

Smith, T. D. 1977. Calculation of apparent increases in the Antarctic sei whale population between 1930 and 1960. Rep. Int. Whaling Comm. (Special Issue 1), 337-342.

Smith, T. (ed.) 1979. Report of the status of Porpoise Stocks Workshop, August 27-31, 1979. Southwest Fish. Ctr. Admin. Rep. LJ-79-41, Nat. Mar. Fish. Serv., La Jolla, Calif.

Smith, T. D., and T. Polacheck. 1979. Analysis of a simple model for estimating historical population sizes. Fish. Bull. 76:771-779.

Experiences in the Study of Large Mammals

RICHARD M. LAWS ————————————————————————

INTRODUCTION

This chapter describes research involving a variety of large mammals covering a period of 30 years. It is presented from a personal point of view, but follows significant developments and trends. Attempts are made to draw some general conclusions and relate the experience of this 30 year period to the general field of population dynamics of large mammals.

ANTARCTIC STUDIES

Very early studies of large mammal population dynamics involve the work of the "Discovery" Investigations, financed by the British Colonial Office out of the profits of the whaling industry, at South Georgia in the South Atlantic. A tax on whale products paid the cost of these far-reaching studies, which ran from 1924 to World War II.

Within 5 years of starting the research program, Mackintosh and Wheeler (1929) produced a monograph with the initial quantitative description of the biology of blue and fin whales, *Balaenoptera musculus* and *B. physalus*. Working largely on dead whales, they accumulated considerable biometric data and reproductive material. Their analysis gave the first real insights into the population biology and life history of these species. This work was backed up by a wide ranging program of research on the Southern Ocean that was directed through physical and chemical oceanographic studies, to the phytoplankton and the Antarctic krill, *Euphausia superba*, recognized as the key organism in the food web of the Southern Ocean and the staple food of the baleen whales (Hardy, 1976).

As a result of length-frequency analysis, Mackintosh and Wheeler put the age of sexual maturity of fin and blue whales at 2 years! They were the first to use estimates of the rate of accumulation of corpora albicantia in the ovaries to draw conclusions about the longevity of whales and the age structure of their populations. For the first time the familiar "catch curves" appeared in the whale literature. Other colleagues pursued similar investigations (e.g., Wheeler, 1934; Laurie, 1937; Matthews, 1937). They believed that these whales reached maximum ages of about 30 years, although we now know that the larger whales reached ages of 80 to 100 years. Their work and its subsequent development illustrate two important points — the necessity for reliable methods of age determination and a general tendency for age scales to be stretched as knowlege increases. The very important and fundamental work of the "Discovery" Investigations was written up by Mackintosh during World War II and appeared as a summary paper on the southern stocks of whalebone whales (Mackintosh, 1942).

Two years of personal research (1948-1950) were spent on Signy Island in the South Orkney Islands undertaking work on the southern elephant seal, *Mirounga leonina*, for the Falkland Islands Dependencies Survey (FIDS). Biologists of the "Discovery" Investigations had drawn attention to the summer population of elephant seals in this island group, and in 1947 Gordon Robin, a physicist, had drawn attention to the existence of a small breeding population. During these years at Signy the elephant seals were observed to form discrete harems on the fast ice, affording unusual opportunities for behavioral study compared with the more usual crowded beaches.

The Signy breeding population numbered only a couple of hundred animals. In 1950 these studies were extended to the much larger population (more than a quarter of a million) at South Georgia, where a sealing industry based on blubber oil had been managed since 1910 on a rational basis (Laws, 1953). During the first year of this work a method of aging elephant seals from internal layers in the teeth was developed. Southern elephant seals spend two periods of the year hauled out on land when they fast. One is during the spring breeding season when the bulls spend as long as three months without feeding and the cows spend three to four weeks. The second is in the summer or autumn moult, when they again haul out on land. This is reflected in the teeth by different densities of dentine appearing as concentric rings in cross sections, and layering in the cement (Laws, 1952, 1962b). Studies undertaken a few years later by Australian biologists at Macquarie Island confirmed the validity of the method, which is now widely used, and showed that growth rates and ages at maturity of the stocks differed (Carrick et al., 1962). At South Georgia it was found that female elephant seals produced their first pup at age 3. This was deferred to 4½ to 5 years, at Macquarie, where male maturity was also delayed. The faster growth rates at South Georgia were responsible for precocious puberty and were, in turn, a result of the reduction of the male population by a sealing industry taking adult males. At Macquarie the population was close to carrying capacity, probably determined by food resources at sea near the breeding concentrations.

Seal populations in this area had been overexploited during the nineteenth century, first during the 1820s and later, after some recovery, during the 1870s. In 1910 the Falkland Islands government initiated a farsighted management regime that ranks with the better-known Pribilof Islands industry as a case of good management of a renewable natural resource (Laws, 1953, 1960b); it continued until 1964. Sealing was undertaken as a subsidiary venture by one of the whaling companies. The island's coastline was divided into four divisions, one of which was left unhunted each year. The other three were worked by sealers who traveled around the island on obsolete whale catchers, landing in small boats to take bulls on the beaches. The largest bulls were selected and driven to the water's edge, where they were shot and flensed. The blubber was then taken out to the sealing vessel and back to the whaling station for processing.

Working as a scientist accompanying the sealers in 1951, I was able to make counts of pups and collect material from the harvest. A very simple static model based on the age structure was constructed, and some inferences about the natural age structure and that under the sealing regime were made. The selection of the oldest males had resulted in reductions in male longevity and survivorship. The mean age of the catch was 6.5 years. My research had shown that the bulls reached sexual maturity on the average at 4.5 years, but did not become socially mature until 7.5 years. This finding suggested overexploitation. An examination of the catch per catcher's day's work confirmed that the adult male stock had been declining over the previous two decades (Laws, 1960b).

As a result of this work new regulations were implemented in the 1952 season. A minimum size limit was introduced and autumn sealing was discontinued. The overall quota remained the same, according to the model, but was subdivided proportionately to the estimated stock in each division. The objective was to raise the mean age of the catch to about the age of social maturity and to ensure a sustained yield.

On introducing the new regulations there was an immediate increase in the catch per unit effort, an increased yield of oil, and a shorter sealing season. A system of monitoring the catch for age distribution by means of a 5% sample of teeth was established; the mean age of the catch increased steadily and then stabilized. The rotation of divisions was later ended and the annual catch taken from all the divisional stocks in proportion to their abundance (Laws, 1960b, 1979). This was one of the first pinniped models. It was successful because the exploitation system was simple, and the fact that the mean age of the catch stabilized at 7.7 years suggested it was valid. (Even now it is not easy to validate models by testing predictions.) Experimental adjustment of the catches for further tests would have been interesting and possible, but the industry came to an end in 1964 when the South Georgia whaling industry ended.

WHALE RESEARCH

In 1953 my attention turned more directly to whales in work with the National Institute of Oceanography in Britain. Whaling with factory ships was quite different from the whaling at South Georgia of the early 1950s. During a five-month season the expedition took some 2500 whales. These included sperm, *Physeter catodon*, humpback, *Megaptera novaeangliae*, blue, fin, and sei, *B. borealis*. Individual carcasses were dismembered and processed, to collect ovaries or testes, establish the state of fusion of the vertebrae, and record other information (Laws, 1961).

Despite the early advances made by the "Discovery" Investigations team there was still no perfected method of age determination. The state of the stocks was assessed from changes in mean length of the catches, the percentage of the catch sexually mature, and other similar indices. At that time the fin whale was the major species in the catches. The catch per unit effort (CPUE) for fin whales had been increasing, but at the same time catcher tonnage and efficiency had been improving. Later, the CPUE began to level off, even though catcher efficiency continued to increase, and it was evident that it was going to decline. This was in the late 1950s. The CPUE fell in the succeeding few years, and it became obvious that the stocks were being grossly overexploited (Laws, 1960a, 1962a).

The best method of age determination in the 1950s was from the ridges on the baleen plates, a method pioneered by Johan Ruud in Norway (Ruud, 1945). Because of wear at the tip, the method did not apply to more than the first few year classes and, as we now know, each of the initial baleen groups incorporated several year classes. Thus, the life span of even the longest-lived whales was still thought to be relatively short — perhaps 30 years. At that time, however, the age at maturity was believed to be somewhat higher than 2 years, perhaps 4-5 years. The first indication that whales were limited by food appeared when Ruud's baleen plate data were examined to show that the percentage mature in baleen group III had increased between 1945 and 1955. At the same time the mean length at sexual maturity had remained constant, and the inference was that fin whales were growing faster and reaching maturity earlier. This was probably due to the reduction of the stocks, first of blue whales, followed by fin. The density of available food (krill) in the sea was probably higher, and the whale stocks were adjusting to this change (Laws, 1962a). There was also evidence that the pregnancy rate was increasing (Laws, 1961).

Work on fin whale reproduction confirmed that ovarian corpora albicantia did persist throughout life, accumulating at a constant rate. The age structure of catches could be compared over time within areas and indicated that there had been selective fishing of older animals (Laws, 1961). The general approach of the work and the models used were still somewhat crude, although they represented an advance.

Baleen whales migrate between the winter breeding grounds in the tropics and the subtropics and the summer feeding areas in the Antarctic, where they graze on krill, following the ice edge as it retreats southward. Because the seasonal events are so marked, the reproductive cycle is matched to the annual rhythm. In the baleen whales this involves an acceleration of the growth of the fetus when the mother begins to feed on krill (Laws, 1959). Such an acceleration seems to be unknown in other mammals. Another interesting pattern that emerges is that lengths vary with longitude in a regular way. The largest individuals migrate into the Antarctic earlier in the season than do others, taking up positions on the most productive parts of the feeding grounds. Smaller, and younger individuals, appearing later, are apparently displaced to the east and west. It seems likely, therefore, that there was competition on the feeding grounds (Laws, 1960a).

One of the most important events in whale research was the almost accidental finding that the ear plug could be used to age baleen whales (Purves, 1955). The ear plug had been described by Lillie (1911) as an interesting anatomical feature, characteristic of baleen whales, but not of toothed whales. A presumed moult cycle results in the accumulation of layers of epithelium and wax in the ear passage. Although we postulated an annual accumulation rate of two full layers, later work showed that only one layer forms each year. When ear plug layer counts were plotted against ovarian corpora numbers, the earlier method (for mature females) was validated (Laws and Purves, 1956). Thus corpora albicantia age series from earlier years could be used.

Layers formed in the ear plugs of immature whales are thicker and more irregular than those formed in adults. Lockyer (1972) described how a transition zone occurred and could be used to calculate the age at maturity of different year classes. Through this work she showed that whales were maturing earlier than in former years. Fin whales showed evidence of an advancement from maturation at 10 years of age in the 1930s to 6 years in the 1950s. This must have been associated with an acceleration in the growth rate, because the mean length on attainment of sexual maturity had not changed. This confirmed earlier suggestions and was later extended to the sei and minke, *B. acutorostrata,* whales.

Studies as described above may now seem rather oversimplified, but they provided the basic biological foundation for later work. Accurate age determination is fundamental to investigations of population age structures and to the estimation of age-specific parameters. The details of the reproductive process in whales and knowledge of the ways in which variations in the reproductive rates are achieved are important. Without this knowledge, population models would be far from reality and we would not be as near to understanding large whale population dynamics as we are today.

HIPPOPOTAMUS

Work on hippo provided a very good point of entry to complex tropical terrestrial ecosystems. My work on hippo, *Hippopotamus amphibius,* was carried out with the Nuffield Unit of Tropical Animal Ecology (NUTAE) in Uganda, initially in the Queen Elizabeth National Park in Western Uganda, now called the Rwenzori National Park. It had the highest known biomass density of large mammals in the world at 21,000 kg/km², compared with 19,000 kg/km² for the next highest, in Lake Manyara National Park in Tanzania (Leuthold and Leuthold, 1976). American Fulbright scholars (notably George Petrides, Wendell Swank, and Bill Longhurst) had recommended a management cropping scheme to reduce overgrazing. A cropping program began in 1957 and provided opportunities to collect quantitative information on age structures, growth, reproduction, nutrition, and disease. The hippo is central in the food web, and its population was larger than could be sustained by its resources. An obvious approach was to study its ecological relations and the dynamics of its populations, and to try to determine the optimum grazing density for this area.

The hippo spends the day kneeling or lying in the water, expending little energy, and at night follows trails, marked by fecal deposits, inland to close-grazed areas of short grasses — the hippo lawns. It feeds by plucking grass with its massive lips, as much as 50 cm wide. The grazing area extends about 5 km from the shore on the average, except where there were wallows that enabled it to penetrate further inland and enlarge the range. The highest-density populations in the world were found in the lakes and rivers of the western Rift Valley. Maximum grazing densities were formerly over $31/km^2$, equivalent to a stocking rate of 62 steers/km^2, which is very high indeed. They created an energy sink from land to water, removing organic matter by grazing and fertilizing lakes by defecating in the water (Laws, 1968a). Because of this, the fishery on Lake George had one of the highest fish production rates per unit area of any lake in the world.

An experimental management scheme, aimed at maintaining a range of grazing densities by regular cropping and monitoring by counts, was set up. The maintained grazing densities ranged from 0 to $23/km^2$. Changes in the habitats were followed by studying vegetation transects, quadrats, and exclosures to see how the grasslands were influenced by different hippo grazing pressures. The reduction of the hippo populations was followed by analysis of those cropped to obtain estimates of population parameters such as ages at maturity, pregnancy rates, growth rates, and mortality rates, as well as nutrition and disease. Each year about 1000 hippo were shot in the water, then hauled out on land and examined. A method of age determination based on tooth replacement and wear was developed (Laws, 1968b). The population size was monitored by aerial and boat counts. Unfortunately the method of cropping was very biased, as the larger animals are easier to shoot, but jaws from natural deaths provided an independent indication of the population dynamics. The life span of the hippo is 45 to 50 years.

An instructive example of one of the experimental areas was the Mweya Peninsula ($4.4\ km^2$) where, in 1957, there had been 90 hippo. At that time it was largely bare ground except for some bush thickets. By 1963, following the removal of hippo, the grass had come back. At the same time the numbers of other grazing species had increased substantially (from 40 to 179), and the total standing stock animal biomass had also increased by 7.7%. The buffalo *(Syncerus caffer)* population growth closely followed that of a logistic curve.

The study lasted for about five years and established that in the Queen Elizabeth Park the optimal grazing density for hippopotamus was about $8/km^2$. As the cropping reduced the overall population size, the age of sexual maturity declined from 12.1 to 9.7 years, the pregnancy rate increased slightly, and the proportion of calves increased from 5.9 to 14.0%. (Laws, 1968; Laws and Clough, 1965).

Studies in Murchison Falls National Park (later renamed Kabalega National Park) showed that a hippo population of about $19/km^2$ was associated with substantial erosion along the banks of the River Nile. A similar cropping scheme there (in association with Wildlife Services, Ltd., run by Ian Parker) also produced a mass of high-quality data. Some 4000 hippo were taken on land at night, because the strong currents ruled out shooting in the river.

Unfortunately, owing to the identification in 1964 of more pressing problems related to elephant overpopulation (described in the next section), only part of the results of the hippopotamus studies have been published.

Field and Clough joined in the hippopotamus investigation and worked on feeding habits, digestibility of the vegetation, and the effects of grazing on the vegetation (Field, 1966, 1968a,b) and on reproduction (Clough, 1966). Later Lock undertook a study of the grasslands in relation to soil types and grazing by hippopotamus (Lock, 1967), involving the construction of experimental enclosure plots.

Although the work began on a narrow front with in-depth studies of hippopotamus ecology and management, the ultimate objective was to study interrelationships of the important components of tropical ecosystems as well as the biology of key species. Nearly 70 scientific publications resulted from research undertaken at NUTAE during the first five years. These and publications resulting from subsequent studies are listed in an account of the 10 year program of NUTAE by Beadle (1974). After 1971 the unit became the Uganda Institute of Ecology and is still in existence.

Other research at NUTAE on large-mammal population ecology, concerned elephants (described below), buffalo (Grimsdell, 1968, 1969), the waterbuck (Spinage, 1967a,b; 1969a,b; 1970), and large herbivores in general (Field and Laws, 1970). Aerial counting, immobilization, marking, and telemetry techniques were used, and pathology and parasites of the animals were studied (Cowan et al., 1967; Kangwagye, 1968; Plowright et al., 1964; Thurlbeck, 1965; Thurlbeck et al., 1965). Work on small mammals was initiated by Delany and Neal (Delany, 1964; Delany and Neal, 1969; Neal, 1967, 1970; Neal and Cook, 1969). The significance of these studies in relation to the ecosystem as a whole is summarized by Beadle (1974).

ELEPHANT BIOLOGY

During the 1960s the elephant, *Loxodonta africana*, populations in several regions of Africa were giving cause for concern. Among key populations were those in Murchison (now Kabalega) Falls National Park in North Bunyoro, Uganda and in the Tsavo ecological unit, Kenya, respectively, in high and low rainfall areas. Results from studies in these area are presented in Laws (1969a,b, 1970), Laws and Parker (1968), and Laws et al. (1975), where information presented below is documented.

In 1947, Eggeling, a forester, had been the first scientist to discuss the developing elephant problem in Uganda (Eggeling, 1947). During the early 1950s, Perry (1953) almost single-handedly undertook the first seriously scientific study of elephants in Africa, working under great difficulties. Later, the Fulbright scholars Buss and Buechner worked in Murchison Falls National Park, carrying out aerial counts, vegetation studies, elephant observations, and some limited sampling (see Laws et al. 1975 for review). Further research on elephants began with the collection of jaws representing natural deaths and led to a paper on determining the age of elephants (Laws, 1966). This work permitted quantitative studies on the age structure of elephant populations along with their age-specific growth and reproduction.

The problem in North Bunyoro had its origins in the sleeping sickness and evacuation of local people that took place during the first decade of this century. Elephants moved into the area that had been vacated, and subsequently, as the surrounding human population increased, increasing numbers of elephants were compressed into the area centering on what is now the Kabalega Falls National Park. The main concentration of the elephant range occurred from 1929, when most of Uganda was open to elephant occupancy. By the early 1970s they were virtually confined to the National Parks.

The concentration of elephants in limited areas led to a buildup in their densities, even though absolute population size was decreasing. Much of the park area was woodland, and its destruction was initiated by elephants that debarked and killed trees, opening the leaf canopy for invasion by inflammable grasses. Grass fires thus penetrated the woodland and accelerated its decline. By the 1960s most of the southern half of the park was converted to open grassland; the northern sector was rapidly changing from woodland to grassland. South of the park lay the Pabidi and Budongo Forests, the latter a valuable source of commercial timber, into which the peripheral elephant herds moved during the dry season.

In 1964 the park authorities agreed to proposals for reducing elephant densities by cropping so as to allow the habitats to recover. In two years 2000 elephants were "cropped" and examined in a carefully planned operation involving the Nuffield Unit and Wildlife Services, Ltd. With scientific supervision, the cropping was done by Ian Parker and colleagues. Later, in Uganda, Ron Johnstone worked with us on various aspects of the elephants' relation to forests. This close collaboration led eventually to the amassing and analysis of data on six separate elephant populations in Uganda, Kenya, and Tanzania. Our book published in 1975 is mainly concerned with two of these in Uganda (Laws et al. 1975); the rest still remains to be written up properly.

Before we began elephant cropping in 1965, most people believed it was impossible, because in other hunting schemes the elephants had become very wild. We avoided such problems by taking complete family units. After the initial experiments had perfected the method, two hunters were able to drop a group of up to 29 elephants in under 2 minutes. There were no survivors to alarm the rest. It was repugnant but, we believed, necessary work, carried out humanely. Because the method was nonselective, the population could be reduced without altering its structure, and the samples were much more valuable scientifically than any collected before. African assistants dismembered the huge carcasses with knives and axes, each trained to specialize in the collection of a particular sample. Most of them were Waliangulu, the "elephant people" (Holman, 1967). The animals were sexed, measured, weighed, and aged to give the basic population and growth data. Additional collections related to reproduction, nutrition, physiology, biochemistry, pathology, and tusk growth. In only three years we had collected data on more than 3000 elephants. (In 1966 we took a sample of 300 elephants in Tsavo National Park, Kenya, and in 1967 and 1968 two samples of 300 each from Mkomasi Game Reserve, Tanzania. There was also a small sample from Forest Department control shooting in the Budongo Forest Reserve.)

Such cropping was based on the fact that there are two fundamental social units comprising elephant populations. The first is the family unit, which is a cohesive group consisting usually of a matriarch, her daughters, and their off-spring. The second social unit is the bull herds, which are loose temporary ag-gregations of often unrelated males. Such organization had been suggested by workers such as Buss. With more complete data I was able to construct kinship diagrams showing putative relationships within the family unit, taking into ac-count age, reproductive status, and placental scars, which represent pregnan-cies. These conclusions were later confirmed by Ian Douglas-Hamilton, as a result of behavioral studies of family groups that he was able to recognize from individual natural characters (Hamilton and Hamilton, 1975).

Initially, based on the age structure of the first 400 elephants examined, 200 north and 200 south of the River Nile, several simple models were developed. These indicated that in both populations recruitment was declining, beginning in about 1946 on the South bank and about 1957 on the North bank (Laws and Parker, 1968). Using such approaches we made predictions about future trends. Fowler and Smith (1973), using our data and a Leslie-matrix model, came up with very similar conclusions.

It was shown that the age at sexual maturity in elephants was very plastic and was deferred in unfavorable situations associated with elephant-induced changes in habitat. The mean ages at sexual maturity ranged from 12 to 23 years in several populations; individual animals were reaching maturity at from 8 to 30 years. This is the widest range of variability in this character that has been reported for wild mammals and clearly, if caused by nutritional conditions or population density, it would be an important mechanism in population adjust-ment to changed conditions. As mentioned earlier, baleen whale populations have responded by increased rate of growth and by an advancement in the age at maturity, to increased food availability. Size rather than age determines the at-tainment of maturity. The elephant's social structure and reproductive cycle ap-pear to be very similar to those of the sperm whale. In discussion at the Interna-tional Conference on the Biology of Whales in 1970 (Schevill, 1974) it was sug-gested that the established elephant parameters could be used in constructing models for this species of whale until it is better known.

From the cropping of elephants we also obtained information on other age-specific characters, such as the proportion of females pregnant, lactating, or in anestrus. Placental scars, used for the first time for an ungulate, gave indepen-dent estimates of the pregnancy rate (averaged over the reproductive span) and of the mean age at first parturition for several populations (Laws, 1967; Laws et al., 1975). Mean calving intervals ranged from 3 years in a very productive population to 9 years in the least productive. Individual females had as many as 11 placental scars in their uteri. We found that an increased proportion of females were reproductively inactive above 50 years and in the least productive population, all aged more than 55 years were postreproductive. This is equivalent to the menopause in the human female and the associated selection for survival of grandmothers who help care for the young and, by inference, reduce infant mortality. Although reproductive inactivity at that end of the life span has a negligible effect on natality, it was interesting that those populations in which sexual maturity was deferred also showed the highest proportion of postreproductive animals; the age-specific pregnancy rate was reduced at both ends of the age scale.

Age structures of the sampled populations showed a series of peaks and troughs that appeared to correlate with rainfall cycles (see Chapter 19). It now seems that this pattern may be related to deficiencies in the allocation of precise chronological ages and needs further attention. When lumped into five-year classes, however, these fluctuations largely disappear, and it is possible to estimate adult mortality rates assuming a stable age structure. In fact, the adult survival rates calculated from regressions for the several populations were quite close, at 94 to 96%. Simple models were developed in terms of numbers, biomass, growth increment, and biomass transfer for an initial population, one in early decline and one in advanced decline.

In the South bank population at Murchison Falls National Park it was clear that the bulk of the population biomass was provided by old animals, with relatively high survival rates — age classes that would take some decades to work through and out of the population structure — and recruitment was very greatly reduced.

The Murchison Falls National Park area was about 3900 km², with counts that indicated a population of about 14,000 elephants. Murray Watson carried out aerial photographic transects that clearly showed how the density of mature healthy trees increased toward the edge and outside of the elephant range. Dead trees were at peak densities on the periphery, with largely open grassland over most of the heavily occupied range (Laws et al., 1975). This indicated that the habitat conversion began centrally and spread radially as a zone of damage.

In 1967 work began in cooperation with Watson and Goddard at the newly formed Tsavo Research Project funded by the Ford Foundation in Tsavo National Park, Kenya. In addition to elephants, this involved black rhinoceros ecology (Goddard, 1970). Earlier, in 1964, with the cooperation of David Sheldrick, the park warden, the collection of elephants and rhinoceros jaws from skeletons and carcasses had been initiated. We also had material from 300 elephants cropped in 1966. We marked and began to photograph a series of aerial photographic transects for vegetation studies, and we began a series of aerial reconnaissance flights over the National Park and surrounding areas (about 39,000 km²).

The reconnaissance flights showed that there was a series of 10 regular aggregations totaling about 33,000 to 40,000 elephants, which maintained a fairly constant pattern throughout the period. We concluded that for management purposes these might be considered as unit populations, although we had no indications of genetic isolation. Just as in Murchison Falls National Park, these groups appeared to have different age structures, group sizes, and differing proportions of bulls and young calves. Furthermore, the groups maintained their integrity through wet and dry seasons. From the earlier cropping, an age/length key was constructed. We measured back lengths on aerial photographs and obtained age structures (Laws, 1969b).

The aerial photographic transects and aerial surveys showed that annually 6% of the *Commiphora* bush and 2 to 4% of the long-lived baobab, *Adansonia digitata,* trees were dying. The bush was rapidly subsiding, and wide expanses of grassland developing. Plains game were more easily seen, aggregating in herds that were sometimes quite large. This was taken by some to indicate that the plains game species were increasing, although there was no good evidence for this, and the observed changes could just as likely be due to the opening up of a closed bush habitat without associated changes in population sizes.

It was a period of higher-than-average rainfall, and the park looked generally lush and green. It was pointed out that drier conditions were to be expected, and it was predicted that a serious crisis would develop, probably in the early 1970s with mass mortality of elephants. Critics claimed that the 1960-1961 drought was the worst in living memory, but even so not many elephants had died. They maintained that it was unlikely to recur. When it was proposed, as part of the research program, that up to 1500 elephants be cropped in the park (in scientific samples from five of the putative unit populations), a political fracas developed which made it impossible to continue. In concluding this work, populations in the Mkomasi Game Reserve, Tanzania (part of the Tsavo ecological unit) were analyzed by means of sample cropping. Detailed stratified random aerial transect counts confirmed both the integrity of populations and the estimates of their size on the basis of the reconnaissance survey flights.

The possible existence of 10 unit populations in 1967 has been criticized on the basis of radiotracking work carried out in 1972 (Leuthold and Sale, 1973). This work showed that 10 individuals moved across the boundaries of several of the supposed unit populations. Such work hardly disproves our hypothesis, but it does indicate that at least some minimal interchange may occur among the populations. This may be especially true under conditions described below.

The predicted, massive die-off of elephants occurred during 1970-71 (Corfield, 1973). An extrapolation to the larger area of the ecological unit gives rise to the conclusion that an estimated 10,000 elephants died. The distribution of deaths was rather similar to that in 1960-61 (Glover, 1963), but the densities were much higher, reaching 4 carcasses/km² in some areas.

The highest densities of dead animals were associated with water sources, but the deaths were due to malnutrition, not shortage of water. Year after year since the successful antipoaching campaign in 1957 the elephants became more concentrated along the rivers and around water holes during the dry season (Laws, 1969b). When the poachers were active, hunting with poisoned arrows, they operated in the vicinity of water, and elephants did not concentrate there. However, the browse that elephants need to carry them through the dry season, when there is no other food, was progressively destroyed, and the boundary between relatively natural bush and the zone denuded of bush was pushed further back each season.

The work of Corfield (1973) together with earlier work (Laws, 1969b), throws light on the selectivity of the deaths. Corfield showed that the deaths were mainly young animals, mature females with calves, and old animals. The adult male population was relatively unscathed. The age structure was greatly modified, and in the worst areas all the calves died. This must have influenced the social structure and behavior. There would be pressure to roam more widely in search of food. The family units were broken up, and surviving females were not tied to local areas by the need to remain with their calves. Thus one might expect that the survivors would be more mobile, the home ranges larger, and the supposed integrity of unit populations lost. Contrary to the picture developed by Leuthold and Sale (1973), it seems highly probable that the conditions of 1970-71 and later were abnormal.

Outside the park boundary, conditions were relatively unchanged. Inside, large tracts were converted to desert or semidesert with a pall of dust rising to several thousand feet. When the rains came rapid runoff removed even more topsoil. The predicted catastrophe had materialized. Despite all these events there are those who still maintain that what we are witnessing is part of a natural cycle that has happened before.

It has been suggested that calf mortality is the most likely method of density-dependent population control (Hanks and Mackintosh, 1973). But it is not enough to show that a specific factor has the potential of primary importance in population regulation. What is necessary is to show that it actually operates in natural situations. Evidence indicates that the mortality rates assumed by Hanks and Mackintosh (1973) are probably unrealistic. Their models show that "an annual population increase of 4% would be close to the maximum value." Hall-Martin (1977) has reported on the increase of the population of elephants enclosed in the Addo Elephant National Park where numbers increased from 18 in 1954 to 81 in 1976, an annual rate of 7%.

Under natural conditions population regulation seems more likely to occur through adjustment in ages at first maturity and pregnancy rates, which have been shown to change in other large mammals (Chapter 23). The elephant represents an extreme example of a K-selected species, with a 22 month gestation period and a large parental investment. Thus it seems inconceivable that elephant populations would be regulated by calf mortality, since adult mortality is variable only within narrow limits. In view of the wide range of adjustment possible in mean calving interval (3 to 9 years) and in the attainment of maturity (12 to 23 years), it seems more likely that reproduction is the actual regulatory process. The compression of elephants into parks, caused by human activities, creates artificially high densities, leading to rapid habitat change and malnutrition. The resulting deaths, not being one of the natural regulatory mechanisms, are highly unnatural.

In Murchison Falls National park the change was cushioned by a higher rainfall. The process was similar, but the population crash was slower and less catastrophic than in the dry Tsavo region (Laws et al., 1975).

It is difficult to see how Tsavo can recover rapidly (as suggested in Chapter 15) even though poaching has further reduced elephant densities. Topsoil has been blown away in the dust clouds or washed away in the floods. Future successive droughts are likely to result in further destruction of browse and further deaths. In 1969 a conceptual model based on the logistic equation was used to describe elephants (Laws, 1969b). It showed that if K, the carrying capacity of the environment, were to be reduced (e.g., either by lower rainfall or destruction of the vegetation) at a rate faster than the elephant population could adjust to the change, there could be a phase when the elephant densities were in excess of carrying capacity, when the rate of habitat destruction would accelerate. The very high survival rate of adults contributes to this (see Chapter 16 for a similar pattern among predators). Corfield (1973) suggested that the catastrophic mass deaths associated with the drought would bring the population below K. It seems likely that before the process of dying reduces the population the habitat is changed and the value of K is further reduced. It is highly likely that the next major drought will result in another die-off in Tsavo. Potential salvation lies in the increasing elephant-poaching activities in the 1970s, which took the place of the recommended rational cropping and reduced the elephant population substantially.

Of course, one cannot be totally justified in assuming a stable population. There is evidence from some areas to show that over the past few hundred years elephant populations may have fluctuated, perhaps within a period of 100-200 years or so, quite significantly (Caughley, 1976; Laws et al., 1975). However, over large regions of Africa there is no longer room for elephants to move about and adjust their densities to changing vegetation patterns. Their ranges are being further compressed every year. The situation has reversed from one in which human islands existed in a sea of elephants, to a sea of people with elephant islands. Any such natural cycles have been interrupted and, in the absence of management, one by one the environments in quite large areas have become trapped in the trough of the cycle, which has then been pushed ever deeper by sustained overpopulation.

Paradoxically, the ivory trade may have done some good through wasteful and inhumane destruction of elephant populations. In Uganda, which had accepted the need for wildlife management, the change in government and the disturbances that followed have led to the decrease of the elephant populations in Murchison Falls (Kabalega) National Park, from 14,000 to 2000 (Eltringham and Malpas, 1980).

In Murchison Falls National Park, experimental exclosure plots, constructed in 1965 and 1967, already had 10 m high *Acacia* trees, eight years later, which show up as dark rectangles in the open grassland when viewed from aircraft several miles away. The reduction in elephant densities brought about by poachers has resulted in similar more general changes. If the population reduction can be halted, if the populations and habitats can be monitored, and if further management, including cropping, can be undertaken in the future, it should be possible to restore and maintain diverse habitats. Whether the political situation will permit this is another question. In Tsavo, with its lower rainfall, the environment is more vulnerable and the effects of earlier nonintervention more serious. It may not be possible to restore the former diversity and faunal assemblage because, as is suggested in Chapter 19, the vegetation standing stock may itself be dependent on the herbivore activity.

RETURN TO THE ANTARCTIC

Returning to the Antarctic in 1969, the British Antarctic Survey (BAS), formerly FIDS, provided opportunities for further work on marine mammals. In October 1977, for example, an international expedition was organized by Don Siniff on the R/V *Hero*, involving seven scientists of four nationalities. The study (in the pack ice) was directed toward the breeding behavior and biology of crabeater seals, *Lobodon carcinophagus*. Formerly this species was thought to breed in large aggregations like the harp seal, *Pagophilus groenlandicus*, but work by Oritsland (1970) and Corner (1971) had indicated that the breeding unit was a family group made up of a female, a pup, and an adult male. When the pup is weaned the adults form a mated pair. Recent work has confirmed this through larger samples wherein the species has been studied using modern techniques of immobilization and radiotelemetry.

During earlier studies an aging method for crabeater seals was developed, based on the dentinal annuli of the canine teeth collected at the site of a mass die-off in Prince Gustav Channel, Antarctic Peninsula in 1957 (Laws, 1958). Eighty-five percent of these seals were dead or dying, due, it was thought, to a virus infection (Laws and Taylor, 1957). Later, in 1973 several hundred specimens, killed for dogfood by BAS, were aged using the annual cemetum layers in thin sections of postcanine teeth. Maximum age was 35 years and the "catch curves" showed a very high mortality rate during the first year (Laws, 1977b). Later collections have increased the sample size to about 1500, covering 11 years. Some 85% of the adults bear scars now known to be caused by leopard seals, *Hydrurga leptonyx*. The 1977 expedition confirmed that weaned crabeater pups are killed and eaten by leopard seals in large numbers. Toward the end of the first year the pups begin to develop enough agility to escape the leopard seals, but are scarred in the process (Siniff and Bengtson, 1977).

Comparison of crabeater and Weddell seal, *Leptonychotes weddelli*, strategies shows that in contrast to the crabeater (with its high initial mortality, longevity, and nongregarious behavior), the Weddell seal forms pupping aggregations (the males occupying three-dimensional subice aquatic territories), and experiences low predation but doesn't live long due to tooth wear incurred in keeping open breathing holes in the fast ice (Kaufman et al., 1975; Laws, 1977b; Siniff et al., 1977; Stirling, 1971).

Studies of whales and seals have an important place in the study of the Southern Ocean ecosystem. In particular, as will be shown, development of research on crabeater seal population dynamics is important for monitoring changes in the Antarctic ecosystem. Fuller accounts will be found in Everson (1977), Laws (1977a,b), and in Chapter 17, in which the facts given below are substantiated.

The seas south of the Antarctic Convergence are very productive and constitute an extensive ecosystem containing unique faunal assemblages, characterized by unusually short food chains and large populations of relatively few species compared with other marine areas. The Antarctic continent is surrounded by a belt of pack ice that waxes and wanes seasonally from a maximum area of 22 million km^2 in late winter to a minimum of 4 million km^2 in spring. This has far-reaching effects on the ecosystem, particularly on phytoplankton production, provides a platform for seals and whales, and is a limit to the range of some whale species.

A single zooplankton species, *Euphausia superba,* the Antarctic krill, is the key organism in the Antarctic marine ecosystem, although there are several other species known collectively as krill. The zooplankton standing crop in Antarctic waters is signficantly higher than in tropical and temperate regions, and *Euphausia superba* probably represents half of this. Together with the high summer production, caused by a high growth rate of phytoplankton, the pack-ice contraction leads to a summer increase in krill abundance in ice-free surface waters amounting to perhaps a hundredfold. Although the krill is circumpolar in distribution, there appear to be several areas of concentration that correspond in general to a series of eddies or gyres in the oceanic circulation. Whether or not these krill stocks are separate is of crucial importance to the way in which the resource is managed.

Another important group in this ecosystem is the Cephalopoda, mainly the oceanic squids. There may be about 20 species occurring in large numbers and as major consumers of krill. Very little is known about them. Although some 100 species of fish occur south of the Antarctic Convergence, in contrast to other oceans the Southern Ocean does not contain dense shoals of pelagic fish. Many of the fish feed on krill, but the amount they consume is not known. The breeding colonies of birds are concentrated on ice-free ground on a few small islands, mainly between 50°S and 65°S. The total bird biomass is at least 500,000 tons. They consume at least 20 million tons of krill, 8 million tons of squid, and 8 million tons of fish annually; 90% of this is taken by penguins.

But the seals and whales comprise by far the most important groups of krill consumers. There are six species of seals in Antarctic waters. Two of them, the fur seal, *Arctocephalus gazella,* and the elephant seal, *Mirounga leonina,* which are mainly sub-Antarctic in distribution, were brought to near extinction by sealers in the nineteenth century. The stocks are still recovering and are the subject of research by BAS (Laws, 1960b; McCann, 1980; Payne, 1977). The fur seal at South Georgia has increased from a few hundred in the 1950s to an estimated 350,000 today, at an annual rate of 17%. At this rate it could regain its original numbers of perhaps 3 to 4 million before the end of the century (Payne, 1977). McCann (1980) compared the present situation of the South Georgia elephant seal population with earlier studies during the exploitation phase and with the Macquarie Island situation in the 1950s (long after sealing had ended). He found changes in the time bulls haul out, number of bulls ashore, cow/bull ratio, harem size, and the age of harem bulls. All such changes can be attributed to the ending of exploitation. The structure of the female portion of the herd has not changed appreciably during the same period.

In quantitative terms, however, the fur seal and elephant seal are relatively unimportant in the dynamics of the Southern Ocean ecosystem. The pack-ice seals and the baleen whales are much more significant. The four ice-breeding seals show an adaptive radiation based on feeding habits. One of them, the crabeater seal, numbering at least 15 million, is the most abundant species of seal in the world, comprising half the total world populations of seals and much more than half the biomass. There is a well-defined ecological separation between the species, which tend to haul out on different types of pack ice and to feed at different depths on different food species. The estimated populations of all four seals total 17 million (although this could be a gross underestimate), with a biomass of nearly 4 million tons, eating about 70 million tons of krill (some 50% more than current estimates of consumption by whales), 6 million tons of squid and 8 million tons of fish.

Five species of migratory baleen whales are present in Antarctic waters during the summer, feeding on krill for an average of 120 days. These whales (and the sperm whale which feeds on squid) have been the object of an industry since 1904, when their numbers are estimated to have been about 1.1 million, with a total biomass of about 45 million tons, eating each year an estimated 190 million tons of krill, 12 million tons of squid and nearly 5 million tons of fish. By 1973 their numbers had declined through overhunting to about 0.5 million of mainly smaller species, weighing an estimated 9 million tons and eating about 42 million tons of krill, 5 million tons of squid, and 130,000 tons of fish. In fact these estimated weights of whales and the food they consume are probably too low because growth rates have increased (see below). Even so the enormous reduction of the whales means that possibly 100 million tons of krill formerly eaten by whales have become available to the remaining whales and to other consumers. It has been suggested that some of this "surplus" could be safely harvested by man while allowing the depleted populations of whales to increase to a reasonable level. (An indication of the potential commercial importance of the krill resource is that the total annual yield from all the world's fisheries is about 70 million tons.) Evidence was presented earlier that in the past competition for food limited the numbers and body size of whales (see Chapter 17). As their numbers were reduced by whaling the amount of available food presumably increased and was reflected in increased growth rates, earlier maturity, and increased pregnancy rates.

Just as the whales show responses to increased food availability, so do the birds and seals. Several penguin species, which can easily be counted on their breeding colonies, have shown substantial increases in numbers, particularly at South Georgia, in the Scotia arc and the Antarctic Peninsula.

The crabeater seal, like the baleen whales experienced advancement of the age at maturity from about 4 years in the 1930s to about 2½ years in the 1970s. As in the whales, this increased reproductive rate is presumably the result of increased food availability improved nutritive condition and increased growth rates. It would be expected to lead to an increase in numbers. Further studies, currently under way on the crabeater seal samples collected by BAS since 1966, will establish whether the pregnancy rate has changed as predicted. The 17% recovery rate for the Antarctic fur seal, the females of which feed on krill during the breeding season, is unusually high—for other marine mammals 8 to 10% is the rule (Laws, 1979)—and is probably related to the increased abundance of krill accompanying the reduction of the whale stocks.

The central position of krill in the Antarctic ecosystem is clear and so, too, are the dangers to the natural consumers of krill, such as an uncontrolled commercial exploitation of krill, which has now begun. Catches reached 125,000 tons by 1977/78. Catches are likely to increase rapidly over the next 5 to 10 years, and it is likely that at some point man as a competitor will begin to affect the populations of squid, fish, birds, seals, and whales. If the natural predators have increased to fill the potential gap left by the reduction of the initial whale stock it is at least possible that the available krill is already being taken by them.

If this prediction is correct, then the population regulatory responses of seals and whales discussed earlier should now be undergoing a reversal as density-dependent checks come into play. Growth rates should have begun to slow, age at maturity to be deferred and pregnancy rates to decline. It is no longer possible, however, to sample the whale populations adequately, because most species are now protected. On the basis of present knowledge it seems probable that blue whales, minke whales, and crabeater seals are in a special competitive relationship, because they all feed on krill and their distributional patterns are similar. It has also been suggested that before exploitation by man began the blue whale may have had a competitive advantage over minke whales and crabeater seals. Because it was the initial target of exploitation, the blue whale has been hit harder than other species and its recovery under protection is less sure and very difficult to monitor.

Monitoring of crabeater seal populations for density-dependent responses, using the methods already developed, then becomes important. They mature at a relatively low age compared with the baleen whales; they appear to consume more krill now than the baleen whales as a whole; and so could provide an early warning of changes in the ecosystem. Carefully planned crabeater seal samples (say, $n = 300$) could be taken from a circumpolar range of key localities and analyzed for changes in growth rates, age at maturity and pregnancy rates. Depending on the results of these studies, an appropriate management tool to ensure the recovery of the blue whale might be large scale culling of Antarctic minke whales and crabeater seals.

If a large-scale krill industry develops, the additional consumption of krill by man could drive the krill population down, with very serious consequences. This is probably the greatest single conservation problem which we face. The exploitation of the food base of the Antarctic vertebrates may have unpredictable effects on the recovery rates of the large whales. The developing industry should be managed with due attention being paid to the well-being of the ecosystem as a whole. We must add to knowledge by research programs designed to produce the necessary information. A comprehensive long-term international program of Biological Investigations of Marine Antarctic Systems and Stocks (BIOMASS) has been formulated. Research is beginning. In 1972 the Scientific Committee for Antarctic Research drew up the Annex to the Convention for the Conservation of Antarctic Seals. This Convention came into force in 1978 and represents a turning point in fishery conservation agreements because it was concluded before any industry developed. It also paved the way for a later development. Since 1977 the Antarctic Treaty Nations have held several meetings to draw up for signature another Convention establishing a definitive regime for the conservation of Antarctic marine living resources (most of which are not adequately protected under existing agreements). Most fisheries conventions are based on a few species with little regard for their interactions with other species and with the environment. The Convention on the Conservation of Antarctic Marine Living Resources (1980) provides in its wording for an ecosystem approach and is a significant breakthrough. The problems of achieving a multi- species convention based on an ecosystem approach are formidable, but they must be solved if the unique Antarctic marine ecosystem is to survive in a recognizable form — with its impressive assemblage of birds, seals, and whales — while contributing to the nutritional needs of the human race. Studies of the population dynamics of the whales and seals are essential to an understanding of the problems and to their solution.

CONCLUSIONS

It is clear that in important respects marine ecosystems are fundamentally different from land ecosystems. They are more continuous, with fewer barriers. Whales, for example, can migrate rapidly and easily across 60° of latitude. They have much more equable climates. Nutrients are less limiting. Above all, the phytoplankton exhibit very high productivity, a very short generation time (measured in days) and, therefore, rapid turnover. The standing crop of phytoplankton is often much less than that of its consumers. No mammals depend directly on this trophic level. There are no long-lived plants to destroy so that recovery times at the lower trophic levels should be shorter in marine ecosystems. In terrestrial ecosystems, by contrast, the generation times of even the grasses are much longer and the life span of the baobab tree is longer than that of any animal.

There are basic biological differences between terrestrial and marine mammals. Notably, marine mammals show larger size, large-scale feeding dispersal, breeding concentrations, and apparent lack of territoriality associated with feeding, although they may show strong breeding territoriality. Feeding strategies are very different; only one small group of marine mammals, the Sirenia, are herbivorous, and none of the very large land mammals are carnivorous. But across the terrestrial and marine divisions there are marked similarities in feeding strategies, social behavior, reproduction, and demographic parameters for large mammals.

The demographic characteristics of mammals in general differ greatly. The basic difference is between small and large mammals. The former are short- lived (a few years), with a low biomass density, high production rates, high mortality, rapid population turnover, and high intrinsic rates of increase. Large mammals are long-lived with much energy and materials locked up in their high biomass. They show a low intrinsic rate of increase, low reproductive rates, low mortality and low rates of population turnover. As a group they are characterized by the life history traits of K strategists, exemplified by very large mammals like whales and elephants and large mammals such as seals and antelopes. These strategies can be regarded as a set of coadapted traits, designed to solve particular ecological problems faced by the animal. The adaptations can be viewed as solutions to the problems posed by the environments, solutions that have evolved by natural selection (Block, 1980).

Among such adaptations is the adjustment of lifespan and annual cycle to that of prey or the productivity of food resources. Food resources may be classified in two ways. They may involve species that are short-lived with a high productivity per unit biomass as typical of annual grasses in terrestrial systems and of marine zooplankton. Alternatively they may be long-lived or show low productivity per unit biomass as typical of terrestrial trees and bushes and marine benthos or large invertebrates and vertebrates.

The food supply is a fundamental determinant of mammal population size and carrying capacity. Temporal relationships between the life span and productivity or rate of increase of food and its mammal consumer strongly influence the nature of the interaction and its reversibility. This is especially true in terms of the effect on the environment, and the development of conservation and management problems. But there is no clear set of universal rules relating to the demographic characters of populations, although most are related to body size.

A few examples are set out in Table 1, in which mammals are related to their environments and food organisms. An exception to this classification that springs to mind, is the Arctic reindeer or caribou which is dependent on a long-lived resource (lichens) of low productivity and low biomass. When translocated to a sub-Antarctic environment it can adapt remarkably well to a tussock grass diet (Leader-Williams, 1980). It seems clear from an inspection of this table that *comparisons* of mammals within the two size groups, very large and large, would be very profitable, as would *contrasts* between these size groups.

In general the body size of a mammal tends to be correlated with its longevity, but marine mammals are larger than terrestrial mammals for energetic reasons (there are no small marine mammals, with the possible exception of the sea otter). The problems posed by the environment differ markedly and have been solved in different ways. The baleen whales are large (and long-lived) because they are aquatic and not restricted in size by gravity and because they feed on small swarming food organisms. They need to have a large feeding apparatus for energetic efficiency. Most can feed only in the summer months in cold waters and lay down a fat reserve for the winter in the form of blubber. Because of the highly seasonal changes in the environment their gestation period is less than a year and their fetal growth rate is exponential during the second half of pregnancy. The larger baleen whales tend to have the shorter gestation periods (although the minke whale is an exception). This rapid growth extends into postnatal growth, and weaning occurs at about 6 to 7 months, so the mean calving interval is about 2 years.

In contrast the sperm whale feeds on larger prey, the females, socially mature males and sexually immature males in warmer waters; the gestation period is 15 months and the slower growth rate continues into later growth with weaning at 16 months or longer and a mean calving interval of about 4 years. The social structure appears to be more advanced and more permanent than in the baleen whales, and polygyny is a feature of it. The hippopotamus is closer to the baleen whales in some respects than to other large terrestrial mammals, with a similarly short gestation period and suckling period. It is dependent on a food resource (grass) with characteristics similar in some respects to the marine zooplankton, including marked seasonality and high productivity. Its social structure appears to be simple. Elephants resemble the sperm whale in their life history and social organization, having a gestation period of 22 months, weaning at 2 years or longer and with a calving interval of about 4 years. Males are expelled from the family unit at puberty, and longevity is 60 to 70 years. In all these species adult survival appears to be fixed within fairly narrow limits and appears not to vary as much as growth, age at first reproduction, fecundity, and juvenile survival.

Tooth wear can be an important factor in limiting longevity and is usually related to the nature of the food base. It may be significant that the longest-lived mammals, the baleen whales, have replaced teeth by baleen plates. Wear in sperm whale teeth, which take soft-bodied prey, should be less than in the elephant which takes substantial amounts of siliceous grasses, but has evolved a dentition that grows and is replaced throughout life; even so tooth wear probably sets an upper limit to longevity in elephants, hippo, rhinoceros, and many other ungulates. Tooth wear is thought to be a factor limiting longevity of Weddell seals (although not caused by feeding), but not in other seal or whale species.

Predation is a more important cause of mortality in small than in larger mammals.

Eberhardt (1977) and Eberhardt and Siniff (1977) have made a first attempt to draw some general conclusions for marine mammals and Fowler (1981) describes a general pattern in the population dynamics of large mammals. It is clear that a comparative approach will be profitable as is argued in Chapter 23. First, the basic demographic data need to be obtained; these are generally lacking and, because of the longevity of large mammals, difficult to obtain within the compass of the usual research project lasting three to five years.

Table 1　Suggested Relationships Between Mammal Sizes and Longevity and Their Food Resources

Controlling Food Resource	Small, Short-lived Units, High Productivity, Low Biomass		Large, Long-lived Units, Low Productivity, High Biomass	
Environment	Terrestrial	Marine	Terrestrial	Marine
Food organisms	Grass	Plankton Sea grass	Trees Bushes	Large invertebrates Vertebrates
Life span (approximate)	years	days/years	decades/centuries	years/decades
Very large mammals	Hippopotamus White rhino	Baleen whales Dugong	Elephant Giraffe Black rhino	Sperm whale Killer whale
Life span (approximate range)	30-50 yr	40-100 yr	40-70 yr	30-60 yr
Large mammals	Antelopes Deer Vicuna Buffalo	Crabeater seal Ringed seal Antarctic fur seal	Lesser kudu Gerenuk	Dolphins Weddell seal Gray seal Elephant seal Sea otter[a]
Life span (approximate range)	15-20 yr	20-40 yr	15-20 yr	20-40 yr

[a]The sea otter is included, although it is not strictly a large mammal.

Knowledge of the status of the populations studied is vital because the important parameters change in a density-dependent manner, usually in relation to the limiting food resources; examples of agonistic interactions due to crowding are few. Very few large mammals appear to be predator limited, although the crabeater seal may be one example. It is best to study single populations over time to determine the nature of density dependent changes and then compare them with other populations and other species, both terrestrial and marine.

The regulatory mechanisms of large mammals that have been identified (Fowler et al., 1980, Chapter 23) can be listed as reproductive rate of adult females, age of first reproduction, immature mortality rates, and adult mortality rates. There is good evidence for the operation of the first three, but less evidence for density dependent changes in adult mortality. This is partly due to the difficulty of accurately estimating adult mortality, which varies within rather narrow limits (see Chapter 20). Within these limits small percentage changes can produce large effects. Changes in juvenile mortality are more marked and more easily established.

Large mammals (except for some large terrestrial carnivores) exhibit a marked lack of plasticity in one reproductive character, litter size. This is almost invariably fixed at one in marine mammals and at one to two in large land mammals (and at one in the larger ones). The age at maturity is quite variable for most species. It may vary between the ages of 9 and 30 years in the African elephant, for example. That is 14 to 46% of the life span. In the crabeater seal the range is from 2 to 7 years or 6 to 20% of the life span. Pregnancy rates appear to be less variable among antelopes and seals, for example, than among elephants and whales. It is important to determine the magnitude of the density-dependent variations in these parameters, and the order in which they come into play. In terms of evolution and natural selection, one would expect this sequence (with increasing population density) to be births, age at maturity, juvenile mortality and adult mortality, (see Chapters 3 and 17) but mortality (and especially adult mortality) may be more important than the present evidence indicates (see Chapter 22).

Comparative studies of the rates of increase following depletion could be illuminating (where immigration can be ruled out) because they are the resultant of birth and death (see, e.g., Chapter 14). Laws (1979) gave values for the intrinsic rate of increase in seals of 0.06 to 0.15 and in whales 0.02 to 0.13. Experimental manipulation by population reductions (decreased densities) or reducing the range available (increased densities) could also be instructive. There is also scope for studies of social interactions in relation to population density.

It must now be apparent that there is a real need for long-term studies. If we had started some of these studies a few decades earlier, and given them more support, we would certainly have been in a much better position to resolve such issues as the decline of the whale stocks, the Tsavo elephant problem, and, now, to respond to the challenge that the Southern Ocean represents.

Marking and sustained field observations of living mammals are very important tools in research aimed at providing these data, but owing to the longevity of large mammals, necessarily require the assurance of long-term research programs. This cannot always be guaranteed, at least at the beginnning of a project.

Consequently, studies of the population dynamics of living animals need to be based on at least some studies of dead animals. After having developed methods of aging whales, seals, hippo, elephants, antelopes, and other groups one can turn to studies of age structures of living mammals by means of age/length keys. Such studies of living mammals should be extended, where possible, by using remote sensing so as to obtain real time data for monitoring and feedback in management programs.

Models are very important at this stage of development of our subject and they receive much attention in this book. Models are an important tool, but their limitations must be recognized. Models can often be manipulated, consciously or unconsciously, to bias resulting conclusions. They are not a substitute for thought, but an aid to interpretation and prediction. By changing the juvenile mortality rates of some models one can produce increases or decreases in population size that are much more important than changes produced by varying reproductive parameters. This does not mean that juvenile mortality is the most important contributor to population change in the real world. It is very important that modeling be based very firmly on empirical biological studies of populations and their natural history.

In our studies we build on the work that has gone before. Population counts or estimates of abundance by capture-recapture methods often show trends as more effort is put into the investigations. These trends are not necessarily, indeed usually are not, related to actual changes in the populations being studied—but rather are an indication that we are tackling the problems more effectively.

Thirty years ago, my research in Signy Island began with primitive equipment. Now BAS has 30 to 40 biologists and assistants working in the Antarctic, with computers, automatic electronic data recorders, chemical laboratories, vehicles, boats, diving equipment, and teleprinter communication with other workers. This alone illustrates the great change there has been in the approach to field studies, including population studies. The continuing developmnent of computers, telemetry, and other specialized equipment gives us a tremendous opportunity to advance our subject. The prospects for the next 30 years seem limitless.

LITERATURE CITED

Beadle, L. C. 1974. The Nuffield Unit of Tropical Animal Ecology (1961-1971). J. Zool. London 173:539-548.

Block, W. (1980). Survival strategies in polar terrestrial arthropods. Biol. J. Linn. Soc. (in press).

Carrick, R., S. E. Csordas, and S. E. Ingham. 1962. Studies on the southern elephant seal, *Mirounga leonina* (L.). 4. Breeding and development. Commonw. Sci. Ind. Res. Organ. Wildl. Res. 7:161-197.

Caughley, C. 1976. The elephant problem—An alternative hypothesis. E. Afr. Wildl. J. 14:265-284.

Clough, G. 1966. Reproduction in the hippopotamus (*Hippopotamus amphibius* Linn.). Ph.D. Thesis, University of Cambridge.

Corfield, T. F. 1973. Elephant mortality in Tsavo National Park, Kenya. E. Afr. Wildl. J. 11:339-368.

Corner, R. W. M. 1971. Observations on a small crabeater seal breeding group. Br. Antarct. Surv. Bull. 30:104-106.

Cowan, D. F., W. M. Thurlbeck, and R. M. Laws. 1967. Some diseases of *Hippopotamus* in Uganda. Pathol. Vet. 4:553-567.

Delany, M. J. 1964. An ecological study of the small mammals in the Queen Elizabeth Park, Uganda. Rev. Zool. Bot. Afr. 70:149-229.

Delany, M. J., and B. R. Neal. 1969. Breeding seasons in rodents in Uganda. J. Reprod. Fert. Suppl. 6:229-235.

Eberhardt, L. L. 1977. Optimal policies for conservation of large mammals, with special reference to marine ecosystems. Environ. Conserv. 4:205-212.

Eberhardt, L. L., and D. B. Siniff. 1977. Population dynamics and marine mammal management policies. J. Fish. Res. Board Can. 34:183-190.

Eggeling, W. J. 1947. Observations on the ecology of the Budongo rain forest, Uganda. J. Ecol. 34:20-87.

Eltringham, S. K., and R. C. Malpas. (1980). The decline in elephant numbers in Rwenzori and Kabalega Falls National Parks, Uganda, between 1973 and 1975. E. Afr. Wildl. J. (in press).

Everson, I. 1977. The living resources of the Southern Ocean. FAO, UNEP. Southern Ocean Fisheries Survey Program. GLC/SO/77/1, 155pp.

Field, C. R. 1966. A comparative study of the food habits of some wild ungulates in the Queen Elizabeth Park, Uganda. Symp. Zool. Soc. London. 21:135-151.

Field, C. R. 1968a. Methods of studying the food habits of some wild ungulates in Uganda. Proc. Nutr. Soc. 27:172-177.

Field, C. R. 1968b. The food habits of some wild ungulates in Uganda. Ph.D. Thesis, University of Cambridge.

Field, C. R., and Laws, R. M. 1970. The distribution of the larger herbivores in the Queen Elizabeth National Park, Uganda. J. Appl. Ecol. 7:273-294.

Fowler, C. W. (1981). Density dependence as related to life history strategy. Ecology. (in press).

Fowler, C. W., and T. Smith. 1973. Characterizing stable populations: An application to the African elephant population. J. Wild. Manage. 37:513-523.

Fowler, C. W., W. T. Bunderson, R. J. Ryel, and B. B. Steele. 1980. A preliminary review of density dependent reproduction and survival in large mammals. Appendix B *in:* C. W. Fowler, W. T. Bunderson, M. B. Cherry, R. J. Ryel, and B. B. Steele. Comparative population dynamics of large mammals: A Search for management criteria. Report to the U.S. Marine Mammal Commission Contract #MM7AC013. NTIS #PB80-178627.

Glover, J. 1963. The elephant problem at Tsavo. E. Afr. Wildl. J. 1:30-39.

Goddard, J. 1970. Age criteria and vital statistics of a black rhinoceros population. E. Afr. Wildl. J. 8:105-121.

Grimsdell, J. J. R. 1968. Breeding in certain large mammals in Uganda, with special reference to the African buffalo, *Syncerus caffer*. J. Anim. Ecol. 37:7P.

Grimsdell, J. J. R. 1969. Ecology of the buffalo *Syncerus caffer* in Western Uganda. Ph.D. Thesis, University of Cambridge.

Hall-Martin, A. 1977. *in:* WWF/IUCN Elephant Survey and Conservation Program, Newsletter No. 2, p. 3.

Hamilton, I. Douglas-, and O. Douglas- 1975. Among the Elephants. Viking Press, New York, 285 pp.

Hanks, J., and J. E. A. Mackintosh. 1973. Population dynamics of the African elephant *(Loxodonta africana)*. J. Zool. London 169:29-38.

Hardy, Sir Alistair. 1967. Great Waters. Collins, London. 542 pp.

Holman, D. 1967. The Elephant People. John Murray, London. 226 pp.

Kangwagye, T. N. 1968. The food habits of some wild ungulates in Uganda. Ph.D. Thesis, University of East Africa.

Kaufman, G. W., D. B. Siniff, and R. Reichle. 1975. Colony behavior of Weddell seals, *Leptonychotes weddelli*, at Hutton Cliffs, Antarctica. Rapp. P. V. Reun. Cons. Int. Explor. Mer. 169:228-246.

Laurie, A. H. 1937. The age of female blue whales and the effect of whaling on the stock. Discovery Rep. 15:223-284.

Laws, R. M. 1952. A new method of age determination for mammals. Nature (London) 169:972-973.

Laws, R. M. 1953. The elephant seal industry at South Georgia. Polar Rec. 6:746-754.

Laws, R. M. 1958. Growth rates and ages of crabeater seals, *London carcinophagus*, Jacquinot and Pucheran. Proc. Zool. Soc. London 130:275-288.

Laws, R. M. 1959. The foetal growth rates of whales, with special reference to the Fin whale, *Balaenoptera physalus* Linn. Discovery Rep. 29:281-308.

Laws, R. M. 1960a. Problems of whale conservation. Trans. N. Am. Wildl. Conf. 25:304-319.

Laws, R. M. 1960b. The southern elephant seal *(Mirounga leonina* Linn.) at South Georgia. Nor. Hvalfangsttid. 49:466-476, 520-542.

Laws, R. M. 1961. Reproduction, growth and age of southern fin whales. Discovery Rep. 31:327-486.

Laws, R. M. 1962a. Some effects of whaling on the southern stocks of baleen whales. Pp. 137-158 *in:* E. D. Le Cren and M. W. Holdgate (eds.). The Exploitation of Natural Animal Populations. Blackwell, Oxford.

Laws, R. M. 1962b. Age determination of pinnipeds with special reference to growth layers in the teeth. Z. Saugetierk. 27:129-146.

Laws, R. M. 1966. Age criteria for the African elephant *(Loxodonta a. africana)*. E. Afr. Wildl. J. 4:1-37.

Laws, R. M. 1967. Occurrence of placental scars in the uterus of the African elephant *(Loxodonta africana)*. J. Reprod. Fert. 14:445-449.

Laws, R. M. 1968a. Interactions between elephants and hippopotamus populations and their environments. E. Afr. Agric. For. J. 33:140-147.

Laws, R. M. 1968b. Dentition and ageing of the hippopotamus. E. Afr. Wildl. J. 6:19-52.

Laws, R. M. 1969a. Aspects of reproduction in the African elephant, *Loxodonta africana.* J. Reprod. Fert. Suppl. 6:193-217.

Laws, R. M. 1969b. The Tsavo Research Project. J. Reprod. Fert. Suppl. 6:495-531.

Laws, R. M. 1970. Elephants as agents of habitat and landscape change in East Africa. Oikos 21:1-15.

Laws. R. M. 1977a. Seals and whales in the Southern Ocean. Philos. Trans. R. Soc. London Ser. B 279:81-96.

Laws, R. M. 1977b. The significance of vertebrates in the Antarctic marine ecosystem. Pp. 441-438 *in:* G. A. Llano, (ed.). Adaptations within Antarctic Ecosystems. Gulf Publishing Co., Houston.

Laws, R. M. 1979. Monitoring whale and seal populations. Pp. 115-140 *in:* D. Nichols (ed.). Monitoring the Marine environment. Institute of Biology, London.

Laws, R. M., and G. Clough. 1965. Observations on reproduction in the hippopotamus, *Hippopotamus amphibius.* Symp. Zool. Soc. London 15:117-140.

Laws, R. M., and I. S. C. Parker. 1968. Recent studies on elephant populations in East Africa. *in:* Comparative nutrition of wild animals. Symp. Zool. Soc. London 21:319-359.

Laws, R. M., I. S. C. Parker, and R. C. B. Johnstone. 1975. Elephants and Their Habitats: The Ecology of Elephants in North Bunyoro, Uganda. Oxford University Press, London. Pp. xxi + 376.

Laws, R. M., and R. J. F. Taylor. 1957. A mass dying of crabeater seals, *Lobodon carcinophagus* (Gray). Proc. Zool. Soc. London 129:315-324.

Leader-Williams, N. (1980). Population ecology of reindeer on South Georgia. Proc. Second Int. Reindeer/Caribou Symp. Røros, Norway, 1979.

Leuthold, W., and B. M. Leuthold. 1976. Density and biomass of ungulates in Tsavo East National Park, Kenya. E. Afr. Wildl. J. 14:49-58.

Leuthold, W., and J. B. Sale. 1973. Movements and patterns of habitat utilization of elephants in Tsavo National Park, Kenya. E. Afr. Wildl. J. 11:369-384.

Lillie, D. G. 1910. Observations on the anatomy and general biology of some members of the larger Cetacea. Proc. Zool. Soc. London 1910:769-792.

Lock, J. M. 1967. Vegetation in relation to grazing and soils in the Queen Elizabeth Park, Uganda. Ph.D. Thesis, University of Cambridge.

Lockyer, C. H. 1972. The age at sexual maturity of the southern fin whale *(Balaenoptera physalus)* using annual layer counts in the ear plug. J. Cons. Perma. Int. Explor. Mer 34:276-294.

Mackintosh, N. A. 1942. The southern stocks of whalebone whales. Discovery Rep. 22:197-300.

Mackintosh, N. A., and J. F. G. Wheeler. 1929. Southern blue and fin whales. Discovery Rep. 1:257-540.

Matthews, L. H. 1937. The humpback whale, *Megaptera nodosa.* Discovery Rep. 17:7-92.

McCann, T. S. (1980). Population structure and social organization of southern elephant seals *Mirounga leonina* (L.). *Biol. J. Linn. Soc.* (in press).

Neal, B. R. 1967. The ecology of small rodents in the grassland community of the Queen Elizabeth Park, Uganda. Ph.D. Thesis, University of Southampton.

Neal, B. R. 1970. The habitat distribution and activity of a rodent population in Western Uganda, with particular reference to the effects of burning. *Rev. Zool. Bot. Afr.* 81:29-50.

Neal. B. R., and A. G. Cook. 1969. An analysis of the selection of small African Mammals by two break-back traps. J. Zool. London 158:335-340.

Øritsland, T. 1970. Sealing and seal research in the south-west Atlantic pack ice Sept.-Oct. 1964. Pp. 367-376 *in:* M. W. Holdgate (ed.). Antarctic Ecology, Vol. 1. Academic Press, London.

Payne, M. R. 1977. Growth of a fur seal population. Philos. Trans. R. Soc. London Ser. B 270:67-79.

Perry, J. S. 1953. The reproduction of the African elephant *Loxodonta africana.* Philos. Trans. R. Soc. London Ser. B 237:93-149.

Plowright, W., R. M. Laws, and C. W. Rampton. 1964. Serological evidence for the susceptibility of the hippopotamus *(H. amphibius* L.) to natural infection with rinderpest virus. *J. Hyg. Camb.* 62:329-336.

Purves, P. E. 1955. The wax plug in the external auditory meatus of Mysticeti. Discovery Rep. 27:293-302.

Ruud, J. T. 1945. Further studies on the structure of the baleen plates and their application to age determination. Hvalradets Skr. 29:1-69.

Schevill, W. E. (ed.). 1974. The Whale Problem: A Status Report. Harvard University Press, Cambridge, Mass. 419 pp.

Siniff, D. B., and J. L. Bengtson. 1977. Observations and hypothesis concerning the interactions among crabeater seals, leopard seals, and killer whales. J. Mammal. 58:414-416.

Siniff, D. B., D. P. DeMaster, R. J. Hofman, and L. L. Eberhardt. 1977. An analysis of the dynamics of a Weddell seal population. Ecol. Monogr. 47:319-335.

Spinage, C. A. 1967a. The autecology of the Uganda waterbuck *Kobus defassa ugandae* with special reference to territoriality and population controls. Ph.D. Thesis, University of London.

Spinage, C. A. 1967b. Ageing the Uganda defassa waterbuck *(Kobus defassa ugandae* Neumann). E. Afr. Wildl. J. 5:1-17.

Spinage, C. A. 1969a. Reproduction in the Uganda defassa waterbuck *Kobus defassa ugandae* Neumann. J. Reprod. Fert. 6:445-457.

Spinage, C. A. 1969b. Territoriality and social organization of the Uganda defassa waterbuck *Kobus defassa ugandae,* J. Zool. London 159:329-361.

Spinage, C. A. 1970. Population dynamics of the Uganda defassa waterbuck *(Kobus defassa ugandae* Neumann) in the Queen Elizabeth Park, Uganda. J. Anim. Ecol. 39:51-78.

Stirling, I. 1971. Population dynamics of the Weddell seal *(Leptonychotes weddelli)* in McMurdo Sound, Antarctica, 1966-1968. Pp. 141-161. *in:* W. H. Burt (ed.). Antarctic Research Series 18, Antarctic Pinnipedia. American Geophysical Union, Washington, D. C.

Thurlbeck, W. M. 1965. Arteriosclerosis in hippopotami. *in:* J. C. Roberts and S. Straus (eds.). Comparative Arteriosclerosis. Harper & Row, New York. 156 pp.

Thurlbeck, W. M., C. A. Butas, E. M. Mankiewicz, and R. M. Laws. 1965. Chronic pulmonary disease in wild buffalo *(Syncerus caffer)* in Uganda. Am. Rev. Resp. Dis. 92:801-805.

Wheeler, J. F. G. 1934. On the stock of whales at South Georgia. Discovery Rep. 9:351-372.

THREE

Characterization of Population Condition

JOHN HANKS

INTRODUCTION

As wildlife management becomes more intensive, the need for objective criteria to describe both animal and habitat condition and trend will increase. The literature on mammalian population dynamics has frequent references to a useful, but often poorly defined concept that attempts to describe the "condition" of a population. Although not usually specified, in most cases "condition" is closely linked to the individual animal's chances of living or dying, and as such is an important factor influencing mortality (Klein and Olsen, 1960; Hirst, 1969; Sinclair, 1970; Klein, 1970; Sinclair and Duncan, 1972; Hanks et al., 1976). It has been indexed by a variety of parameters, including deposited fat reserves, adrenocortical hypertrophy, blood chemistry and hematology, urinary excretion of hydroxyproline, and aspects of body growth. In this chapter, such indices will be considered as a measure of the "physiological condition" of a population and will be linked to the individual animal's chances of living or dying.

Caughley (1971, 1977) favored a demographic approach to the assessment of "condition" and suggested it should be expressed by a single statistic that weighs up and combines the vigor of each age and sex class in the population. He proposed the use of the survival-fecundity rate of increase, symbolized r_s, which he termed "demographic vigor," noting that any change in the environment will usually result in a rapid change in r_s. This concept, which provides an assessment of the dynamics of a population, is a valuable adjunct to "physiological condition," and the two terms are considered independently in this chapter. An index that measures "physiological condition" need not necessarily be equated with, nor predict, "demographic vigor."

Ideally, any index of population condition, whether physiological or demographic, should include a measure of resilience. The single statistic r_s tells us nothing about the recuperative powers of a population subjected to sudden environmental stress. Similarly, a single physiological parameter does not indicate the extent to which an animal can tolerate further deprivations. In both cases, a measure of resilience in a condition index would provide a wildlife manager with an indication of the anticipated response of a population to continued environmental stress or ameliorative management practices.

The main purpose of this study is to review the more important criteria that have been used to describe physiological condition and demographic vigor in populations of large mammals, and to consider the implications of introducing a component to describe "population resilience."

PHYSIOLOGICAL CONDITION

Deposited Fat Reserves

A commonly used criterion for the description of physiological condition in large mammals is the quantification of deposited fat reserves. The removal of all the fat and the expression of this fat as a percentage of the carcass weight is an expensive, tedious, and time-consuming task, although it has been attempted (Ledger and Smith, 1964). A preferable alternative is the kidney fat index.

Kidney Fat Index (KFI)

Riney (1955) postulated that in ruminants, fat can be taken as a direct measure of the animal's condition, "reflecting the metabolic level or goodness of physiologic adjustment of an animal with its environment." He proposed a method whereby an estimate of total body fat was obtained from the amount of fat deposited immediately around the kidney (ignoring the fat extending anteriorly and posteriorly in the kidney mesentery). This fat was used to calculate the kidney fat index, where

$$\text{kidney fat index} = \frac{\text{perinephric fat weight}}{\text{kidney weight}} \times 100$$

Riney's method has subsequently been used extensively in an identical or slightly modified form by Hughes and Mall (1958), Taber et al. (1959), Ransom (1965), Allen (1968), Trout and Thiessen (1968), Caughley (1970a), Laws et al. (1970), Smith (1970), Albl (1971), Bear (1971), Caughley (1971), Huntley (1971), Anderson et al. (1972a), Sinclair and Duncan (1972), Dauphine (1975), Laws et al. (1975), Williamson (1975), Hanks et al. (1976), McNab (1976), Brooks et al. (1977), and Malpas (1977).

Smith (1970) found that in seven species of East African wild ungulates KFI and the total body fat exhibited a significant correlation at the 0.01 level. Smith also examined subcutaneous fat deposition and visceral fat and concluded that of the various methods he used for assessing condition, KFI was the most useful and best fulfilled the needs of a workable technique.

In addition to the seasonal changes in deposited fat reserves associated with plane of nutrition, fluctuations in KFI can also reflect physiological and behavioral events associated with reproduction. This has been demonstrated in several species, including Himalayan thar *(Hemitragus jemlahicus)* by Caughley (1970a) pronghorns *(Antilocarpa americana)* by Bear (1971), mule deer *(Odocoileus hemionus hemionus)* by Anderson et al. (1972a), and impala *(Aepyceros melampus)* by Hanks et al. (1976) (Figure 1).

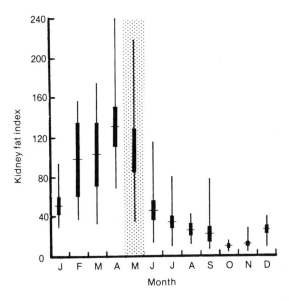

Figure 1 Monthly changes in kidney fat index (KFI) in male impala. (Range, vertical line; mean, crossbar; Standard error, broad portion of line. The shaded area represents the rutting season.)

For any given use of KFI, the kidney is assumed to be a constant function of body size. A recent important study by Dauphine (1975) indicated that the kidneys of caribou *(Rangifer tarandus)* undergo such pronounced seasonal weight fluctuations that they are unsuitable for use as a measure of body size in any index intended to display seasonal changes in another physical attribute. Dauphine suggested that a summer increase in kidney weight is a pattern common to cervids. His work has drawn attention to the need to check kidney weights for seasonal differences and to adjust them if necessary before attempting to use KFI for interseasonal comparisons. A further limitation of the use of KFI occurs with younger animals, where very little body fat is deposited regardless of condition. Hanks et al. (1976) found that impala under 3 years of age had a mean KFI significantly lower than older animals, and all young impala were consequently eliminated from the analysis of seasonal variation in KFI.

Bone Marrow Fat (BMF)

Several studies have used techniques that involve an assessment of BMF (Cheatum, 1949; Bischoff, 1954; Riney, 1955; Ransom, 1965; Allen, 1968; Greer, 1968; Trout and Thiessen, 1968; Neiland, 1970; Bear, 1971; Huntley, 1971; Sinclair and Duncan, 1972; Verme and Holland, 1973; West and Shaw,

1975; Franzmann and Anreson, 1976; Hanks et al., 1976; Brooks et al., 1977). The fat content of the bone marrow is usually expressed as a percentage of the fresh weight of the marrow, the fat being assessed by extraction in a Soxhlet apparatus. As this is a time-consuming technique, several attempts have been made to look for alternatives. Bear (1971) concluded that visual estimates of BMF, using color and consistency, are reliable only for extreme fat values, an observation supported by Franzmann and Anreson (1976), who concluded that marrow fat indices based on color and consistency have subjective error potential. To save time on chemical analyses, Greer (1968) proposed a compression method to indicate BMF in elk *(Cervus canadensis)*, an improvement on the visual estimates when chemical analyses are not feasible.

For the purposes of most studies, a technique should be used that indicates the fat content accurately enough without delay or laboratory costs. Neiland (1970) analyzed femur marrow samples from barren ground caribou *(Rangifer tarandus granti)* and demonstrated that the marrow was a three-component system comprised of fat, water, and nonfat residue. He came to the useful conclusion that the dry weight of the marrow, corrected for nonfat residue, gives an adequate measure of femur fat content. This technique has been used subsequently by Sinclair and Duncan (1972), Hanks et al. (1976), and Brooks et al. (1977), who reported that the dry weight of bone marrow, expressed as a percentage of its fresh weight, was a good and easy-to-use quantitative estimate of its fat content. Brooks et al. (1977) used the general formula

$$\% \text{ marrow fat} = \% \text{ dry mass} - 7$$

to estimate BMF in eight African ungulates, and it is quite probable that a similar relationship exists for many other species.

Hanks et al. (1976) reported that BMF is of limited use as a measure of condition in younger animals. In impala under 2 years of age, hemopoietic tissue extends throughout most of each limb bone, and the marrow has a very opaque gelatinous appearance. In these cases, the bone marrow is still very active in red blood cell formation, and the red color is not necessarily an indication of poor condition. Limb bones examined for BMF in the absence of age criteria could give an erroneous estimate of deposited fat reserves.

Sequence of Fat Mobilization

Harris (1945) was one of the first to describe a sequence of fat mobilization. In deer, rump fat is the first to disappear, followed by subcutaneous fat, visceral fat, and marrow fat. Riney (1955) came to a similar conclusion, giving the sequence as subcutaneous, abdominal, perinephric and, finally, bone marrow, an observation subsequently confirmed and described in more detail by Ransom (1965), Sinclair and Duncan (1972), and Brooks et al. (1977). In most species, KFI decreases to an index of about 40 without appreciable changes in BMF, but with further fat mobilization, BMF declines sharply (Figure 2). As a useful field guide, if KFI is above 40, there is little to be gained by carrying out an assessment of BMF.

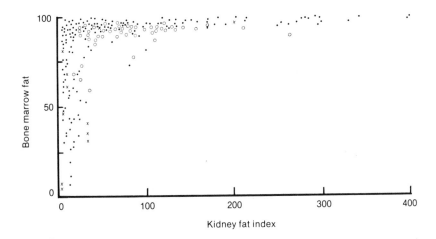

Figure 2 The relationship between bone marrow fat (BMF) and kidney fat index (KFI) in impala ●, buffalo ○, and eland ×. [(From S. Afr. J. Wildl. Res. 7(2):61-66 (1977).]

A further and very important aspect concerning the sequence of fat mobilization was described by Brooks et al. (1977). These investigators showed that in at least four species there was a sequence of BMF mobilization in the limb bones that was initiated in the femur and humerus (Table 1). They concluded that for more sensitive comparative studies of condition these two bones should be collected in preference to the metatarsus and metacarpus.

Table 1 The Percent Fat Content in the Bone Marrow of Complete Impala Limb Bones, Arranged in Order of Increasing Fat Content of the Humerus and Femur, Respectively[a]

Forelimbs									
Humerus	10.3	15.5	16.1	22.2	28.1	30.2	46.6	86.8	90.1
Radius	94.7	36.2	36.9	89.1	91.6	80.8	93.9	91.1	91.1
Metacarpus	93.9	16.3	14.9	65.0	90.9	89.9	90.0	90.1	90.7

Hindlimbs										
Femur	1.3	16.5	19.0	21.3	27.5	36.5	67.1	77.3	89.0	90.3
Tibia	0.7	45.9	38.3	86.2	94.0	93.8	89.9	95.1	91.4	91.0
Metatarsus	2.2	21.0	18.5	90.5	87.8	91.1	81.6	93.3	89.0	90.3

[a]From S. Afr. J. Wildl. Res. 7(2):61-66 (1977).

Although it may be assumed that the sequence of fat deposition occurs in the reverse sequence to that described above, further studies are required. In particular, information is needed on the extent to which lipogenesis is delayed after a marked improvement in the nutritional plane.

Adrenocortical Hypertrophy

It is generally accepted that adrenal hypertrophy and hyperplasia are reactions of the body to stress, and that an increase in adrenocortical tissue has a direct relationship to adrenal weight (Christian, 1955). It is also Christian's (1975) contention that with an increase in population density, there is a concurrent intensification in social aggressive interactions due to crowding. In these cases, a stressful stimulus results in endocrine responses in the form of increased pituitary-adrenocortical activity. Although an animal stressed by a long-term chronic stimulus, such as a high population density, should show adrenocortical hypertrophy, it is important to appreciate that a variety of other factors unrelated to the quantity and quality of social interactions will influence adrenal weight; and these include low temperature, sexual activity, photoperiod, and diet (Southwick, 1957; McKeever, 1959; Quay, 1960; Chitty, 1961; Bronson and Eleftheriou, 1965; Christian and Davis, 1964; Eleftheriou, 1964; Christian and Davis, 1966; Andrews, 1968; Pinter, 1968; Andrews, 1970a,b,; Bujalska, 1971; To and Tamarin, 1977). All these studies were made on small mammals, but they do nevertheless introduce some valuable concepts in connection with condition assessment. (I have had to select only some of the many examples available, because space is short.) In contrast, fewer studies have analyzed adrenal weights of the larger mammals, and most of these have been of a more superficial nature (Browman and Sears, 1956; Hughes and Mall, 1958; Taber et al., 1959; Christian et al., 1960; Welch, 1962; Hoffman and Robinson, 1966; Krumrey and Buss, 1969; Flook, 1970; Smith, 1970; Anderson et al., 1971; Hanks, 1972; McKinney and Dunbar, 1976).

Several attempts have been made with larger mammals to relate adrenal size to physiological condition. Hughes and Mall (1958) correlated adrenal weight of California deer with condition, taking the amount of perinephric fat as the measure of condition. They concluded that in the deer sampled, adrenocortical size, most easily measured as adrenal weight, may be considered as a condition factor. However, Smith (1970) found that in only one of the ten species of East African ungulates he studied was there a significant correlation between adrenal weight and total fat. More recent studies on large mammals have been cautious in the interpretation of adrenal weight changes in relation to physiological condition. Anderson et al. (1971), in their examination of adrenal weight in mule deer, concluded that significant seasonal changes in the mean relative adrenal weights of each sex *appeared* to be a function of variation in fat reserves. Welch (1962) found that adrenal size in white-tailed deer *(Odocoileus virginianus)* was related to population density, and although he suggested that trends in mean adrenal weight may furnish a useful supplementary index of condition, he emphasized the need for additional information before using the technique for management purposes.

Quite clearly, controlled experimental work is required in large mammals to show that adrenal hypertrophy is caused by social stress associated with high density per se, and further studies are needed to establish the relationships between adrenal size and deposited fat reserves.

Blood Chemistry and Hematology

In recent years, an increasing number of papers have referred to aspects of blood chemistry and hematology of large wild mammals, mostly reporting baseline values for a number of parameters (Rosen and Bischoff, 1952; Bandy et al., 1957; Taber et al., 1959; Erickson and Youatt, 1961; Stewart et al., 1964; Ullrey et al., 1964; Youatt et al., 1965; Herin, 1968; McEwan, 1968; Johnson et al., 1968; Tumbleson et al., 1968; Houston, 1969; McEwan and Whitehead, 1969; Seal and Erickson, 1969; Tumbleson et al., 1970; Franzmann, 1971; Anderson et al., 1972b; Halloran and Pearson, 1972; Seal et al., 1972; McCullagh, 1973; Heidt and Hargraves, 1974; Coblentz, 1975; Cooper et al., 1975; Pedersen and Pedersen, 1975; Seal et al., 1975; Thompson et al., 1977; deCalesta et al., 1977; Lee et al., 1977; Malpas, 1977; Wilson and Hirst, 1977; Brown et al., 1978). Unfortunately, with few exceptions, the blood came from animals with undefined histories, and it is usually impossible to relate the blood parameters to environmental variables, physiological condition, or demographic vigor. Furthermore, few studies have recognized the possibility of differences in blood characteristics associated with sex and age, yet these differences can often be significant (Dacie and Lewis, 1968). An additional complication was noted by Seal et al. (1972) and Franzmann and Thorne (1970), who reported that drug immobilization and the stress associated with handling have been shown to cause aberrant values in blood levels. In many studies, these problems appear to have been ignored or overlooked.

The use of blood constituents to describe physiological condition has met with mixed success. For example, Rosen and Bischoff (1952) related body weight to red blood cells, packed cell volume (PCV), and hemoglobin in California deer, and found no correlation. In contrast, Wilson and Hirst (1977) found that in sable (Hippotragus niger) KFI was very significantly related to live weights of animals, total plasma proteins, PCV, and serum albumin.

Nevertheless, certain aspects of blood chemistry and hematology do appear to have potential as indices of physiological condition, and the more promising are reviewed breifly below. No attempt has been made to review those aspects associated with a pathological state or impaired organ function. The parameters are not arranged in any order of priority or potential value.

Serum Cholesterol

Cholesterol is synthesized in the liver and other tissues from various amino acids, carbohydrates, and fatty acids when they are supplied in excess of metabolic needs. Thus a reduction in cholesterol levels should indicate a reduction in the overall quality of the diet. Coblentz (1975) found that in the white-tailed deer serum cholesterol levels fell with the approach of winter, reflecting a reduction in the quantity and quality of the animals' diet. He suggested that cholesterol levels may have significant value as an index of "nutritional condition." As an animal experiencing a marked decline in "nutritional condition" would be less resilient to environmental stress than an animal on a much higher plane of nutrition, for the purposes of the characterization of population condition, "nutritional condition" could be regarded as synonymous with physiological condition as defined in the introduction, although the term should be used with caution.

Other studies have emphasized the relationship between cholesterol levels and body weight. Colby et al. (1950) reported that cholesterol was significantly correlated with the rate of gain in beef cattle bulls, whereas Stewart et al. (1964) found that a rise in plasma cholesterol in black-tailed deer *(Odocoileus hemionus columbianus)* was associated with the point of maximum gain in body weight.

Although serum cholesterol levels appear to be a useful index of physiological condition, much more information is required on the nature of the relationship between food quality and serum cholesterol and the extent to which body growth is delayed after a marked improvement in the nutritional plane.

Serum Protein

Serum or plasma protein levels are one of the most commonly measured blood parameters in North American cervids and more recently in African ungulates, and yet it is probably one of the least useful for the assessment of physiological condition. Malpas (1977) studied serum protein levels in the elephant *(Loxodonta africana)*, and concluded that there were no significant differences between areas and seasons. Although levels of serum protein can and do reflect the quality and quantity of the dietary protein, it appears that these levels are more slowly affected than other relevant blood parameters, and their value lies more in the reflection of the severity of protein deprivation (Kumar et al., 1972). However, as serum protein levels can be decreased significantly by several other factors, such as poor absorption, wounds, renal failure, pregnancy, and lactation, and increased by shock, infections, and dehydration (Benjamin, 1961), serum protein levels should be used with caution as an index of physiological condition.

Blood Urea Nitrogen (BUN)

BUN levels depend on the quality and quantity of dietary protein, and this easily measured characteristic appears to be the best early test of reduced protein intake (Berrier, 1968; Eggum, 1970; Kumar et al., 1972; Eskeland et al., 1974; Kirkpatrick et al., 1975; Malpas, 1977; Brown et al., 1978). When food intake or food quality is extremely low and a mammal is approaching starvation, it initiates the catabolism of muscle protein to sustain levels of energy metabolites

(Young and Scrimshaw, 1971; deCalesta et al., 1977). Under such circumstances, BUN levels may be elevated (Ullrey et al., 1968). Clearly, BUN levels alone could give a false impression of physiological condition.

Packed Cell Volume (PCV)

Although Wilson and Hirst (1977) found that KFI was very significantly related to PCV in sable, they did not consider the physiological basis of this relationship. PCV (hematocrit) values are useful in the diagnosis of various pathological conditions, but until information is available on the precise relationship between deposited fat reserves, PCV and plane of nutrition, PCV values seem to be of limited use as an index of physiological condition. Further studies are also required on the extent to which the erythropoietic capability of the bone marrow and spleen will be influenced by the plane of nutrition.

Plasma Glucose

Frankel et al. (1970) and Ravel (1973) have shown that plasma glucose levels can vary with diet, age, physical activity, environmental conditions, and time of day. Unless all these variables are taken into consideration, plasma glucose levels have a limited value as an index of physiological condition.

Other Blood Parameters (of Doubtful Value)

The three widespread intracellular enzymes LDH (lactic dehydrogenase), GOT (glutamic oxalocetic transaminase), and GPT (glutamic pyruvic transaminase) escape into the circulation when cells are injured. Lee et al. (1977) found significant elevations in serum levels of LDH and GOT in polar bears (Ursus maritimus) caught in snares, probably due to the snare causing trauma in striated muscles. Seal et al. (1972) found a 200-fold increase in serum levels of both LDH and GOT in white-tailed deer 24 hours after they had been physically restrained. Although there is no doubt as to the relationship between cell damage and circulating levels of the three intracellular enzymes, it is unlikely that these values will prove useful as indices of physiological condition.

Franzmann (1971) used the serum albumin/globulin (A/G) ratio to detect differences in protein intake in bighorn sheep (Ovis canadensis). He suggested that the high A/G ratios result from high serum albumin levels, and he considered that these in turn might reflect an increase in certain dietary proteins. Malpas (1977) was unable to find any significant difference in A/G ratios between areas and seasons in the elephant populations he examined in Uganda. Although further studies are required, it seems reasonable to conclude that the difficulties associated with the assessment of serum protein levels will also apply to the assessment of serum A/G ratios.

There are certain fatty acids, the essential fatty acids, (EFA), which animal tissues cannot synthesize and which must be supplied in the diet. Animals fed on artificial diets devoid of EFA fail to thrive, and develop a scaly dermatitis (Passmore, 1968), but it is unlikely that such deficiencies occur naturally. The EFA deficiency described by McCullagh (1973) in the elephant did not appear to cause either dermatitis or an obvious loss in physiological condition, but he did present evidence to suggest that excessive tree damage by elephants may be a natural response to an inadequate EFA intake.

Urinary Excretion of Hydroxyproline

Hydroxyproline is an amino acid derived from the breakdown of collagen. Its excretion is related to the rate of collagen metabolism, and hence also to the rate of growth (Smiley and Ziff, 1964). A low excretion of hydroxyproline is associated with malnutrition. Whitehead (1965) showed that the measurement of total urinary hydroxyproline was of value in assessing the nutritional status of communities of children, because those living on diets deficient in either protein or total calories excreted subnormal amounts. Howells and Whitehead (1967) subsequently developed the hydroxyproline-creatinine index (HCI) in which hydroxyproline in a random sample of urine is related to the concentration of creatinine:

$$\text{hydroxyproline-creatinine index} = \frac{\text{m}M\text{-hydroxyproline/liter}}{\text{m}M\text{-creatinine/liter/kg body weight}}$$

The level of HCI was considered to measure essentially the rate of growth. Mc-Cullagh (1969) and Malpas (1977) both found that HCI in elephants ran parallel to the seasonal variation in rainfall and vegetation growth, and they concluded that growth rates were higher in the west season than in the dry. McCullagh went on to suggest that the importance of the ratio lies in the fact that it measures growth at the instant of collecting the sample, whereas measurements of body size represent the sum of growth throughout the animal's life.

In terms of the value of HCI in the assessment of physiological condition, it is assumed that a high HCI can be equated with good condition as manifested by the rate of growth. However, in a recent study Woodall (1977) has cast doubt on this concept. Studying the ecology and nutrition of the water vole *(Arvicola terrestris)* he concluded that urinary hydroxyproline excretion appeared to be related to both the digestibility and nitrogen content of the food, and HCI did not provide an index of the instantaneous growth rate as suggested by McCullagh (1969) and Malpas (1977). Woodall suggested that HCI should be regarded as providing an indication of the animal's nitrogen balance.

This potentially promising technique requires further study before its true value can be assessed.

Body Growth

An analysis of an animals growth in weight, height, and length can provide objective criteria for assessing physiological condition, based on the concept that a reduction in weight at age or a reduction in growth rates can be equated with poor condition. Using body weight as the criterion of condition, Park and Day (1942) suggested an average dressed weight for various age groups of white-tailed deer as a simplified measure of condition in relation to available forage.

Taber and Dassmann (1958) used body weight corrected for skeletal size as a means of comparing black-tailed deer from different habitat types. Brand et al. (1975) used a cubic relationship between weight and hindfoot length to obtain a "condition index" for snowshoe hares *(Lepus americanus)* similar to a cubic relationship between weight and total length in cottontails *(Sylvilagus floridanus)* used by Bailey (1968). In these cases, the term "condition index" was used to describe a measure of the slenderness or heaviness of an animal. Klein (1968) has provided one of the best examples of the value of body weight measurements as an index of physiological condition. Describing the introduction of reindeer to St. Matthew Island, Klein reported that in 1957, when there were 1350 reindeer on the island, their body weights were found to exceed those of reindeer in domestic herds by 24 to 53% among females and 46 to 61% among males. By 1963, the population had increased to 6000, and average body weights had declined from 1957 values by 38% for adult females and 48% for adult males. This reduced body growth was almost certainly related to qualitative and quantitative changes in the food supply. A massive population crash followed soon after the determination of the 1963 values, and Klein concluded that food supply, through interaction with climatic factors, was the dominant population regulating mechanism for reindeer on the island.

The value of body weight as an index of physiological condition becomes particularly apparent if mean weight loss is related to death from starvation, and deCalesta et al. (1975) indicated the potential of this line of research. They showed that in mule deer weight loss is linearly related to length of starvation, and was significantly faster in fawns than in does. The mean weight loss by four fawns succumbing to starvation was 30%, the doe dying during starvation losing 37% of prestarvation body weight. They made the important observation that adult deer originally in good physical condition can be starved for varying lengths of time up to 64 days and still be refed successfully. The point at which starving deer cannot be refed successfully is associated with a physiological state at which approximately 30% body weight has been lost, at which point there are changes in the levels of blood parameters suggestive of near exhaustion of fat reserves. Fawns succumbed sooner to starvation than did does, probably due to the exhaustion of their more meager endogenous energy reserves. In a review of the literature, Klein (1970) concluded that body weight is one of the more universally used criteria for reflecting the growth rate and nutritive status of deer, and he quoted numerous examples of large body size being correlated with good forage quality.

Laws and Parker (1968) used the von Bertalanffy (1938) growth equation to measure growth in height and weight in elephant populations in East Africa. They suggested that an elephant population that has become overcrowded and undernourished within the last few years could experience a consequent reduction in growth. They proposed that in future comparative work on elephant populations, two of the von Bertalanffy coefficients, K and H_∞ would be useful parameters to measure. However, in a detailed analysis of the derivation of the three coefficients in the equation, Hanks (1972) concluded that in animals that have a long life span the von Bertalanffy equation serves as a purely empirical representation of weight (or height) at age data, and there is little biological significance in the parameters it contains.

In an attempt to facilitate the recording of physiological condition in live deer and several other ungulates, Riney (1960) described a field technique in which animals were classified and placed in one of three condition classes—good, fair, or poor. These classes were based on one of the many characteristics of inanition, the tendency of an animal to become thin as its condition drops. The method was used subsequently by Child et al. (1972) with some success with tsessebe *(Damaliscus lunatus lunatus)*, and in a modified form by Albl (1971) with elephants. However, both Huntley (1971) and Hanks et al. (1976) have drawn attention to the fact that considerable changes can occur in deposited fat reserves without expressing themselves in the external appearance of animals. As an example, Hanks et al. (1976) described how two impala could have identical weight and girth measurements (and look identical in the field), and yet at the same time they could differ substantially in their deposited fat reserves as demonstrated by KFI and BMF. It is only after a further mobilization of fat reserves that the body weight falls significantly and the girth measurement is reduced (a reclassification resulting in terms of Riney's field technique). This was confirmed by the fact that although the relationship

$$W = a + bLG^2$$

[where W is total body weight (kg), L is total length (cm), and G is girth (cm)] was highly significant, the relationship between the same linear measurements and KFI gave a much lower correlation.

DEMOGRAPHIC VIGOR

Caughley (1977) has defined several concepts of rate of increase that are relevant to this review:

$\bar{r}$ = observed rate of increase

r_s = survival-fecundity rate of increase, the rate implied by prevailing schedules of survival and fecundity

r_m = intrinsic rate of increase, the rate achieved in the absence of crowding and of shortage of resources

These three rates are mutually dependent and occasionally congruent. Caughley has pointed out that each of these rates is designated by the generic symbol $\bar{r}$, and subscripts are used only when a particular interpretation of $\bar{r}$ must be specified. Thus r_s, the demographic vigor of a population (as defined by

Caughley), weighs up and combines the vigor of each sex and age class of the population under the environmental conditions it actually faces. The concept of demographic vigor is a valuable adjunct to physiological condition. Whereas the latter is related to an individual animal's chances of living or dying, the demographic vigor of a population is its level of well-being, describing the average reaction of all members of a population to the collective action of all environmental variables.

Unfortunately, there are many difficulties and tautologies associated with the measurement of r_s; these have been described in detail by Caughley (1977) and need not be repeated here. Whereas r_s is difficult to determine, $\bar{r}$, a much more general measure of rate of increase, can be calculated relatively easily (Grimsdell and Bell, 1972) by regression analysis from two or more estimates of population size (Figure 3). A disadvantage of using $\bar{r}$ as an index of demographic vigor is that it is averaged over a period of time, in contrast to r_s, which is calculated from age-specific survival and fecundity schedules under conditions that a population actually faces at a given point in time.

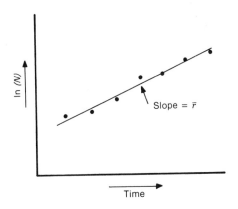

Figure 3   The estimate of $\bar{r}$ when natural logarithms of animal numbers are plotted against time.

Whatever single statistic is used to index demographic vigor, none can predict possible future population trends. The introduced reindeer population on St. Matthew Island increased from 1350 in 1957 to 6000 in 1963. At any one time between those two dates, both $\bar{r}$ and r_s would have had high positive values, indicating that the population was in "good condition" in terms of demographic vigor. Yet, soon after 1963 the population experienced a spectacular crash. If the index of demographic vigor had had, as an adjunct, an index of physiological condition, the fact that the majority of animals in the population were close to death although not dying, could have been detected, and possible population projections made accordingly. Quite clearly, a single statistic of population condition based on rate of increase, can at times be quite misleading. Nevertheless, certain demographic parameters can be used rather more profitably than rate of increase per se, providing that the relative importance of each parameter is specified.

In the assessment of physiological condition, it was noted that certain events take place in sequence as an individual declines in condition. As demographic vigor declines it is assumed that the death rate increases and the birthrate declines. Should it be possible to attach a sequence and index of relative importance to these events, the wildlife manager might have some very useful and relatively easily determined field criteria with which to assess demographic vigor. There is now a good deal of evidence to suggest that an increase in juvenile mortality is not only the proximate expression of downward trend in rate of increase in a wide range of mammalian species, but that it is also the major component influencing the rate of increase (Klein and Olson, 1960; Richens, 1967; Caughley, 1970b; Hanks and McIntosh, 1973; Corfield, 1973; Eberhardt, 1977). A possible sequence of events in a population of large mammals as demographic vigor declines could be as follows (modified after Eberhardt, 1977):

1   Juvenile mortality rate increases.

2   Age at first reproduction increases.

3   Fecundity (m_x) declines.

4   Adult mortality rate increases.

It seems reasonable to conclude that in large mammals, all of which are long-lived species, security of the adult population is the key to the persistence of the species. The last parameter to give way as a population is subjected to environmental stress is adult survival, in particular survival of adult females. It also seems reasonable to conclude that the juvenile mortality rate is a vital key factor in the analysis of demographic vigor.

In an important series of studies on the relationship between reproductive success and nutritional levels in white-tailed deer, Verme (1963, 1967, 1969, 1977) concluded that a doe's nutritional plane, especially during the latter part of pregnancy, greatly influences the growth of her fetus, and, therefore, its chances of survival at birth. Early neonatal death occurred because stunted fawns were too weak to stand or too small to reach the teats, their mother had abandoned them, or lactation was delayed or absent. Verme was able to demonstrate that fawn survival can be appraised by fetal body weight analysis, relatively low fetal weights being correlated with impending natal loss. Similar conclusions have been reached in domestic ruminants by Tassell (1967) and Everitt (1968). Variations in fetal development yield an integrated measure of the sum of stresses imposed on the doe throughout pregnancy. Thus a measurement of subnormal fetal development could indicate an impending increase in neonatal mortality, and as such could have value as an index of demographic vigor.

Several studies have related reproductive homeostasis to either the nutritional plane or population density, and have called attention to the importance of variations in fertility (Cheatum and Severinghaus, 1950; Dahlberg and Guettinger, 1956; Taber, 1956; Barick, 1958; Myers and Poole, 1962; Severinghaus and Tanck, 1964; Verme, 1969; Sadleir, 1969; Klein, 1970; Sinclair, 1974; Laws et al., 1975). However, a comparison with other studies leads to the conclusion that although these variations can and do influence rate of increase, they are far less important as indices of demographic vigor than are variations in fetal development and juvenile mortality.

BEHAVIORAL ATTRIBUTES AS AN INDEX OF POPULATION CONDITION

Several studies of population condition have indicated that a decline in either physiological condition or demographic vigor can be correlated with certain behavioral attributes of the population. While it is beyond the scope of this review to consider these behavioral attributes in any detail, attention should be drawn to their potential value as either additional indices of population condition, or as a possible "early-warning" system of adverse population trends. Ideally, the relevant behavioral characteristics should be included in all studies of population condition.

The behavioral attributes of interest can be grouped into three very broad categories: (1) rate and quality of social interaction, (2) population density, and (3) feeding strategy.

An excellent review of the majority of the important behavioral attributes of relevance, including such aspects as dominance, spacing behavior, and aggression in relationship to population limitation in vertebrates, was made by Watson and Moss (1970). They reached the important conclusion that although there are many papers dealing with territorial behavior and with other forms of dominance and spacing behavior in vertebrates and many on population studies, as well as much work on nutrition, very few studies consider any two of these aspects together, and practically none considers all three. Consequently, there is an inordinate amount of speculation in the literature, but there are relatively few data about the relationships among these aspects. Watson and Moss also concluded that dominance and spacing behavior are usually affected by a change in the plane of nutrition before animals are actually killed by starvation. This last observation by Watson and Moss points to the value of regular monitoring of the rate and quality of social interaction, coupled with the determination of population density, in all studies of population condition.

The term "feeding strategy" is taken in this review to include all components of an animal's feeding habits, including qualitative and quantitative aspects of the diet. The value of this behavioral attribute in a study of population condition is illustrated by the work of Klein (1970). Describing the feeding strategy of deer in North America, he found that deer are selective in their feeding habits, usually choosing plants, or parts of plants, which are of the highest nutrititive quality. Where food becomes limited, either through plant ecological changes or increases in population density, the preferred plant species of high forage quality are reduced and often eliminated by the deer. Klein also concluded that growth of North American deer is more directly related to the qualitative and quantitative aspects of food supply and may not necessarily be related to the density of the population.

The regular monitoring of feeding strategies should clearly become an integral part of all determinations of population condition.

THE IMPORTANCE OF THE ASSESSMENT OF HABITAT CONDITION AND TREND

All condition measurements should be related to the composition, condition and trend of available food resources. Unfortunately, very few studies of population condition have attempted to relate the animal to its habitat, and as a consequence both physiological and demographic parameters come from populations with undefined nutritional histories.

Primary production itself must not be equated simply with food. Sinclair (1975) pointed out that in most grassland ecosystems there are times during the growing season when there is a considerable excess of food, but there are also times when growth ceases and there is a shortage of food. He stresses that what is overlooked is that in the nongrowing season, the previous excess of food is no longer available to the animals because it changes in nature, both structurally and chemically, and becomes unsuitable as forage. In grasslands there is a transport of proteins and carbohydrates to the roots, where it is unavailable to most herbivores, leaving a highly lignified and indigestible remnant with a low nutrient content. It is now generally known that grass with less than 4% crude protein is of too low a quality for ruminants to maintain their body weight, and hence it ceases to be food for them.

In Serengeti, Sinclair (1975) found that green grass, with its higher protein content, is the limiting resource, the quantity of which was affected by the size of the grazing populations. The more conspicuous quantities of dry grass were of little value as food. Within three grassland systems, there was on average, a short period of one to four months during the dry season when available food was lower than the requirements of the herbivore trophic level. Such a shortage would be sufficient to limit the herbivore population.

Although several authors have attempted to use a variety of physiological parameters to evaluate habitat condition, the significant seasonal variations in food quality described by Sinclair (1975) suggest that no measure of physiological condition can ever be a satisfactory substitute for a direct evaluation of the habitat itself. Figure 4 illustrates a hypothetical relationship between diet quality and deposited fat reserves, as indexed by KFI, in two populations of animals. In situation A, the population has a seasonally stable diet, and KFI does not fluctuate in any significant manner. Sampling the population at the points indicated by the arrows would yield identical KFI values, leading to the possible conclusion that the condition of the habitat, as indicated by diet quality, was the same in both cases. In situation B, diet quality undergoes marked fluctuations during the year (a common occurrence in subtropical ecosystems).

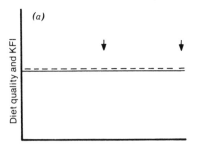

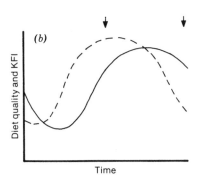

Figure 4 Hypothatical relationship between diet quality (broken line) and kidney fat index (KFI) (solid line). Arrows indicate time of sampling. (For explanation, see text.)

Under such circumstances, it is reasonable to anticipate a delay in lipogenesis as diet quality improves, coupled with a delay in the mobilization of fat reserves as diet quality falls. In this case, as in situation A, sampling the population at the points indicated by the arrows would yield identical KFI values, once again leading to the possible, but nevertheless incorrect conclusion, that the habitat was identical in terms of diet quality. The true situation was that at the first point of sampling population B, KFI was increasing, as was diet quality, but at the second point of sampling, KFI was decreasing with a very low diet quality.

This example is, of course, deliberately oversimplified for the purposes of discussion, but it does serve to illustrate the importance of a direct evaluation of the habitat itself in relation to the herbivore populations. Particularly useful recent studies in this context have been those of Ferrar and Walker (1974), Mentis and Duke (1976), Walker (1976), and Mentis (1977).

POPULATION CONDITION AND RESILIENCE

My interpretation of resilience differs from that of Holling (1973), who used the terms "resilience" and "stability" to describe ecological systems. Holling regarded "stability" as "the ability of a system to return to an equilibrium state after a temporary disturbance," whereas "resilience" determines "the persistence of relationships within a system." I have used "resilience" to indicate the extent to which an individual or a population can recover from "displacement" of any nature, and this must indicate its ability to absorb "displacement." "Stability," as I understand it, implies an immovability, which is unacceptable within this context.

This discussion of physiological condition and demographic vigor deals with systems profoundly affected by changes external to them. In an individual animal, the physiological system has been affected by gradual variations in food quantity and quality, or by environmental perturbations. If a measure of these physiological changes is to be of any practical value, it must be linked to an animal's chances of living or dying. An animal in "good" physiological condition should be much further removed from death than one in "poor" condition. Ideally, any index of physiological condition should include a measure of resilience. Thus, an individual animal in "good" condition should be not only far removed from death, but should be sufficiently resilient to absorb either a progressive decline in food resources or sudden environmental stress. Similarly, an individual animal in "poor" condition should be not only close to death, but it should also be so low in resilience that it is unable to tolerate any further deprivations.

In this review, five broad methods have been used to index physiological condition, namely, deposited fat reserves, adrenocortical hypertrophy, blood chemistry and hematology, urinary excretion of hydroxyproline, and body weight (linked with a "visual" classification system). With the exception of the work by deCalesta et al. (1975) relating body weight loss to starvation and eventual death, no attempt was made in any of the papers reviewed to equate a particular physiological parameter to the animal's chances of living or dying, although in many cases this was indirectly assumed by ill-defined references to "good" and "poor" condition. As an example, it is assumed that an animal experiencing adrenocortical hypertrophy is in "poor" condition, yet in most cases no evidence is led to support this assertion. A population biologist or wildlife manager should be able to evaluate the significance of a "poor" condition index associated with adrenocortical hypertrophy, and unless such an index can be related to the capacity for the animal to absorb further environmental stress, a measure of hypertrophy per se will be of little use.

The characterization of population condition, based on an assessment of physiological condition, requires a great deal of further study to elucidate the details of the complex relationships between the various physiological parameters concerned, coupled with a study of the significance of these parameters in terms of an individual animal living or dying. Furthermore, each parameter within the "physiological system" needs to be much more carefully examined in relation to the "systems" response to all the relevant environmental variables. Finally, to enhance the practical value of a physiological condition index, each parameter should be assessed in terms of its resilience.

At the present state of our knowledge it is difficult to recommend the best technique to measure physiological condition. So many of the techniques reviewed are subject to errors of interpretation in the absence of stringent methods of field collection, and are further complicated by the fact that individual animals and populations have undefined nutritional histories. As a preliminary recommendation, deposited fat reserves, as indexed by KFI and BMF, appear to be the most useful and least contentious methods for assessing physiological condition. With a KFI value of 80 or above, an individual animal is in "good" condition, and is still very resilient to deprivations. As long as KFI remains above 40, there is little to be gained by carrying out an assessment of BMF, as very little fat is mobilized from that source. However, once KFI falls below that value, BMF declines rapidly. As BMF is the last of the major fat reserves to be mobilized, any individual animal experiencing a reduction in BMF moves into the "poor" condition category. Once fat has been mobilized from the metatarsus and metacarpus, the reserves are exhausted. Resilience is very low, and the individual animal is in "very poor" condition.

The use of a single statistic to index demographic vigor tells us nothing about the recuperative powers of a population subjected to sudden environmental stress. Figure 5 shows the hypothetical responses of two populations subjected to similar environmental perturbations. In population A, prior to the perturbation, the rate of increase was high, and r_s was equal to r_m. The population was unable to tolerate the environmental pressures, and experienced a drastic population crash and slow recovery, very similar to the reindeer on St. Matthew Island described by Klein (1968). In contrast, population B was stationary, and in this case, $r_s = 0$. Although environmental pressures resulted in a decline, the population soon recovered. If demographic vigor alone had been equated with population condition, population A would have received a much higher rating than population B, yet such a statistic could be very misleading to a wildlife manager. Population A was low in resilience, because most of the animals had a low physiological condition index. In contrast, population B was able to absorb stress and deprivation, and recovered rapidly without declining drastically. It was high in resilience.

The problems associated with the measurement of demographic vigor have already been discussed and need not be repeated here. As a useful field guide, subnormal fetal development and increases in juvenile mortality are the best indices of declining demographic vigor and should be used in conjunction with a measure of physiological condition.

In summary, the characterization of population condition may be described in terms of physiological condition and demographic vigor. Deposited fat reserves, adrenocortical hypertrophy, blood chemistry and hematology, urinary excretion of hydroxyproline, and body growth may be considered indices of

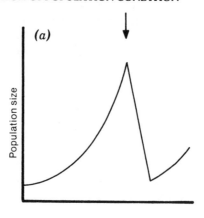

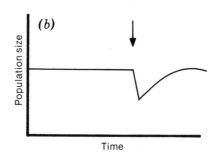

Figure 5 Hypothetical population response to environmental perturbations (arrows). Population (*a*), low resilience; population (*b*), high resilience. (For explanation, see text.)

physiological condition. Deposited fat reserves, as indexed by perinephric and bone marrow fat, are the most useful and least contentious methods for assessing physiological condition. To enhance the practical value of these indices, each should be assessed in terms of its resilience. The use of a single statistic to describe demographic vigor in general presents many problems. As an alternative, subnormal fetal development and increases in juvenile mortality are the best practical indications of declining demographic vigor. Ideally, demographic vigor should be assessed together with physiological condition, and, where possible, both indices should be related to relevant behavioral attributes of the population and to an assessment of habitat condition and trend.

The characterization of population condition is an important aspect of large mammal population dynamics that deserves further study. It should be integrated with a multidisciplinary approach that includes behavioral characteristics and an assessment of habitat condition and trend.

LITERATURE CITED

Albl, P. 1971. Studies on assessment of physical condition in African elephants. Biol. Conserv. 3:134-140.

Allen, E. O. 1968. Range use, foods, condition and productivity of white-tailed deer in Montana. J. Wildl. Manage. 32:130-141.

Anderson, A. E., D. E. Medin, and C. D. Bowden. 1971. Adrenal weight in a Colorado mule deer population. J. Wildl. Manage. 35:689-697.

Anderson, A. E., D. E. Medin, and C. D. Bowden. 1972a. Indices of carcass fat in a Colorado mule deer population. J. Wildl. Manage. 36:579-594.

Anderson, A. E., D. E. Medin, and C. D. Bowden. 1972b. Total serum protein in a population of mule deer. J. Mammal. 53:384-387.

Andrews, R. V. 1968. Daily and seasonal variation in adrenal metabolism of the brown lemming. Physiol. Zool. 41:88-94.

Andrews, R. V. 1970a. Effect of climate and social pressure on the adrenal response of lemmings, voles and mice. Acta Endocrinol. (Copenhagen) 65:639-664.

Andrews, R. V. 1970b. Circadian variations in adrenal secretion of lemmings, voles, and mice. Acta Endocrinol. (Copenhagen) 65:645-654.

Bailey, J. A. 1968. A weight-length relationship for evaluating physical condition of cottontails. J. Wildl. Manage. 32:835-841.

Bandy, P. J., W. D. Kitt, A. J. Wood, and I. McT. Cowan. 1957. The effect of age and the plane of nutrition on the blood chemistry of the Columbian black-tailed deer *(Odocoileus hemionus columbianus)*. B. Blood glucose, non-protein nitrogen, total plasma protein, plasma albumin, globulin and fibrinogen. Can. J. Zool. 35:283-289.

Barick, F. B. 1958. A study of deer productivity. Wildl. N. Carol. 22:6-10.

Bear, G. D. 1971. Seasonal trends in fat levels of pronghorns, *Antilocapra americana*, in Colorado. J. Mammal. 52:583-589.

Benjamin, M. M. 1961. Outline of veterinary clinical pathology. Iowa State University Press, Ames, Iowa.

Berrier, H. H. 1968. Diagnostic aids in the practice of veterinary medicine. Alban Professional Books, St. Louis, Missouri.

Bertalanffy, L. von. 1938. A quantitative theory of organic growth. Hum. Biol. 10:181-213.

Bischoff, A. I. 1954. Limitations of the bone marrow technique in determining malnutrition in deer. Proc. West. Assoc. State Game Fish Comm. 34:205-210.

Brand, C. J., R. H. Vowless, and L. B. Keith. 1975. Snowshoe hare mortality monitored by telemetry. J. Wildl. Manage. 39:741-747.

Bronson, F. H., and B. E. Eleftheriou. 1965. Adrenal response to fighting in mice: separation of physical and psychological causes. Science 147:627-628.

Brooks, P. M., J. Hanks, and J. V. Ludbrook. 1977. Bone marrow as an index of condition in African ungulates. S. Afr. J. Wildl. Res. 7:61-66.

Browman, L. G., and H. S. Sears. 1956. Cyclic variation in the mule deer thymus. Proc. Soc. Exp. Biol. Med. 93:161-162.

Brown, I. R. F., P. T. White, and R. C. Malpas. 1978. Proteins and other nitrogenous constituents in the blood serum of the African elephant, *Loxodonta africana*. Comp. Biochem. Physiol. 59A:267-270.

Bujalska, G. 1971. Self-regulation of reproduction in an island population of *Clethrionomys glareolus* (Schreber, 1780). Ann. Zool. Fenn. 8:90-93.

Caughley, G. 1970a. Fat reserves of Himalayan Thar in New Zealand by season, sex, area and age. N. Z. J. Sci. 13:209-219.

Caughley, G. 1970b. Eruption of ungulate populations, with emphasis on Himalayan Thar in New Zealand. Ecology 51:53-72.

Caughley, G. 1971. Demography, fat reserves and body size of a population of red deer *Cervus elaphus* in New Zealand. Mammalia 35:369-383.

Caughley, G. 1977. Analysis of vertebrate populations. John Wiley and Sons, London.

Cheatum, E. L. 1949. Bone marrow as in index of malnutrition in deer. N. Y. State Conserv. 3:19-22.

Cheatum, E. L., and C. W. Severinghaus. 1950. Variations in fertility of white-tailed deer related to range condition. Trans. N. Am. Wildl. Conf. 15:171-190.

Child, G., H. Robbel, and C. P. Hepburn. 1972. Observations on the biology of tsessebe, *Damaliscus lunatus lunatus* in Northern Botswana. Mammalia 36:342-388.

Chitty, H. 1961. Variations in the weight of the adrenal gland of the field vole, *Microtus agrestis*. J. Endrocrinol. 22:387-397.

Christian, J. J. 1955. The effects of population size on the adrenal glands and reproductive organs of male mice in populations of fixed size. Am. J. Physiol. 182:242-300.

Christian, J. J. 1975. Hormonal control of population growth. Pp. 205-274 *in:* B. E. Eleftheriou and R. L. Sprott (eds.). Hormonal Correlates of Behavior, Vol. 1. Plenum Press, New York.

Christian, J. J., and D. E. Davis. 1964. Endocrines, behavior and population. Science 146:1550-1560.

Christian, J. J., and D. E. Davis. 1966. Adrenal glands in female voles *(Microtus pennsylvanicus)* as related to reproduction and population size. J. Mammal. 47:1-18.

Christian, J. J., V. Flyger, and D. E. Davis. 1960. Factors in the mass mortality of a herd of sika deer, *Cervus nippon*. Chesapeake Sci. 1:79-95.

Coblentz, B. E. 1975. Serum cholesterol level changes in George Reserve deer. J. Wildl. Manage. 39:342-345.

Colby, R. W., J. H. Ware, J. P. Baker, and C. M. Lyman. 1950. The relationship of various blood constituents to rate of gain in beef cattle. J. Anim. Sci. 9:652.

Cooper, A. C. D., G. M. Stuttaford, and I. H. Carmichael. 1975. Studies on the serum proteins of game animals in Botswana. A preliminary report. E. Afr. Wildl. J. 13:145-148.

Corfield, T. F. 1973. Elephant mortality in Tsavo National Park, Kenya. E. Afr. Wildl. J. 11:339-368.

Dacie, J. V., and S. M. Lewis. 1968. Practical Haematology. J. & A. Churchill, London.

Dahlberg, B. L., and R. C. Guettinger. 1956. The white-tailed deer in Wisconsin. Tech. Wildl. Bull. Game Manage. D. V. Wisc. 14:282.

Dauphiné, T. C. 1975. Kidney weight fluctuations affecting the kidney fat index in caribou. J. Wildl. Manage. 39:379-386.

deCalesta, D. S., J. G. Nagy, and J. A. Bailey. 1975. Starving and refeeding mule deer. J. Wildl. Manage. 39:663-669.

deCalesta, D. S., J. G. Nagy, and J. A. Bailey. 1977. Experiments on starvation and recovery of mule deer does. J. Wildl. Manage. 41:81-86.

Eberhardt, L. L. 1977. Optimal policies for conservation of large mammals, with special reference to marine ecosystems. Environ. Conserv. 4:205-212.

Eggum, B. C. 1970. Blood urea measurements as a technique for assessing protein quality. Br. J. Nutr. 24:983-988.

Eleftheriou, B. E. 1964. Bound and free corticosteriod in the plasma of two subspecies of deer mice *(Peromyscus maniculatus)* after exposure to low ambient temperature. J. Endocrinol. 31:75-79.

Erickson, A. W., and W. C. Youatt. 1961. Seasonal variation in the hematology and physiology of black bears. J. Mammal. 42:198-203.

Eskeland, B., W. E. Pfander, and P. L. Preston. 1974. Intravenous energy infusions in lambs; effects on nitrogen retention, plasma free amino acids and plasma urea nitrogen. Br. J. Nutr. 31:201-211.

Everitt, G. C. 1968. Prenatal development of uniparous animals with particular reference to the influence of maternal nutrition in sheep. In: G. A. Lodge and C. E. Lamming (eds.). Growth and Development of Mammals. Butterworth, London.

Ferrar, A. A., and B. H. Walker. 1974. An analysis of herbivore/habitat relationship in Kyle National Park, Rhodesia. J. S. Afr. Wildl. Manage. Assoc. 4:137-147.

Flook, D. R. 1970. Causes and implications of an observed sex differential in the survival of wapiti. Can. Wildl. Serv. Rept. Ser. 11. 71 pp.

Frankel, S., S. Reitman, and A. Sonnenwirth. 1970. Gradwohl's clinical laboratory methods and diagnosis. C. V. Mosby Co., St. Louis, Missouri.

Franzmann, A. W. 1971. Application of physiological values to Bighorn sheep management. Trans. 1st N. Am. Wildl. Sheep Conf. Pp. 87-90.

Franzmann, A. W., and P. D. Anreson. 1976. Marrow fat in Alaskan moose femurs in relation to mortality factors. J. Wildl. Manage. 40:336-339.

Franzmann, A. W., and E. T. Thorne. 1970. Physiologic values in wild bighorn sheep (Ovix canadensis canadensis) at capture, after handling, and after capture. J. Am. Vet. Med. Assoc. 157:647-650.

Greer, K. R. 1968. A compression method indicates fat content of elk (wapiti) femur marrow. J. Wildl. Manage. 32:747-751.

Goldzieher, M. A. 1946. Endocrine aspects of senescence. Geriatrics 1:226-231.

Grimsdell, J. J. R., and R. H. V. Bell. 1972. Population growth of red lechwe (Kobus leche leche Gray) in the Busanga Plain, Zambia. E. Afr. Wildl. J. 10:117-122.

Halloran, D. W., and A. M. Pearson. 1972. Blood chemistry of the brown bear (Ursus arctos) from southwestern Yukon Territory, Canada. Can. J. Zool. 50:827-833.

Hanks, J. 1972. Growth of the African elephant (Loxodonta africana). E. Afr. Wildl. J. 10:251-272.

Hanks, J., D. H. M. Cumming, J. L. Orpen, D. F. Parry, and H. B. Warren. 1976. Growth, condition and reproduction in the impala ram (Aepyceros melampus), J. Zool., Lond. 197:421-435.

Hanks, J., and J. E. A. McIntosh. 1973. Population dynamics of the African elephant (Loxodonta africana). J. Zool., Lond. 169:29-38.

Harris, D. 1945. Symptoms of malnutrition in deer. J. Wildl. Manage. 9:319-322.

Heidt, G. A., and J. Hargraves. 1974. Blood chemistry and hematology of the spotted skunk, Spilogale putorius. J. Mammal. 55:206-208.

Herin, R. A. 1968. Physiological studies in the Rocky Mountain elk. J. Mammal. 49:762-764.

Hirst, S. M. 1969. Predation as a regulating factor of wild ungulate populations in a Transvaal lowveld nature reserv. Zool. Afr. 4:199-231.

Hoffman, R. A., and P. F. Robinson. 1966. Changes in some endocrine glands of white- tailed deer as affected by season, sex and age. J. Mammal. 47:266-280.

Hollings, C. S. 1973. Resilience and stability of ecological systems. Annu. Rev. Ecol. Syst. 4:1-23.

Holter, J. B., and H. H. Hayes. 1977. Growth in white-tailed deer fawns fed varying energy and constant protein. J. Wildl. Manage. 41:506-510.

Houston, D. B. 1969. A note on the blood biochemistry of the Shiras moose. J. Mammal. 50:826.

Howells, G. R., and R. G. Whitehead. 1967. A system for the estimation of the urinary hydroxyproline index. J. Med. Lab. Tech. 24:98-102.

Hughes, E., and R. Mall. 1958. Relation of the adrenal cortex to condition of deer. Calif. Fish Game. 44:191-196.

Huntley, B. J. 1971. Seasonal variation in the physical condition of mature male blesbok and kudu. J. S. Afr. Wildl. Manage. Assoc. 1:17-19.

Johnson, H. E., W. G. Youatt, L. D. Fay, H. D. Harte, and D. E. Ullrey. 1968. Hematological values of Michigan white-tailed deer. J. Mammal. 49:749-754.

Kirkpatrick, R. L., D. E. Buckland, W. A. Abler, P. F. Scanlon, J. B. Whelan, and H. E. Burkhart. 1975. Energy and protein influences on blood urea nitrogen of white-tailed deer fawns. J. Wildl. Manage. 39:692-698.

Klein, D. R. 1968. The introduction, increase, and crash of reindeer on St. Matthew Island. J. Wildl. Manage. 32:350-367.

Klein, D. R. 1970. Food selection by North American deer and their response to over- utilization of preferred plant species. Pp. 25-46 in: A. Watson (ed.). Animal Populations in Relation to Their Food Resources. Blackwell, Oxford.

Klein, D. R., and S. T. Olson. 1960. Natural mortality patterns of deer in southeastern Alaska. J. Wildl. Manage. 24:80-88.

Klein, D. R., and H. Strandgaard. 1972. Factors affecting growth and body size of roe deer. J. Wildl. Manage. 36:64-79.

Krumrey, W. A., and I. O. Buss. 1969. Observations on the adrenal gland of the African elephant. J. Mammal. 50:90-101.

Kumar, V., H. P. Chase, K. Hammond, and D. O'Brien. 1972. Alterations in blood biochemical tests in progressive protein malnutrition. Pediatrics 49:736-743.

Laws, R. M., and I. S. C. Parker. 1968. Recent studies on elephant populations in East Africa. Symp. Zool. Soc. London 21:319-359.

Laws, R. M., I. S. C. Parker, and R. C. B. Johnstone. 1970. Elephants and habitats in North Bunyoro, Uganda. E. Afr. Wildl. J. 8:163-180.

Laws, R. M., I. S. C. Parker, and R. C. B. Johnstone. 1975. Elephants and Their Habitats. The Ecology of Elephants in North Bunyoro, Uganda. Clarendon Press, Oxford.

Ledger, H. P., and N. S. Smith. 1964. The carcass and body composition of the Uganda kob. J. Wildl. Manage. 28:825-839.

Lee, J., K. Ronald, and N. A. Oritsland. 1977. Some blood values of wild polar bears. J. Wildl. Manage. 41:520-526.

Malpas, R. C. 1977. Diet and the condition and growth of elephants in Uganda. J. Appl. Ecol. 14:489-504.

McCullagh, K. G. 1969. The growth and nutrition of the African elephant. 1. Seasonal variations in the rate of growth and the urinary excretion of hydroxyproline. E. Afr. Wildl. J. 7:85-90.

McCullagh, K. G. 1973. Are African elephants deficient in essential fatty acids? Nature 242:267-268.

McEwan, E. H. 1968. Hematological studies of barren-ground caribou. Can. J. Zool. 46:1031-1036.

McEwan, E. H., and P. E. Whitehead. 1969. Changes in the blood constituents of reindeer and caribou occurring with age. Can. J. Zool. 47:557-562.

McKeever, S. 1959. Effects of reproductive activity on the weight of adrenal glands in *Microtus montanus*. Anat. Rec. 135:1-5.

McKinney, T. D., D. M. Baldwin, and R. H. Giles. 1970. Effects of differential grouping on adrenal catecholamines in the cottontail rabbit. Physiol. Zool. 43:55-59.

McKinney, T. D., and M. R. Dunbar. 1976. Weight of adrenal glands in the bobcat *(Lynx rufus)*. J. Mammal. 57:378-380.

McNab, B. K. 1976. Seasonal fat reserves of bats in two tropical environments. Ecology 57:332-338.

Mentis, M. T. 1977. Stocking rates and carrying capacities for ungulates on African rangelands. S. Afr. J. Wildl. Res. 7:89-98.

Mentis, M. T., and R. R. Duke. 1976. Carrying capacities of natural veld in Natal for large wild herbivores. S. Afr. J. Wildl. Res. 7:65-74.

Myers, K., and W. E. Poole. 1962. A study of the biology of the wild rabbit, *Oryctolagus cumiculus* (L) in confined populations. III. Reproduction. Aust. J. Zool. 10:225-267.

Neiland, K. A. 1970. Weight of dried marrow as indicators of fat in caribou femurs. J. Wildl. Manage. 34:904-907.

Park, B. C., and B. B. Day. 1942. A simplified method for determining the condition of white-tailed deer herds in relation to available forage. U.S.D.A. Tech. Bull. 840:1-60.

Passmore, R. 1968. Essential materials and waste. *In*: R. Passmore and J. S. Robson (eds.). A Companion to Medical Studies, Vol. 1. Blackwell, Oxford.

Pedersen, R. J., and A. A. Pedersen. 1975. Blood chemistry and hematology of elk. J. Wildl. Manage. 39:617-620.

Pinter, A. J. 1968. Effects of diet and light on growth, maturation, and adrenal size of *Microtus montanus*. Am. J. Physiol. 215:461-466.

Quay, W. B. 1960. The reproductive organs of the collared lemming under diverse temperature and light conditions. J. Mammal. 41:74-80.

Ransom, A. B. 1965. Kidney and marrow fat as indicators of white-tailed deer condition. J. Wildl. Manage. 29:397-398.

Ravel, R. 1973. Clinical Laboratory Medicine, 2nd ed. Year Book Med. Publ. Inc., Chicago, Illinois.

Richens, V. G. 1967. Characteristics of mule deer herds and their range in north eastern Utah. J. Wildl. Manage. 31:651-666.

Riney, T. 1955. Evaluating condition of free-ranging red deer *(Cervus elaphus)*, with special reference to New Zealand. N. Z. J. Sci. Tech. Sect. B. 36:429-463.

Riney, T. 1960. A field technique for assessing physical condition of some ungulates. J. Wildl. Manage. 24:92-94.

Rosen, M. N., and A. I. Bischoff. 1952. The relation of hematology to condition in California deer. Trans. N. Am. Wildl. Conf. 17:482-496.

Sadleir, R. M. F. S. 1969. The Ecology of Reproduction in Wild and Domestic Mammals. Methuen, London.

Schalm, O. W. 1965. Veterinary Hematology. Lea & Febiger, Philadelphia.

Seal, U. S., L. D. Mech, and V. Van Ballenberghe. 1975. Blood analysis of wolf pups and their ecological and metabolic interpretation. J. Mammal. 56:64-75.

Seal, U. S., and A. W. Erickson. 1969. Hematology, blood chemistry, and protein polymorphisms in the white-tailed deer *(Odocoileus virginianus)*. Comp. Biochem. Physiol. 30:695-713.

Seal, U. S., J. J. Ozoga, A. W. Erickson, and L. J. Verme. 1972. Effects of immobilization on blood analysis of white-tailed deer. J. Wildl. Manage. 36:1034-1040.

Severinghaus, C. W., and J. E. Tanck. 1964. Productivity and growth of white-tailed deer from the Adirondack region of New York. N.Y. Fish Game J. 11:13-27.

Sinclair, A. R. E. 1970. Studies of the ecology of the East African buffalo. Ph.D. Thesis, Oxford University.

Sinclair, A. R. E. 1970. The natural regulation of buffalo populations in East Africa. II. Reproduction, recruitment and growth. E. Afr. Wildl. J. 12:169-183.

Sinclair, A. R. E. 1975. The resource limitation of trophic levels in tropical grassland ecosystems. J. Anim. Ecol. 44:497-520.

Sinclair, A. R. E., and P. Duncan. 1972. Indices of condition in tropical ruminants. E. Afr. Wildl. J. 10:143-149.

Smiley, J. D., and M. Ziff. 1964. Urinary hydroxyproline excretion and growth. Physiol. Rev. 44:30-44.

Smith, N. S. 1970. Appraisal of condition estimation methods for East African ungulates. E. Afr. Wildl. J. 8:123-129.

Southwick, C. H. 1957. The population dynamics of confined house mice supplied with unlimited food. Ecology 36:212-225.

Stewart, S. F., H. A. Norden, A. J. Wood, and I. McT. Cowan. 1964. Changes in the plasma lipids in the black-tailed deer throughout the year. Proc. Int. Congr. Zool. 2:46.

Taber, R. D. 1956. Deer nutrition and population dynamics in the north Coast Range of California. Trans. N. Am. Wildl. Conf. 21:159-172.

Taber, R. D., and R. F. Dasmann. 1958. The black-tailed deer of the chaparral: Its life history and management in the north Coast Range of California. Calif. Fish Game Bull. 8:1-163.

Taber, R. D., K. L. White, and N. S. Smith. 1959. The annual cycle of condition in the Rattlesnake, Montana, mule deer. Proc. Montana Acad. Sci. 19:72-79.

Tassell, R. 1967. The effects of diet on reproduction in pigs, sheep and cattle. V. Plane of nutrition in cattle. Br. Vet. J. 123:459-463.

Thompson, R. D., D. J. Elias, and G. C. Mitchell. 1977. Effects of vampire bat control on bovine milk production. J. Wildl. Manage. 41:736-739.

To, L. P., and R. H. Tamarin. 1977. The relation of population density and adrenal gland weight in cycling and noncycling voles (Microtus). Ecology 58:928-934.

Trout, L. E., and J. L. Thiessen. 1968. Food habits and condition of mule deer in Owyhee County. Proc. West. Assoc. State Game Fish Comm. 48:188-200.

Tumbleson, M. E., J. D. Cuneio, and D. A. Murphy. 1970. Serum biochemical and hematological parameters of captive white-tailed fawns. Can. J. Comp. Med. 34:66-71.

Tumbleson, M. E., J. W. Ticer, A. R. Dommert, D. A. Murphy, and L. J. Korschgen. 1968. Serum proteins in white-tailed deer in Missouri. Am. J. Vet. Clin. Pathol. 2:127-131.

Ullrey, D. E., W. G. Youatt, H. E. Johnson, L. D. Fay, B. E. Brent, and K. E. Kemp. 1968. Digestibility of cedar and balsam fir browse for the white-tailed deer. J. Wildl. Manage. 32:152-171.

Ullrey, D. E., W. G. Youatt, H. E. Johnson, P. K. Ku, and L. D. Fay. 1964. Digestibility of cedar and aspen browse for the white-tailed deer. J. Wildl. Manage. 28:791-797.

Verme, L. J. 1963. Effects of nutrition on growth of white-tailed deer fawns. Trans. N. Am. Wildl. Nat. Resour. Conf. 28:431-433.

Verme, L. J. 1967. Influence of experimental diets on white-tailed deer reproduction. Trans. N. Am. Wildl. Nat. Resour. Conf. 32:405-420.

Verme, L. J. 1969. Reproductive patterns of white-tailed deer related to nutritional plane. J. Wildl. Manage. 33:881-887.

Verme, L. J. 1977. Assessments of natal mortality in Upper Michigan deer. J. Wildl. Manage. 41:700-708.

Verme, L. J., and J. C. Holland. 1973. Reagent-dry assay of marrow fat in white-tailed deer. J. Wildl. Manage. 37:103-105.

Walker, B. H. 1976. An approach to the monitoring of changes in the composition and utilization of woodland and savanna vegetation. S. Afr. J. Wildl. Res. 6:1-32.

Watson, A., and R. Moss. 1970. Dominance, spacing behavior and aggression in relation to population limitation in vertebrates. In: A. Watson (ed.). Animal Populations in Relation to Their Food Resources. Blackwell, Oxford.

Welch, B. L. 1962. Adrenals of deer as indicators of population conditions for purposes of management. Proc. Natl. White-tailed Deer Dis. Symp. 1:94-108.

West, G. C., and D. L. Shaw. 1975. Fatty acid composition of dall sheep bone marrow. Comp. Biochem. Physiol. 50B:599-601.

Whitehead, R. G. 1965. Hydroxyproline creatinine ratio as an index of nutritional status and rate of growth. Lancet 2:567-570.

Williamson, B. R. 1975. The condition and nutrition of elephants in Wankie National Park. Arnoldia (Rhod.) 7:1-20.

Wilson, D. E., and S. M. Hirst. 1977. Ecology and factors limiting roan and sable antelope populations in South Africa. Wildl. Monogr. No. 54.

Woodall, P. L. 1977. Aspects of the ecology and nutrition of the water vole *Arvicola terrestris* (L). Ph.D. Thesis, University of Oxford.

Youatt, W. G., L. G. Verme, and D. E. Ullrey. 1965. Composition of milk and blood in nursing white-tailed does and blood composition of their fawns. J. Wildl. Manage. 29:79-84.

Young, V. R., and N. S. Scrimshaw. 1971. The physiology of starvation. Sci. Am. 225:14-21.

FOUR

Population Dynamics of Bears
—Implications

F. L. BUNNELL

D. E. N. TAIT

INTRODUCTION

Ursavus elmensis, the small "dawn bear" of about 20 million years ago, occurred in the (then) subtropical climate of southern Germany. Its living descendants are members of the *Ursus* line, the brown and grizzly bears (*Ursus arctos* Linne.), black bears *U. americanus* (Pallas) and *U. thibetanus* (Cuvier), and polar bears *U. maritimus* (Phipps). Over the intervening milennia, these bears have evolved different shapes and sets of tactics. Many forms failed, most recently the cave bear *(U. spelaeus)*, only a few thousand years ago. The living forms now extend over the northern hemisphere from about 25° to 88°N.

Here, we examine the tactics of current members of the genus *Ursus* as employed in their pursuit of survival. Of the living forms, only *U. thibetanus*, for which data are very sparse, is not considered. We examine first the size of living forms, for body weight shows close relationships with reproductive behavior. We then consider observed patterns in reproductive rates and mortality. Phenomena influencing these patterns are considered in association with population regulation. Finally, implications of these patterns are addressed.

SIZE OF BEARS

Common conceptions of the size of bears are inaccurate (Burghardt et al., 1972). Black bears *(U. americanus)* are the smallest; mean weights of adult males range from about 80 to 150 kg, depending on location, and are usually 60 to 70% heavier than adult females (Table 1). The largest recorded weights are from the eastern United States [e.g., 272 kg from New York (Black, 1958), 264 kg from New York (Harlow, 1961)]. Brown and grizzly bear weights are most inflated in "conventional wisdom." Typically, adult males average about 200 kg and may be 20 to 80% heavier than females in the same population (Table 1). Most older records of massive brown bears are suspect (see, e.g., Holzworth, 1930). Recent records of very large grizzly or brown bears are 386 kg from coastal British Columbia (D. Hebert, personal communication, 1978), 443 kg from the Alaskan Peninsula (Glenn, 1973), and 500-685 kg from Kamchatka

Table 1   Weights of Bears of the Genus *Ursus*

Location	Source	Adult Male	Adult Female	Ratio	Yearlings Male	Yearlings Female	Cubs Male	Cubs Female
Ursus americanus								
Western populations								
Alberta	Nagy and Russell, 1978	82 (20)[b]	74 (16)	1.11				
Saskatchewan	Miller, 1963				12 (3)			
Washington	Poelker and Hartwell, 1973	87 (18)	58 (17)	1.61	34 (3)	17 (6)	9 (2)	7 (2)
California	Piekielek and Burton, 1975	98 (30)	58 (11)	1.69	28 (3)	18 (2)	13	
Montana	Jonkel and Cowan, 1971	102 (5)	68 (8)	1.50	20 (37)		11 (22)	
Eastern populations								
New York	Harlow, 1961	147 (49)	91 (19)	1.62				
New York	Black, 1958	165 (25)	99 (16)	1.67	44 (17)	40 (19)	17 (13)	19 (11)
New York	Sauer, 1975	136 (43)	85 (24)	1.60	49 (10)	38 (19)	15 (3)	13 (5)
New Hampshire	Harlow, 1961	120 (19)	83 (11)	1.45				
Florida	Harlow, 1961	139 (16)	86 (12)	1.62				
Michigan	Erickson and Nellor, 1964	124 (4)	95 (8)	1.31	45 (2)	—	29 (4)	—
Pennsylvania	Matson, 1954						32-41	
Wisconsin	Knudsen, 1961						43	
Wisconsin	Bersing, 1956				46 (2)			
Minnesota	Rogers, 1977				39 (13)		17 (57)	
Ursus thibetanus japonicus	Hirasaka, 1954	50-120	45-70					
Ursus arctos								
Interior populations								
Yukon	Pearson, 1975	139 (40)	95 (21)	1.46	40	28	12	
Finland	Pulliainen, 1972	165 (84)	135 (23)	1.22	(6)	(1)	(1)	
							21	20
Alaska	Reynolds, 1976	180 (17)	109 (18)	1.65	63	43	(3)	(1)
Alaska	Crook, 1971	217 (12)	147 (11)	1.48	(2)	(2)		
Alberta	Nagy and Russell, 1978	218 (2)	178 (3)	1.22	57	41		23
Alberta	Mundy and Flook, 1973	237 (5)	128 (3)	1.85	(3) 32 (4)	(3)		(1)
Coastal populations								
Alaska	Wood, 1976	159 (9)					31	33
Alaska	Glenn, 1973	225-405	205				(2)	(2)
Kamchatka	Kistchinski, 1972	150-250						
British Columbia	Lloyd, 1978	250-386 (2)	122 (3)					
Ursus maritimus								
Minitoba	Stirling et al., 1977a	276 (19)	176 (23)	1.57	114 (23)	98 (31)	54 (21)	47 (12)
Western Canadian Arctic	Stirling et al., 1975	450-550	180-270	2.0-2.5				
Svalbard	Lønø, 1970	343 (5)	180 (3)	1.91			113 (3)	61 (3)
Franz Josefs Land	Parovshchikov, 1964 ("large bears")	463	251					
Greenland	Pedersen, 1945 ("large bears")	400-450	350-380					
Ursus spelaeus								
Fossils	Kurten, 1967	410-440[c]						

[a] Fall (September-November) weights employed to facilitate comparability. [b] Numbers in parentheses are sample sizes. [c] Estimated from cross section of the femur.

(Novikov, 1969). Uncommonly heavy and well-fatted polar bears attain similar weights [464 kg from the Franz Joseph Archipelago (Parovchshikov, 1964); 550 kg from Svalbard and Canada (Lønø, 1970; Stirling et al., 1975)]. Pedersen (1945) reported a male bear weighing 800 kg. Generally, mean weights of polar bears are heavier than for brown or grizzly bears (Table 1). Adult males average about 280 to 350 kg and may be 60 to 100% larger than adult females (Table 1).

REPRODUCTIVE PATTERNS

Background

Members of the genus *Ursus* share a reproductive syndrome; collectively they exhibit some of the lowest reproductive rates among terrestrial mammals. Periodically, most of the population foregoes reproduction entirely (for polar bears, see data of Stirling et al., 1975; for black bears, Jonkel and Cowan, 1971; for brown bears, Martinka, 1974). The low reproductive rate occurs in spite of features that should encourage reproductive success. All North American *Ursus* are induced ovulators (Erickson and Nellor, 1964; Lønø, 1970; Craighead et al., 1969). Induced ovulation normally encourages higher rates of fertilization (Brambell, 1948). All *Ursus* hibernate, although among polar bears, the phenomenon is largely restricted to parturient females. Lord (1960) noted that hibernators generally have smaller litter sizes than do nonhibernators and suggested that the smaller litter size was associated with higher initial survival rates for the young of hibernators. Compared with other mammals of equivalent size, bear litters are not small (Table 2). Nevertheless, realized rates of reproduction are low, primarily attributable to a late age of first reproduction and a long period (here termed *breeding interval*) between litters (Table 3). We believe the periodic failure of reproduction is adaptive and that it is assisted by another characteristic shared among *Ursus*, that is, delayed implantation (Dittrich and Kronberger, 1963; Wimsatt, 1963; Craighead et al., 1969). Delayed implantation likely provides an energetically efficient means of "aborting" the young before the demands of late gestation and lactation occur.

The denning habit, induced ovulation, and delayed implantation are common among members of *Ursus*. Demographic features that modify reproductive rates and that vary among the species are age of first reproduction, breeding interval, and litter size (see also Chapter 9). Black bears, the most primitive of the genus, most frequently have a mean age of first reproduction of 4 to 5 years, although it may be as late as 7 to 8 years (Table 3). Captive animals show somewhat earlier ages of first reproduction ranging from 3 years to 7 years (Rausch, 1961; Stickley, 1961; Rogers, 1976). Mean litter sizes in the wild range from 1.32 to 2.4 or more, with a mean of 2.25 (n = 516 family groups). Data are sparser for breeding interval. Eastern populations with energy-rich mast and berries as forage may breed every two years (Erickson, 1964; Free and McCaffrey, 1972; Pelton and Beeman, 1975), while western populations appear to have a mean breeding interval of slightly more than three years (Table 3).

Table 2 Mean Litter Sizes of Several Species of the Genus *Ursus* at Various Locations[a]

Ursus americanus

Location and Source	½	1½	2½
Western populations			
Idaho			
(Rust, 1946)		1.32[d] (19)[d]	
Lowell, Idaho			
(Beecham, 1980)	1.65 (23)		
California			
(Piekielek and Burton, 1975)	1.67 (6)		
Montana			
(Jonkel and Cowan, 1971)	1.7 (38)	1.6 (23)	
Alaska			
(Hatler, 1967)	1.73		
Council, Idaho			
(Reynolds and Beecham, 1980)	1.94 (16)		
Alaska			
(Erickson and Nellor, 1964)	1.96 (23)		
Eastern populations			
North Carolina			
(Collins, 1974)	1.80[e] (30)		
Michigan			
(Erickson and Nellor, 1964)	2.15 (20)		
Florida			
(Harlow, 1961)	2.2 (10)		
Alberta			
(Nagy and Russell, 1978)	2.2 (5)		
Maine			
(Spencer, 1955)			2.4 (38)
Wisconsin			
(Schorger, 1949)			2.4 (264)
Virginia			
(Stickley, 1961)	2.63[c] (19)		
Minnesota			
(Rogers, 1976)	2.74[b] (35)		
Captive			
(Baker, 1912)	2.43 (28)		
(Dittrich and Kronberger, 1963)	2.25 (516)		
Weighted mean of wild litters			

Ursus arctos

Location and Source	½	1½	2½
Interior populations			
SW Yukon			
(Pearson, 1972)		1.58 (12)	
SW Yukon			
(Pearson, 1975)	1.7 (11)	1.5 (11)	
Glacier N. Park, Montana			
(Martinka, 1974)	1.7 (35)	1.8 (30)	
Brooks Range, Alaska			
(Reynolds, 1976)	1.77 (13)	2.0 (7)	1.5 (2)
Italy			
(Zunino and Herrero, 1972)		1.9 (42)	
Glacier N. Park, Canada			
(Mundy and Flook, 1973)	2.0 (108)	2.0 (92)	1.8 (25)
Yellowstone N. Park			
(Craighead et al., 1974)	2.13 (213)		
Coastal populations			
Kamchatka			
(Averin, 1948)	2.1 (65)		
Kamchatka			
(Novikov et al., 1969)	2.1 (45)	1.2 (17)	
Admiralty Island, Alaska			
(Perensovich, 1966)	2.0 (10)	1.67 (33)	
Kamchatka			
(Markov, 1969)	2.2 (213)	1.56 (58)	
Admiralty, Baranof, Chichagof Islands, Alaska			
(Klein, 1958)	2.19 (79)	1.9 (115)	
Alaskan Peninsula			
(Klein, 1958)	2.2	2.1	
Kodiak, Alaska			
(Hensel et al., 1969)	2.23 (98)	2.0 (103)	
Kodiak, Alaska			
(Klein, 1958)	2.3 (52)	2.3 (41)	
Kodiak, Alaska			
(Troyer and Hensel, 1964)	2.36 (92)	2.17 (58)	
Captive			
(Dittrich and Kronberger, 1963)	2.05 (213)		
Weighted mean of wild litters	2.12 (1042)		

Ursus maritimus

Location and Source	½	1½	2½
E. James Bay			
(Jonkel et al., 1976)	1.5 (68)		1.7 (26)
Wrangel Island			
(Uspenski and Chernyavski, 1965)	1.5 (14)		
Alaska			
(Lentfer et al., 1980)	1.58 (38)	1.65 (77)	1.47 (57)
Canadian Central and High Arctic			
(Stirling et al., 1977b)	1.58 (43)	1.37 (32)	1.10 (10)
Svalbard			
(Løno 1970)	1.67 (24)		
E Beaufort			
(Stirling et al., 1975)	1.68 (19)	1.65 (23)	1.50 (18)
Alaska			
(Lentfer, 1976)	1.70 (10)		
NE Greenland			
(Manniche, 1910)	1.71[b] (35)		
Pelly Bay			
(van de Velde, 1957)	1.71[b] (56)		
Wrangel Island			
(Uspenski and Kistchinkski, 1972)	1.75[b] (51)		
Canadian Arctic			
(Harrington, 1968)	1.76[b] (136)		
Franz Joseph Arch.			
(Paravshchikov, 1964)		1.77 (141)	
Hudson Bay, Manitoba			
(Jonkel et al., 1972)		1.84 (145)	
Owl River, Manitoba			
(Stirling et al., 1977a)		2.0 (124)	
Hudson Bay, Ontario			
(Kolenosky 1974, 1975)		2.25 (12)	
Captive			
(Harrington 1968, Afonskaja and Krumina 1958)	1.65 (62)		
Weighted mean of wild litters	1.76 (916)		

[a] Sample size is shown in parentheses. [b] Counts of cubs before leaving den. [c] Embryos and cubs. [d] Number of family groups. [e] Corpora lutea.

Table 3 Major Reproductive Features among the Genus *Ursus*

Location and Source	Mean Litter Size of Cubs	Mean Age at First Litter (years)	Birth Interval (years)	Litter Size/ Birth Interval
Ursus americanus				
Montana				
(Jonkel and Cowan, 1971)	1.7	7-8	3.1[a]	0.55[a]
Council, Idaho				
(Reynolds and Beecham, 1980)	1.94	4.76[a]	3.23[a]	0.60[a]
Lowell, Idaho				
(Beecham, 1980)	1.65	4.75[a]	2	0.83[a]
North Carolina				
(Collins, 1974)	1.8[a]	4.2[a]	2	0.90[a]
Minnesota				
(Rogers 1976, 1977)				
abundant forage	3.1	4.5[a]	2.1[a]	1.48[a]
scarce forage	1.96	6.5[a]	3.5[a]	0.56[a]
Captive				
(Baker, 1912)	2.4	4	2[b]	1.20[a]
Ursus arctos				
SW Yukon				
(Pearson, 1972, 1975)	1.59	7.8[a]	⩾ 3.1[a]	⩽ 0.51[a]
Interior, Alaska				
(Reynolds, 1976)	1.77	9.9[a]	> 3	< 0.59[a]
Yellowstone Park				
(Craighead et al., 1974)	2.24	5.8[a]	3.4	0.66
Glacier Park, Canada				
(Mundy and Flook, 1973)	2.0	⩾ 5	> 2.8[a]	⩽ 0.71[a]
Kodiak, Alaska				
(Hensel et al., 1969)	2.23	4-5	⩾ 3	⩽ 0.74[a]
McNeil River, Alaska				
(Glenn et al., 1976)	2.5	6[a]	3.6[a]	0.70[a]
Captive				
(Dittrich and Kronberger, 1963)	2.05[a]	3-4	2[b]	1.02[a]
Ursus maritimus				
Alaska				
(Lentfer et al., 1980)	1.58	5.44[a]	3.6	0.44[a]
Eastern Beaufort Sea				
(Stirling et al., 1975)	1.68	⩾ 4.96[a]	3.03[a]	0.55[a]
Svalbard				
(Lønø, 1970)	1.67	4.0	> 2.18[a]	⩽ 0.77[a]
Captive				
(Kost' jan, 1954; Afonskaja and Krumina, 1958; Volf, 1963; Harrington, 1968)	1.64	4-6	2.1[a,b]	0.78[a]

[a] Our calculations from data provided.

[b] Cubs taken from mother.

The larger grizzly matures more slowly, and mean age of first reproduction is more commonly one year later, at age 5 to 6 years. The northernmost populations in the interior of the Yukon and Alaska mature still later, at 7 to 8 years or older (Pearson, 1972; Reynolds, 1976). Modal litter size among grizzlies ranges from about 1.8 to 2.2 (mean 2.12, $n = 1042$ family groups), and mean breeding interval appears to be about 3.5 to 4 years (Tables 2 and 3). Although much larger than the black bear, female polar bears become sexually mature at about the same age (Table 3). Mean breeding interval appears to be from 3 to 3.5 years, but litter sizes are considerably smaller (mean 1.76, $n = 916$ family groups).

Model

Together, the age of first reproduction, litter size, and breeding interval determine the population's reproductive rate. Combining these appropriately into a simple model generates the maximum mortality that a population could sustain. The model assumes constant mortality rates and that cubs die only when the mother dies. The mortality rate essential to generate a stationary (nondeclining) population is balanced against the natality rate (see Bunnell and Tait, 1980). Figure 1 illustrates isoclines of the maximum sustainable mortality for populations having different average natality rates and ages of first reproduction. The natality rate is for reproductive females and is computed by dividing average litter size by the mean interval between reproduction, the breeding interval. This formulation facilitates comparison of black, brown, and polar bears.

For example, a brown bear population, in which females first breed at age 6.5, first reproduce at age 7, have a mean litter size of 1.5, and breed every three years (natality rate 0.5 cubs/year), can sustain no greater mortality than 10.7%/year (Figure 1).

Discussion

As a result of their higher natality rates, black bears can sustain the greatest mortality (Figure 1). The highest rates, 22 to 24%, are those computed from data for captive bears (Baker, 1912) from which cubs were removed and lactation-induced anestrus terminated, and for wild bears enjoying abundant forage (Rogers, 1976, 1977). Although natality rates for grizzly bears differ little from those of polar bears, their later age of sexual maturity generates a lower sustainable rate of mortality. The results have clear implications to managers establishing harvest policy, for they represent optimistic estimates of maximum sustainable mortality. They are also congruent with expectations derived from natural mortality patterns in *Ursus*. All bears are subject to mortality from conspecifics and man. In addition, black bears are killed by grizzlies (Jonkel and Cowan, 1971), by wolves (Schorger, 1949; Rogers, 1977), and even by coyotes (Boyer, 1948). Grizzlies seem to have no predators beyond man and conspecifics

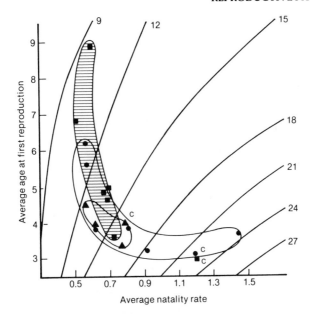

Figure 1   Isoclines of maximum sustainable mortality (%/year) as a function of average age at first reproduction and average natality rate (litter size/years between litters). Grizzly bears, ■; polar bears, ▲; black bears, ●; captive populations, c.

(see chapter 9), although Ognev (1931), Averin (1948), and Pulliainen (1963, 1972) have reported instances of wolves attacking brown bears in Eurasia, and Murie (1944) and G. C. Haber (personal communication, 1978) reported wolf attacks on grizzlies in Alaska. Polar bears have no predators beyond conspecifics, but have been killed by their intended prey, both walrus *(Odobenus rosmarus)* and hooded seal *(Cystophora cristata)* (Lønø, 1970). As predators, polar bears sometimes experience a serious reduction in prey availability; for example, when ice conditions are unfavorable for hunting seals (Stirling et al., 1975). Thus, we expect and observe higher sustainable rates of mortality than for grizzly bears.

Mean litter size of polar bears is smaller than for grizzlies (Table 2), as we would expect given their obligate, carnivorous nature. The higher realized natality rate is achieved by a lower mean age of first reproduction, despite the fact that the polar bear is larger than the grizzly (Table 1). Female polar bears, however, attain adult weight by age 3.5 years (Lønø, 1970; Stirling et al., 1977a), while female grizzlies require 6 to 7.5 years (Glenn, 1973; Reynolds, 1976).

The rather tight crescent of realized natality shown in Figure 1 suggests that both age of first reproduction and litter size are jointly influenced. Litter sizes are never large when age of first reproduction is late (despite potential advantages), nor is the reverse true. Nutrition is implicated and is discussed in association with population regulation. It is noteworthy that the obligate carnivore, polar bear, exhibits the least flexibility in realized reproductive rates.

MORTALITY PATTERNS

All *Ursus* appear to have similar life spans in the wild. The oldest female polar bears observed have been 25 years old (Stirling et al., 1975; Lentfer et al., 1980). Brown and grizzly females may have a somewhat greater longevity, 28 to 30+ years (Couturier, 1954; Storer and Tevis, 1955; Pearson, 1975). The oldest specimens of free-ranging black bear females are equally aged (a 27-year-old reported by Poelker and Hartwell (1973), and a 32.5-year-old reported by Sauer (1975). Females maintain their reproductive competency almost throughout their life span. The oldest reproductively active females reported in the literature are 21 years (Lentfer et al., 1980) and 23 years (Stirling et al., 1977b) for polar bears; 22 years (Nagy and Russell, 1978) and 24.5 years (Pearson, 1975) for brown bears; and 27 years (Poelker and Hartwell, 1973) and 32.5 years (Sauer, 1975) for black bears. Despite the similarity in maximum longevity between species, realized natality rates (Table 3) suggest broad patterns in the mortality rates of *Ursus*. Specifically, brown bear females should experience less mortality than do black or polar bear females (Figure 1).

We consider three stages in the life history of *Ursus*: cub, subadult, and adult. The cub phase lasts while the animal is under the protective care of its mother. This period varies within and between species (Table 3). Generally, black bears are self-sufficient at 1.5 years of age (range: 0.5 to 2.5 years), brown bears at 2.5 years (range: 1.5 to 4.5 years), and polar bears at 2.5 years (range 1.5 to 3+ years) (see review of Bunnell and Tait, 1981). Cubs generally are not hunted and their mortality rate can be estimated from changes in numbers through time without invoking the impacts of hunting. Bunnell and Tait (1981) reported that mortality over the first year of life was higher than had been estimated from changes in litter sizes (an inappropriate method of estimating cub mortality). Specifically, estimated mortality rates for black bear cubs were 25 to 30% and 30 to 40% for brown bear cubs. Assumptions necessary to calculate mortality rates have not held in any of the polar bear populations that have been well studied.

During the subadult stage, animals disperse from the population—an unacceptable violation of mark and recapture assumptions—hence useful estimates of subadult mortality are rare. Bunnell and Tait (1981) noted that the few data available indicated rates of subadult mortality of about 15 to 35% annually—higher than indicated by data derived from live adults.

Estimates of mortality rates of both subadult and adult bears are confused by the impacts of hunting. Cowan (1972) suggested that most of the surviving bear populations were limited by their major predator, man. In their review of the mortality rates of North American *Ursus*, Bunnell and Tait (1981) treated age structures of 30 populations, only one of which was not hunted. Estimated mortality rates of the hunted black and brown bear populations were two to three times greater than for the single, unhunted brown bear population. Reproductive behavior of bears also modifies their age structure. Reproduction can become synchronized in alternate years (Free and McCaffrey, 1972; Collins,

1974; Rogers, 1976; Lindzey and Meslow, 1977b). Studies cited earlier indicated that among all *Ursus*, most females may forego reproduction during a particular year. Bunnell and Tait (1981) found that the most effective approach to accommodate the resultant erratic age structures of bears was an algebraic transformation of the Chapman-Robson statistic to yield an estimate of the mean mortality rate per year.

Considering only the capture data sampling live, hunted populations, it was found that the female mortality rates were consistently lower than those for males, ranging from 9.8 to 24%/year. Weighted averages were 17.7, 16.6, and 19.4%/year for polar, black, and brown bear females, respectively. Thus, there is broad agreement with the relative rates suggested in Figure 1. On average, estimated mortality rates for males are greater than the rates for females; 21.6, 25.9, and 23.4 respectively, for polar, black, and brown bears. Much of the difference between the sexes is a function of their differential vulnerabilities to hunting, as are the differences between mortality rates derived from live and dead samples (Bunnell and Tait, 1981). Since hunting appears to represent the dominant proportion of bear mortality (whether mortality rates are derived from live or dead samples), interpretation of mortality rates is inextricably related to an understanding of the predator-prey interaction between man and bear. Models of this interaction were proposed by Bunnell and Tait (1980) and evaluated by Bunnell and Tait (1981).

POPULATION REGULATION

Background

Bears have few enemies other than man and conspecifics. The omnivorous food habits of most species frequently have been assumed to ensure an adequate food supply. Therefore, by the process of elimination, it often has been deduced that numbers must be self-limited by social factors.

We suggest that within genus *Ursus* a broad but consistent pattern to population regulation is evident. Nutritional condition dominates the reproductive rate, and social mechanisms facilitate access to, or exclusion from, sources of nutrition. Imposed on these fundamental constraints is the bear's apparent desire and ability to impose a haremlike structure on the mating system.

In presenting this suggestion, we treat reproductive rates and mortality rates separately. Furthermore, we consider that reproductive features are largely density independent, while influences of mortality are invoked at high densities (compare to pattern described in Chapter 23) largely through dispersal of subadults.

Regulation of Reproductive Rate

Age of first reproduction, litter size, and breeding interval all appear to be dominated by nutritional condition. Mean age of first reproduction within specific bear populations is generally the age at which females attain adult weight, or slightly later. This phenomenon itself generates broad regional differences in realized natality rates. For example, females among the northern interior grizzly attain adult weights at age 6 to 8 years, and first give birth at about age 7 to 9 years (Pearson, 1975; Reynolds, 1976), while coastal populations grow more quickly and reproduce earlier (see data of Hensel et al., 1969; Glenn, 1973; Glenn et al., 1976). Similar comparisons are possible among black bear populations; for example, compare Jonkel and Cowan (1971) with Rogers (1976, 1977). The carnivorous polar bear more consistently has a rapid growth rate.

Among black and brown bears, females that do not gain sufficient weight prior to denning generally fail to produce cubs. Rogers (1976) reported that none of 16 adult female black bears weighing less than 67 kg on October 1 produced cubs, while 28 of 30 females weighing more than 80 kg and without cubs the previous season did produce cubs. It is likely that delayed implantation facilitates effective "abortion" of the pregnancy. Among polar bears, underweight females appear more often to give birth and then lose the litter (Stirling et al., 1976). Striking evidence is available among polar bears in the Canadian Arctic that experienced ice conditions unfavorable to seal populations and to polar bear hunting of seals. Comparing "before" and "during" unfavorable conditions, we note that the total population declined 33%, the percentage of females with cubs of any age declined from 82.2 to 54.7, and yearlings as a proportion of the population declined from 15 to 3.6% (data of Stirling *et al.*, 1975, 1977b). Observed litter sizes of cubs, however, changed insignificantly, from 1.69 ± 0.05 to 1.61 ± 0.08. During periods of low food abundance, polar bears dispersed widely, seeking more favorable hunting grounds; many females postponed breeding, yearling mortality increased, and entire litters apparently were lost.

A broader relationship is evident between mean litter size and midlatitude of the denning area (Figure 2). Harestad and Bunnell (1978) utilized latitude as a surrogate variable for productivity. These workers noted that regardless of weight or trophic proclivity, mammalian species showed larger home ranges at greater latitude, implicating lower primary productivity with increasing latitude. In all but the most northern regions, female polar bears and their young actively forage for vegetation upon emergence from the maternal den (e.g., Stirling et al., 1977a). At more southerly latitudes, the vegetation and amounts of energy plus nutrients are more abundant upon emergence than in the north. Latitude of denning and litter size show a clear statistical relationship ($r = 0.77$; p of zero slope < 0.003) for the 12 studies summarized in Figure 2:

$$L(x) = 3.05 - 0.02x \qquad (1)$$

where L denotes mean litter size and x denotes latitude, degrees north (°N).

Similar broad relationships are evident among black and brown bears. In western North America, black bears typically dwell in coniferous forests, their

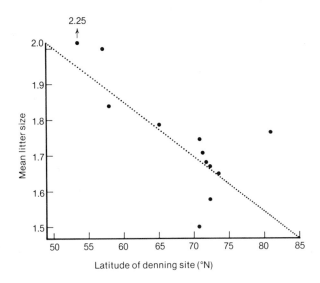

Figure 2   Relationship between mean litter size and midlatitude in degrees north of the denning area among polar bears.

mean litter sizes range from 1.32 to 1.96 with a mean of 1.71 (Table 2). In eastern North America, black bears more frequently inhabit deciduous or mixed forests, where berries and mast commonly provide energy and nutrient-rich forage. Mean litter sizes of black bears east of 95°W, range from 2.15 to 2.74 with a mean of 2.42. Other attributes of their life histories also differ. Mean weights of adult black bears from the west (Washington, California, Montana, and Alberta) are about 93.5 kg for males and 64.5 kg for females (Table 1). Eastern black bears are about 40% larger. Growth rates are also faster in the east, and ages of first reproduction correspondingly younger; 5 to 8 years in the west, about 4 years in the east (Table 3). In eastern populations, the breeding interval is more commonly two years (Erickson, 1964; Free and McCaffrey, 1972; Rogers, 1976), while in the west it usually exceeds 3 years (Table 3).

There is also evidence of the importance of nutrition to black bear populations within a given area. Jonkel and Cowan (1971) reported that reproduction of black bear in Montana approached zero when huckleberries (*Vaccinium* spp.) were scarce. Rogers et al. (1976) noted that mean litter sizes observed near sources of garbage in Michigan ($\bar{x} = 3.1$, $n = 7$) were significantly larger ($p \leqslant 0.01$) than those observed in the same area but away from sources of garbage ($\bar{x} = 1.99$, $n = 129$ (Erickson et al., 1964). In Minnesota, captive black bears receiving rich diets developed more rapidly than did wild bears, even when captive bears were caged together with larger bears that dominated them (Rogers, 1976). Captive females produced their first cubs at 3 years of age, wild females with ready access to garbage at 4 to 5 years ($\bar{x} = 4.4$), and those with little or no access to garbage, still later at age 4 to 7 years ($\bar{x} = 5.6$) (Rogers, 1976). Rausch (1976) reported similar findings for black bear in Alaska. Rogers (1976) further noted that "only 33% (14/43) of the females 5 years of age or older were accompanied by cubs following years of scarce food, whereas 44% (17/39) were with cubs following years of moderately abundant food, and 79% (23/29) were with cubs following years of exceptionally abundant food."

The eastern forms of the grizzly have been extirpated so potential east-west differences cannot be examined. There are, however, differences between interior forms and the coastal forms, that have access to salmon. Omitting the Yellowstone population, which until recently frequently exploited garbage dumps, the weighted mean of cub litters is 1.88 for interior populations (2.00 when Yellowstone bears are included). For coastal populations, the comparable estimate is 2.22 (Table 2). Data are sparser for age of first reproduction. Hensel et al. (1969) estimated the mean age of first reproduction for the coastal population on Kodiak Island at 4 to 5 years. The northern interior populations (Yukon and Alaska) breed much later, at 7 to 10 years, while southern interior populations are only a year or more delayed, behind coastal forms (Table 3).

The association between adult weight and reproductive rates is consistent with that observed in black bears. Adult female grizzlies of interior populations weigh about 100 to 140 kg, whereas along the coast they weigh 120 to 200 kg (Table 1). When they finally reproduce, female grizzlies of the northern interior are only slightly heavier than female black bears from eastern North America.

Together, these observations imply strong nutritional control on all reproductive parameters. Age of first reproduction lags slightly behind the age at which adult body weight is attained. Both litter size and breeding interval are correlated with health and weight of the mother.

If nutritional condition is indeed dominant in regulating populations, one might anticipate a territorial spacing mechanism. In theory, territoriality would be the optimal spacing mechanism where resources were plentiful and evenly distributed, or accessible and predictable. Conversely, it would not pay to defend areas in which food was patchy or unpredictable in abundance (Horn, 1968; Wiens, 1976). Limited overlap of home ranges has been interpreted as evidence of territoriality. Among black and brown bears, reported overlap of home ranges within sex classes has ranged from slight (Jonkel and Cowan, 1971; Poelker and Hartwell, 1973; Pearson, 1975; Rogers, 1976) to extensive (Lindzey and Meslow, 1977a; Reynolds and Beecham, 1980; Craighead, 1976; Berns et al., 1980). Available data seldom permit evaluation of the spacing mechanism relative to forage abundance. Jonkel and Cowan (1971) felt that the great diversity of topography, climate, and vegetation on their study area permitted bears to occupy small areas of limited overlap, which these investigators interpreted as territories. Presumably, the diversity provided continuously renewing food resources that were uniformly distributed. Reynolds and Beecham (1980) noted extensive overlap, which they attributed to patchy and unpredictable forage production. Where overlap of home ranges is low, it is more often lower for females than for males, again implicating the importance of abundant forage for successful reproduction.

The erection of dominance hierarchies at concentrated food sources, such as berry patches, salmon streams, and garbage dumps (e.g., Stonorov and Stokes, 1962; Frame, 1974; Rogers, 1976; Egbert and Stokes, 1976), indicates both the lability of social organization among *Ursus* and the importance of social organization in allocating food resources. Generally, it appears that access to food is not greatly restricted by social factors. Where forage is abundant, populations reproduce well. Dominance hierarchies at concentrated food sources appear to operate primarily in mediating communal access to food by an often solitary and aggressive animal. We conclude that reproductive rate is nutritionally regulated in a largely density-independent fashion.

Regulation of Mortality

Nutritional factors appear to regulate the rate at which a population can grow by controlling fecundity, generally in a density independent fashion. Upper limits to bear abundance appear to be determined by density-dependent mortality factors. Analyses by Bunnell and Tait (1981) support the suggestion of Cowan (1972) that the major cause of mortality in modern bears is hunting by man. Of the 23 studies on 30 populations reviewed by these workers, only one examined an unhunted population. Therefore, discussion of "natural" mortality unassociated with man is necessarily speculative. We noted earlier that conspecifics were the bear's next most important mortality agent after man. Here, we argue that adult males regulate population density by killing or evicting younger males (see also Chapter 9). The manner in which male aggression enhances male fitness is discussed subsequently. For the present, we note that male bears enhance their individual fitness by reducing the density of other adult males. Evidence for such density-dependent regulation is sometimes indirect, but it is plentiful and logically consistent.

Examples of conflict among adult males resulting in severe wounding or death occur in all North American *Ursus*: for polar bear, see Lønø (1970); brown bears, Pearson (1975); and black bears, Beecham (1980). Although this form of conflict could serve to reduce adult male density, it represents a high degree of risk for both participants. Males are also aggressive toward younger bears. "Murder" of subadults and cubs has been observed in all species [polar bear: Nansen (1897), Parovshchikov (1964), Lønø (1970); brown bear: Craighead and Craighead (1967), Pearson (1975), Egbert and Luque (1975); black bear: Erickson (1957), Jonkel and Cowan (1971), Beecham (1980)]. Subadults may simply be driven from the area, as with brown bears (Craighead and Craighead, 1967), or, as with polar bears, driven from their freshly killed prey (Pedersen, 1945; Lønø, 1970; Stirling, 1974; Stirling and Archibald, 1977). Furthermore, subadults frequently show greater rates of dispersal and larger home ranges than do adults (Stickley, 1961; Erickson and Petrides, 1964; Reynolds and Beecham, 1980). Jonkel and Cowan (1971) observed large losses of subadults from their black bear population. Enforced eviction of subadults from more productive habitat could help regulate population numbers. Stokes (1970) suggested that dispersal may result from social intolerance and that social intolerance increases rapidly with density. In brief, there is direct and indirect evidence that older individuals—primarily adult males—act to reduce the survivorship or density of subadults. (The analysis by McCullough (Chapter 9) indicates that this is the case for the Yellowstone grizzly population.) Rogers (1976, 1977) documented that, in black bears, the interaction is principally between adult males and subadult males. We suggest that among Ursidae, adult males have evolved an aggressive disposition directed toward subadult males. The resulting eviction or "voluntary" evacuation from the area by subadult males results in low recruitment of adult males and a disporportionate adult sex ratio in the population. We note subsequently that such behavior, even if directed against one's own offspring, would serve to increase fitness of individual adult males.

Over evolutionary time, these subadult male vagabonds would either experience a high mortality rate or settle new areas. In recent times these subadults have become the principal victims of man as predator.

The greater apparent mortality rates of males over females indicate the greater vulnerability of males to hunting. Among black and polar bears, the generally greater apparent mortality rate of killed males over captured males indicates a biased selection in the kill toward younger over older males. Among grizzly bears, the picture is ambiguous. Data of Nagy and Russell (1978) suggest hunter selection for young males, but represent a small sample; data of Glenn (1975), Wood (1976), and Johnson (1980) suggest hunter bias for older males (Bunnell and Tait, 1981).

Together, these observations suggest that evidence of density dependence should be present in the age structures and dynamics of bear populations. Density dependence can be incorporated into a simple model used to develop the isoclines (see Figure 1). The assumption evaluated is that older individuals, primarily adult males, act to reduce the survivorship or density of subadult bears. The older bears may be either the direct source of mortality or an indirect source of reduction by driving the younger bears away. The model assumes that the rates of mortality and dispersal of male and female subadults increase with increasing numbers of males.

Simulated age structures are compared with data from an unhunted population of grizzlies in Yellowstone and a hunted population from northern Canada (Figure 3). Several features are apparent here. First, the age structure differs between the hunted and unhunted populations in the manner anticipated if density dependence was an important regulating mechanism. Second, the explicit formulation of the model closely approximates field measures. The major difference between simulation and data for the unhunted population is among the oldest age classes. The weakest assumption of the model, a high rate of mortality for old adults, was suggested by examining the data of Craighead et al. (1974). However, the review of mortality patterns by Caughley (1966) suggests that increasing mortality with age is general among mammals. The scale in

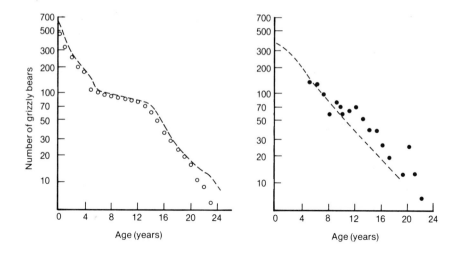

Figure 3 Simulated (dashed line) and sampled (○, Craighead, et al., 1974; ●, Pearson, 1975) age distributions of hunted and unhunted populations of grizzly bears. Sample age distributions have been scaled to a common basis for comparison.

Figure 3 is logarithmic, and the number of bears older than 15 years does not represent a significant segment of the population. The principal effect of hunting is to increase the mortality rate of adult animals. In both simulation and data, the slope of the curve through adult age classes steepens and the characteristic kink in age structure separating subadults from adults is removed.

Despite its simplicity and common-sense appeal, the notion of a specific form of density dependence is nontrivial and its rigorous evaluation is important (see Chapter 9). Whereas density-dependent regulation in wildlife population appears to be general (Chapter 23), it need not be resident in any specific portion of the population and, in many ungulates, it appears to be diffuse within the population. There may be value in concentrating the mechanism in the most stable portion of the population. Older male bears are subject to very few forms of mortality and thus provide an ideal regulatory mechanism. If, as field observations and simulation suggest, older males are the regulatory mechanism, there are important implications to harvest and control. Removal of the older males represents an unnatural, or at least unusual, form of mortality, and one that greatly reduces the effectiveness of intrinsic control. We know too little about bear behavior to speculate on how this removal may affect mating systems in sparse populations, but it seems clear that those populations having a higher reproductive rate (black bears) will be difficult to control by hunting, a practice that selects adult males.

Figure 4 illustrates the change in total size of a bear population simulated using the model that generated the close approximations to field data (Figure 3).

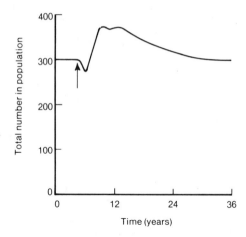

Figure 4 Simulated changes in total size of a population with density dependence regulation from which 45 adult males and 5 adult females were removed.

The projection simulates removal of 45 adult males and five adult females from an otherwise unhunted stable population. The removal abnormally affects that segment of the population that controls total population size with resulting "overcompensation." Far more than 50 young bears "escape" into the adult segment of the population. The effect is not dampened quickly and the population does not return to its original level until after about 25 years (one generation) (Figure 4).

Data of Kemp (1972, 1976) indicate that the simulated pattern occurs in natural populations as well. Kemp experimentally removed adult male black bears from a population in Alberta. After 26 large adult males were removed from the population of about 80 animals, the population doubled. The increase in the population was attributable primarily to a demonstrated increase in numbers of subadult males, and possibly to increased subadult survivorship. The magnitude of immigration could not be evaluated. We conclude that natural mortality of bears is density dependent with control resident in older males, in basic agreement with McCullough (Chapter 9).

FITNESS

Background

The mortality factor dominating the population dynamics of bears, namely, human predation, is evolutionarily recent. The greater vulnerability of subadult and adult males to this mortality, although fortuitous, must be the consequence of other selection pressures. Despite apparent value in regulating population density, these selection pressures should be considered within the context of individual fitness.

Our concept of fitness is that of Orians's (1969) *average reproductive success*—that is, an animal increases its fitness if it increases it genetic contribution to future generations. In a nonexpanding population with equally competitive offspring, fitness can be closely approximated by the total number of offspring produced.

Males

Male bears can improve their individual fitness simply by mating successfully with more females. The process of successful mating by the male can be considered in two phases: the search phase and the encounter phase. The search phase for black bears has been discussed by Rogers (1977). Male bears roam widely, eventually locating females, possibly by following scent trails. On the average, less than one-third of adult females will be in estrus each year (Table 3). Male bears can increase their search success by increasing their home range size to include a number of female home ranges. On the average, home ranges of males are 5.4 and 3.6 times larger than those of females for black and brown bears, respectively (Bunnell and Tait, 1981). Rogers (1977) reported that the home range of each nature male included the "territories" of 7 to 15 females. Jonkel and Cowan (1971) noted extensive overlap in home ranges between adult black bears of opposite sex, but little overlap of adult bears of the same sex. In the Yukon, Pearson (1975); observed extensive home-range overlap between male and female brown bears.

The empirical relationships between size of home range and body weight among mammals, noted by McNab (1963), have been extended by Harestad and Bunnell (1978). The home range of omnivorous mammals is proportionate to body weight to the 0.92 power. Sexual dimorphism does not appear to confer advantages by partitioning forage resources among large omnivores. We suggest that the larger body size of male Ursidae (Table 1) represents in part an adaptation that favors the establishment of larger home ranges and, thus, access to an increased number of potential mates.

The second phase of successful mating, the encounter phase, includes copulation. In its simplest form, a lone male meets and copulates with a lone female. An unusual example among brown bear is provided by Herrero and Hamer (1977). A male brown bear actively herded and confined a female to the peak of a mountain for two weeks. Copulation did not occur until the end of the confinement period. More often, however, the encounter phase is complicated by the presence or arrival of another male. Then, copulatory rights are established by contest between the males or through recognition of a dominance hierarchy. Hornocker (1962) observed that the Yellowstone grizzly established a dominance hierarchy that presumably reduced the incidence of actual fighting. Pearson (1975) observed fresh wounds on most large male grizzlies in the Yukon during the breeding season. Large size confers selective advantage by promoting success in encounters. Encounters between male black bears "were often settled by large males simply chasing away smaller ones" (Rogers, 1977).

But it is not sufficient to achieve success in both searching and encountering an estrus female. Successful mating requires that procreation follow copulation. The promiscuous behavior of the female grizzly (Craighead et al., 1969) and black bear (Rogers, 1977), coupled with induced ovulation, leaves the resolution of successful mating indeterminate. The probability of a bear siring a cub is likely the inverse of the number of males that copulated with the mother. Litter mates need not have a common father. A male's fitness would increase hyperbolically with decreasing promiscuity of the female. Males could remain with the female after copulation and fend off other amorous males. They would then lose their own promiscuous opportunities, might not be able to fend off a large male, and might not have been first to copulate with the female. For whatever reason, it is sufficient to note that males do not generally stay with the female after copulation (Craighead et al., 1969; Herrero and Hamer, 1977 for grizzly; Rogers, 1977 for black). Thus, the adult male's reproductive fitness critically depends on, and is sensitive to, changes in the relative density of adult males to adult females. Anything an adult male does to decrease the abundance of adult males increases his own individual fitness. Thus, we find the density dependence described earlier; male aggression enhances individual fitness.

Evaluation of the presence of density dependence was discussed in association with Figures 3 and 4. We have suggested that adult males reduce numbers of subadult males to generate an adult sex ratio biased toward females. A simple test of this hypothesis would be to examine the adult sex ratio of a "natural" population. The population must be unhunted; park populations with access to garbage dumps might be inadequate tests. Dumps may provide refuges preserving subadult males through to adulthood. Among eastern black bears, males represented 80% of those at garbage dumps (Erickson et al., 1964), and 67% (Rogers et al., 1976) of those feeding at garbage dumps, campgrounds, or residential areas. Rogers et al. (1976) also noted that 72% of those males were age 3 years. McNeil Falls, with its concentrated salmon run, may not be broadly representative for similar reasons.

Females

Circumstances differ for female *Ursus*. With mate selection largely the province of males, females can devote their efforts to rearing of young. From Figure 1 it is apparent that female bears have two broad options whereby they can increase their fitness — to reproduce at an earlier age or to produce more young per unit time after becoming reproductive. The latter option involves shortening the breeding interval or increasing the litter size. All three aspects are under strong nutritional control. Studies have confirmed that territoriality or exclusiveness of home ranges is more strongly expressed in females than in males (Jonkel and Cowan, 1971; Pearson, 1975; Rogers, 1976). The option of territoriality is available to females, as their smaller body size more frequently allows them to establish smaller, exclusive home ranges. It is also likely that territoriality is more often expressed in the females because they have a greater, and longer-term investment, at stake. Although adult females appear to be more tolerant toward their offspring (Pearson, 1975; Rogers, 1977), female territoriality may ultimately limit female density with the resulting dispersal of subadult females.

Having established access to forage, there is relatively little a female bear can do to enhance fitness. Age of first reproduction and litter size will likely be established by the forage base and are beyond her control. All she can modify to increase her fitness is the breeding interval, and this only by abandonment of small litters to terminate lactational anestrus. Tait (1980) has documented the value of abandonment to grizzlies, which can be extended to black and polar bears as well.

LITERATURE CITED

Afonskaja, R. I., and M. K. Krumina. 1958. Observations of polar bears (in Russian). Moscow Zoopark II:56-63.

Averin, Y. V. 1948. Land vertebrates of eastern Kamchatka (in Russian). Proc. Kronotsky State Reserve, No. 1, Moscow: 1-222. As cited in Kistchinski (1972).

Baker, A. B. 1912. Further notes on the breeding of the American black bear in captivity. Smithson. Misc. Collect. 59:1-4.

Beecham, J. 1980. Some population characteristics of two black bear populations in Idaho. Pp. 201-204 in: C. J. Martinka, and K. L. McArthur (eds.). Bears—Their Biology and Management. Bear Biol. Assoc. Series No. 4, 375 pp.

Berns, V. D., G. C. Atwell, and D. L. Boone. 1980. Brown bear movements and habitat use at Karluk Lake, Kodiak Island. Pp. 293-296 in: C. J. Martinka, and K. L. McArthur (eds.). Bears—Their Biology and Management. Bear Biol. Assoc. Series No. 4, 375 pp.

Bersing, O. S. 1956. Wisconsin black bear. Wisc. Conserv. Bull. 21:25-28.

Black, H. C. 1958. Black bear research in New York. Trans. N. Am. Wildl. Conf. 23:443-461.

Boyer, R. H. 1948. Mountain coyotes kill yearling black bear in Sequoia Natl. Park. J. Mammal. 30:75.

Brambell, F. W. R. 1948. Prenatal mortality in mammals. Biol. Rev. 23:370-407.

Bunnell, F. L., and D. E. N. Tait. 1980. Bears in models and in reality—Implications to management. Pp. 15-23 in: C. J. Martinka, and K. L. McArthur (eds.). Bears—Their Biology and Management. Bear Biol. Assoc. Series No. 4, 375 pp.

Bunnell, F. L., and D. E. N. Tait. 1981. Mortality rates in bears. J. Wildl. Manage. (Submitted.)

Burghardt, G. M., R. O. Hietala, and M. R. Pelton. 1972. Knowledge and attitudes concerning black bears by users of the Great Smoky Mountains National Park. Pp. 255-273 in: S. Herrero (ed.). Bears—Their biology and management. IUCN Publ. New Ser. No. 23, Morges, Switzerland.

Caughley, G. 1966. Mortality patterns in mammals. Ecology 47:906-918.

Collins, J. M. 1974. Some aspects of reproduction and age structures in the black bear in North Carolina. Proc. 27th Annu. Conf. Southeast Assoc. Game Fish Comm. pp. 163-170.

Couturier, M. A. J. 1954. The Brown Bear Ursus arctos (in French). L'imprimerie Allier, Grenoble. 504 pp.

Cowan, I. McT. 1972. The status and conservation of bears (Ursidae) of the world—1970. Pp. 343-367 *in*: S. Herrero (ed.). Bears—Their biology and management. IUCN Publ. New Ser. No. 23, Morges, Switzerland.

Craighead, F. C., Jr. 1976. Grizzly bear ranges and movement as determined by radiotracking. Pp. 97-109 *in*: M. R. Pelton, J. W. Lentfer, and G. E. Folk (eds.). Bears—Their biology and management. IUCN Publ. New Ser. No. 40, Morges, Switzerland.

Craighead, J. J., and F. C. Craighead, Jr. 1967. Management of bears in Yellowstone National Park. Environ. Res. Inst. Montana Coop. Wildl. Res. Unit Rep. 113 pp. (multilith). As cited in Stokes (1970) and Martinka (1974).

Craighead, J. J., M. G. Hornocker, and F. C. Craighead, Jr. 1969. Reproductive biology of young female grizzly bears. J. Reprod. Fert. Suppl. 6:447-475.

Craighead, J. J., J. R. Varney, and F. C. Craighead, Jr. 1974. A population analysis of the Yellowstone grizzly bears. Montana For. Conserv. Exp. Sta. Bull. 40:1-20.

Crook, J. L. 1971. Determination of abundance and distribution of brown bear *(Ursus arctos)* north of the Brooks Range, Alaska. M.Sc. Thesis, University of Alaska, College. 78 pp. As cited by Reynolds (1976).

Dittrich, L., and H. Kronberger. 1963. Biological-anatomical research on the reproductive biology of the brown bear (*Ursus arctos* L.) and other Ursids in captivity (in German). Z. Saugetierk. 28:129-155.

Egbert, A. L., and M. H. Luque. 1975. Among Alaska's brown bears. Natl. Geogr. Mag. 148:428-442.

Egbert, A. L., and A. L. Stokes. 1976. The social behavior of brown bears on an Alaskan salmon stream. Pp. 41-56 *in*: M. R. Pelton, J. W. Lentfer, and G. E. Stokes (eds.). Bears—Their biology and management. IUCN Publ. New Ser. No. 40, Morges, Switzerland.

Erickson, A. W. 1957. Techniques for live trapping and handling black bears. Trans. N. Am. Wildl. Conf. 22:520-543.

Erickson, A. W. 1964. Breeding biology and ecology of the black bear in Michigan. Ph.D. Thesis, Michigan State University, East Lansing. 311 pp.

Erickson, A. W., and J. E. Nellor. 1964. Breeding biology of the black bear. Pp. 1-45 *in*: Erickson et al. (eds.). The black bear in Michigan. Mich. State Univ. Res. Bull. 4.

Erickson, A. W., and G. A. Petrides. 1964. Population structure, movements, and mortality of tagged black bears in Michigan. Pp. 46-67 *in*: A. W. Erickson et al. (eds.). The black bear in Michigan. Mich. State Univ. Res. Bull. 4.

Erickson, A. W., J. E. Nellor, and G. A. Petrides, eds. 1964. The black bear in Michigan. Mich. State Univ. Res. Bull. 4:1-102.

Frame, G. W. 1974. Black bear predation on salmon at Olsen Creek, Alaska. Z. Tierpsychol. 35:23-28.

Free, S. L., and E. McCaffrey. 1972. Reproductive synchrony in the female black bear. Pp. 199-206 *in*: S. Herrero (ed.). Bears—Their biology and management. IUCN Publ. New Ser. No. 23, Morges, Switzerland.

Glenn, L. P. 1973. Report on 1972 brown bear studies. Alaska Dept. Fish Game Proj. Prog. Rep. Fed. Aid Wildl. Rest. Proj. W-17-4 and W-17-5. 16 pp.

Glenn, L. P. 1975. Report on 1974 brown bear studies. Alaska Dept. Fish Game Proj. Rep. Fed. Aid Wildl. Rest. Proj. W-17-6 and W-17-7. 10 pp. + App.

Glenn, L. P., J. W. Lentfer, J. B. Faro, and L. H. Miller. 1976. Reproductive biology of female brown bears, *Ursus arctos*, McNeil River, Alaska. Pp. 381-390 *in*: M. R. Pelton, J. W. Lentfer, and G. E. Folk (eds.), Bears—Their biology and management. IUCN Publ. New Ser. No. 40, Morges, Switzerland.

Harestad, A. S., and F. L. Bunnell. 1978. Home range and body weight—A reevaluation. Ecology 60:385-403.

Harlow, R. F. 1961. Characteristics and status of Florida black bear. Trans. N. Am. Wildl. Nat. Resour. Conf. 26:481-495.

Harrington, C. R. 1968. Denning habits of the polar bear (*Ursus maritimus* Phipps). Can. Wildl. Serv. Rep. Ser. No. 5:1-30.

Hatler, D. F. 1967. Some aspects in the ecology of the black bear, *Ursus americanus*, in interior Alaska. M.Sc. Thesis, University of Alaska, College. 111 pp.

Hensel, R. J., W. A. Troyer, and A. W. Erickson. 1969. Reproduction in the female brown bear. J. Wildl. Manag. 33:357-365.

Herrero, S., and D. Hamer. 1977. Courtship and copulation of a pair of grizzly bears, with comments on reproductive plasticity and strategy. J. Mammal. 58:441-444.

Hirasaka, K. 1954. Basking habit of the japanese bear. J. Mammal. 35:128.

Holzworth, J. M. 1930. The Wild Grizzlies of Alaska. G. P. Putnam's Sons, New York.

Horn, H. S. 1968. The adaptive significance of colonial nesting in the Brewer's blackbird (*Euphagus cyanocephalus*). Ecology 49:682-694.

Hornocker, M. G. 1962. Population characteristics and social and reproductive behavior of the grizzly bear in Yellowstone National Park. M.Sc. Thesis, Montana State University, Bozeman. 94 pp. As cited in Pearson (1975).

Johnson, L. 1980. Brown bear management in southeastern Alaska. Pp. 263-270 *in*: C. J. Martinka, and K. L. McArthur (eds.). Bears — Their Biology and Management. Bear Biol. Assoc. Series No. 4, 375 pp.

Jonkel, C. J. and I. McT. Cowan. 1971. The black bear in the spruce-fir forest. Wildl. Monogr. 27:1-57.

Jonkel, C. J., G. B. Kolenosky, R. J. Robertson, and R. H. Russell. 1972. Further notes on polar bear denning habits. Pp. 142-158 *in*: S. Herrero (ed.). Bears — Their biology and management. IUCN Publ. New Ser. No. 23, Morges, Switzerland.

Jonkel, C. J., P. Smith, I. Stirling, and G. B. Kolenosky. 1976. The present status of the polar bear in the James Bay and Belcher Islands area. Can. Wildl. Serv. Occ. Pap. No. 26:1-42.

Kemp, G. A. 1972. Black bear population dynamics at Cold Lake, Alberta, 1968-1970. Pp. 26-31 *in*: S. Herrero (ed.). Bears — Their biology and management. IUCN Publ. New Ser. No. 23, Morges, Switzerland.

Kenp, G. A. 1976. The dynamics and regulation of black bear, *Ursus americanus*, populations in northern Alberta. Pp. 191-197 *in*: M. R. Pelton, J. W. Lentfer, and G. E. Folk (eds.). Bears — Their biology and management. IUCN Publ. New Ser. No. 40, Morges, Switzerland.

Kistchinski, A. A. 1972. Life history of the brown bear (*Ursus arctos* L.) in north-east Siberia. Pp. 67-73 *in*: S. Herrero (ed.). Bears — Their biology and management. IUCN Publ. New Ser. No. 23, Morges, Switzerland.

Klein, D. R. 1958. Alaska brown bear studies. Alaska Dept. Fish Game Fed. Aid Wildl. Rest. Proj. W-3-R-13, Job 1.

Knudsen, G. J. 1961. We learn about bears. Conserv. Bull. 26:13-15.

Kolenosky, G. B. 1974. Polar bears in Ontario — maternity denning and cub production, 1974. Fish Wildl. Res. Branch, Ont. Min. Nat. Res., Maple, Ontario. 26 pp.

Kolenosky, G. B. 1974. Polar bears in Ontario — Maternity denning and cub production, 1975. Fish Wildl. Res. Branch, Ont. Min. Nat. Res., Maple, Ont. 31 pp.

Kost'jan, E. Ja. 1954. (New data on the breeding of polar bears). Zool. Zh. 33:207-215.

Kurten, B. 1968. Pleistocene Mammals of Europe. Weidenfeld and Nicolson, London.

Kurten, B. 1976. The Cave Bear Story. Life and Death of a Vanished Animal. Columbia University Press, New York.

Lentfer, J. W. 1976. Polar bear reproductive biology and denning. Alaska Dept. Fish Game Final Rep. Fed. Aid Wildl. Rest. Proj. W-17-3 and W-17-4.

Lentfer, J. W., R. J. Hensel, J. R. Gilbert, and F. E. Sorensen. 1980. Population characteristics of Alaskan polar bears. Pp. 109-115 *in*: C. J. Martinka, and K. L. McArthur (eds.). Bears—Their Biology and Management. Bear Biol. Assoc. Series No. 4, 375 pp.

Lindzey, F. G., and E. C. Meslow. 1977a. Population characteristics of black bears on an island in Washington. J. Wildl. Manage. 41:408-412.

Lindzey, F. G., and E. C. Meslow. 1977b. Home range and habitat use by black bears in southwestern Washington. J. Wildl. Manage. 41:413-425.

Lloyd, K. A. 1978. Aspects of the ecology of black and grizzly bears in coastal British Columbia. M.Sc. Thesis, University of British Columbia, Vancouver, B.C. 135 pp.

Lønø, O. 1970. The polar bear (*Ursus maritimus* Phipps) in the Svalbard area. Nor. Polarinst. Skr. 149:1-103.

Lord, R. D. 1960. Litter size and latitude in North American mammals. Am. Midl. Nat. 64:488-499.

McNab, B. K. 1963. Bioenergetics and the determination of home range size. Am. Nat. 97:133-140.

Manniche, A. L. V. 1910. The terrestrial mammals and birds of northeast Greenland. Denmark-Ekspeditionen til Gronlands Nordostkyst 1906-08. Vol. 5, No. 1, 200 pp.

Markov, V. I. 1969. Biology of the Kamchatkan brown bear *(Ursus arctos beringianus)* and principles regulating its hunting. 9th Int. Congr. Game Biol. Moscow, pp. 88-90.

Martinka, C. J. 1974. Population characteristics of grizzly bears in Glacier National Park, Montana. J. Mammal. 55:21-29.

Matson, J. R. 1954. Observations on the dormant phase of a female black bear. J. Mammal. 35:28-35.

Miller, R. S. 1963. Weights and color phases of black bear cubs. J. Mammal. 44:129.

Mundy, K. R. D., and D. R. Flook. 1973. Background for managing grizzly bears in the National Parks of Canada. Can. Wildl. Serv. Rep. Ser. No. 22:1-34.

Murie, A. 1944. The wolves of Mt. McKinley. Fauna Series No. 5. U.S. Dept. of Interior. 238 pp.

Nagy, J. A., and R. H. Russell. 1978. Ecological studies of the boreal forest grizzly bear (*Ursus arctos* L.) Annu. Rep. 1977, Can. Wildl. Serv. 85 pp.

Nansen, F. 1897. Fram over polhavet. Den norske Polarfaerd 1893-96 med et Tillaeg af Otto Sverdrup, I-II. Oslo.

Novikov, B. 1969. A giant requires protection (in Russian). Hunting Game Manag. No. 10:9. As cited in Kistchinski (1972).

Novikov, G. A., A. E. Airapetjants, Yu. B. Pukinsky, E. K. Timofeeva, and I. M. Fokin. 1969. (Some peculiarities of populations of brown bears in the Leningrad district.) Zool. Zh. 48:885-901.

Ognev, S. I. 1931. Wild Animals of Eastern Europe and Northeastern Asia (in Russian), Vol. II. Moscow. 766 pp.

Orians, G. H. 1969. On the evolution of mating systems in birds and mammals. Am. Nat. 103:589-603.

Parovshchikov, V. Ja. 1964. Of a study of the population of the polar bear *Ursus (Thalarctos) maritimus* Phipps, of the Franz Joseph Archipelago (in Russian). Vestn. Cesk. Spol. Zool. 28:167-177.

Pearson, A. M. 1972. Population characteristics of the northern interior grizzly in the Yukon Territory, Canada. Pp. 32-35 *in*: S. Herrero, (ed.). Bears—Their biology and management. IUCN Publ. New Ser. No. 23, Morges, Switzerland.

Pearson, A. M. 1975. The northern interior grizzly bear *Ursus arctos* L. Can. Wildl. Serv. Rep. Ser. 34:1-84.

Pedersen, A. 1945. Der Eisbar (*Thalarctos maritimus* Phipps), Verbreitung und Lebensweise. [The Polar Bear (*Thalarctos maritimus* Phipps), Distribution and Habits.] Bruun and Co., Copenhagen.

Pelton, M. R., and L. E. Beeman. 1975. A synopsis of the black bear in the Great Smoky Mountains National Park. Proc. Southern Reg. Zoo Workshop, pp. 43-48.

Perensovich, M. 1966. Brown bear studies, 1960-1966. U.S. For. Serv. Completion Rep. 38 pp.

Piekielek, W., and T. S. Burton. 1975. A black bear population study in California. Calif. Fish Game 61:4-25.

Poelker, R. J. and H. D. Hartwell. 1973. The black bear of Washington. Wash. State Game Dep. Biol. Bull. 14:1-180.

Pulliainen, E. 1963. [Food and feeding habits of the wolf (*Canis lupus*) in Finland] (in Finnish). Suom. Riista 16:136-150.

Pulliainen, E. 1972. Distribution and population structure of the bear (*Ursus arctos* L.) in Finland. Ann. Zool. Fenn. 9:199-207.

Rausch, R. L. 1961. Notes on the black bear, *Ursus americanus* Pallas in Alaska, with particular reference to dentition and growth. Z. Saugetierk. 26:77-107.

Reynolds, D. G., and J. J. Beecham. 1980. Home range activities and reproduction of black bears in west-central Idaho. Pp. 181-190 *in*: C. J. Martinka, and K. L. McArthur (eds.). Bears—Their Biology and Management. Bear Biol. Assoc. Series No. 4, 375 pp.

Reynolds, H. 1976. North slope grizzly bear studies. Alaska Dep. Fish Game Final Rep., Fed. Aid Wildl. Rest. Proj. W-17-6 and W-17-7. 14 pp. + App.

Rogers, L. 1976. Effects of mast and berry crop failures on survival, growth, and reproductive success of black bears. Trans. N. Am. Wildl. Nat. Resour. Conf. 41:431-438.

Rogers, L. L. 1977. Social relationships, movements, and population dynamics of black bears in northeastern Minnesota. Ph.D. Thesis. University of Minnesota, Minneapolis. 194 pp.

Rogers, L. L., D. W. Luehn, A. W. Erickson, E. M. Harger, L. J. Verme, and J. J. Ozoga. 1976. Characteristics and management of bears that feed in garbage dumps, campgrounds, or residential areas. Pp. 169-175 *in*: M. R. Pelton, J. W. Lentfer, and G. E. Polk (eds.). Bears—Their biology and management. IUCN Publ. New Ser. No. 40, Morges, Switzerland.

Rust, H. J. 1946. Mammals of northern Idaho. J. Mammal. 27:308-327.

Sauer, P. R. 1975. Relationship of growth characteristics to sex and age for black bears from the Adirondack region of New York. N. Y. Fish Game J. 22:81-113.

Schorger, A. W. 1949. The black bear in early Wisconsin. Trans. Wis. Acad. Sci. Arts Lett. 39:151-194.

Spencer, H. E. 1955. The black bear and its status in Maine. Maine Dep. Inland Fish Game Bull. 4:1-55.

Stickley, A. R. 1961. A black bear tagging study in Virginia. Proc. Annu. Conf. Southeast Assoc. Game Fish Comm. 15:43-54.

Stirling, I. 1974. Mid-summer observations on the behaviour of wild polar bears (*Ursus maritimus*). Can. J. Zool. 52:1191-1198.

Stirling, I., and W. R. Archibald. 1977. Aspects of predation of seals by polar bears. J. Fish. Res. Board Can. 34:1126-1129.

Stirling, I., D. Andriashek, P. Latour, and W. Calvert. 1975. Distribution and abundance of polar bears in the Eastern Beaufort Sea. Beaufort Sea Tech. Rep. No. 2. 59 pp.

Stirling, I., A. M. Pearson, and F. L. Bunnell. 1976. Population ecology of polar and grizzly bears in Canada. Trans. N. Am. Wildl. Nat. Resour. Conf. 41:421-429.

Stirling, I., C. Jonkel, P. Smith, R. Robertson, and D. Cross. 1977a. The ecology of the polar bear (*Ursus maritimus*) along the western coast of Hudson Bay. Can. Wildl. Serv. Occ. Pap. No. 33:1-64.

Stirling, I., R. E. Schweinsburg, W. Calvert, and H. P. L. Kiliaan. 1977b. Population ecology of the polar bear along the proposed Arctic Islands gas pipeline route. Prog. Rep. to Environ. Manage. Serv., Dept. of Environ., Edmonton, Alberta. 69 pp.

Stokes, A. W. 1970. An ethologist's views on managing grizzly bears. Bioscience 20:1154-1157.

Stonorov, D., and A. W. Stokes. 1972. Social behavior of the Alaska brown bear. Pp. 232-242 in: S. Herrero (ed.). Bears—Their biology and management. IUCN Publ. New Ser. No. 23, Morges, Switzerland.

Storer, T. I., and L. P. Tevis, Jr. 1955. California Grizzly. University of California Press, Berkeley.

Tait, D. E. N. 1980. Abandonment as a reproductive tactic—The example of grizzly bears. Am. Nat. 115:800-808.

Troyer, W. A., and R. J. Hensel. 1964. Structure and distribution of a Kodiak bear population. Wildl. Monogr. 28:769-772.

Uspenski, S. M., and F. B. Chernyavski. 1965. Data on the ecology, distribution, and conservation of the polar bear in the Soviet Arctic (in Russian). Pp. 215-228 in: The Game Mammals, No. 1. Rosselkhosizdat Publ. House, Moscow.

Uspenski, S. M., and A. A. Kistchinski. 1972. New data on the winter ecology of the polar bear (Ursus maritimus) on Wrangel Island. Pp. 181-197 in: S. Herrero (ed.). Bears—Their biology and management. IUCN Publ. New Ser. No. 23. Morges, Switzerland.

Van de Velde, F. 1957. Nanuk, king of the arctic beasts. Eskimo 45:4-15.

Volf, J. 1963. Observations on the reproductive biology of the polar bear, Thalarctos maritimus (Phipps), in captivity. Z. Saugentierkd. 28:163-166.

Wiens, J. A. 1976. Population responses to patchy environments. Annu. Rev. Ecol. Syst. 7:81-120.

Wimsatt, W. A. 1963. Delayed implantation in the Ursidae, with particular reference to the black bear (Ursus americanus Pallas). In: A. C. Enders (ed.). Delayed Implantation. University of Chicago Press, Chicago, Illinois.

Wood, R. E. 1976. Movement and populations of brown bears in the Hood Bay drainage of Admiralty Island. Alaska Dep. Fish Game, Final Rep. Fed. Aid Wildl. Rest. Proj. W-17-5, W-17-7. 10 pp.

Zunino, F., and S. Herrero. 1972. The status of the brown bear (Ursus arctos) in Abruzzo National Park, Italy 1971. Biol. Conserv. 4:263-272.

FIVE

Reexamination of the Life Table
for Northern Fur Seals with Implications
about Population Regulatory Mechanisms

TIM SMITH

TOM POLACHECK

INTRODUCTION

Fur seal *(Callorhinus ursinus)* populations in the north Pacific have periodically been subjected to intensvie harvesting since the 1700s (Lander and Kajimura, 1976). Several breeding areas are currently used, with apparent segregation of the animals into separate breeding populations (Kenyon et al., 1954). The best known of these populations is that occupying the Pribilof Islands in the Bering Sea. This population has been the subject of intensive scientific study and harvest management since 1911, as a result of a treaty between Japan, Canada, U.S.S.R., and the United States (Roppel and Davey, 1965). The population was at an extreme low point in abundance in 1911, when only about 70,000 pups were born annually. Following 1911 the population increased, and currently approximately 300,000 pups are born annually (Lander and Kajimura, 1976). The fur seals occupy beaches on the Pribilof Islands during the summer months, where pupping and breeding occur. Males establish and defend harems. Initially the treaty terminated all harvesting, but since 1918 a large proportion of the males between the ages of 2 and 6 have been harvested each year under the assumption that they are "surplus" in the sense that they are not necessary for the maintenance of the population.

A series of studies since 1954 (Kenyon et al., 1954; Scheffer, 1955; Nagasaki, 1961; Chapman, 1961, 1964, 1973) have described, elaborated on, and tested the hypothesis that this population is regulated in size by density through changes in the survival rate of young animals (see also Chapter 10). In addition, a lesser number of studies (Ichihara, 1971; Bulgakova, 1971) have tested this same hypothesis for breeding populations in the western Pacific Ocean. These studies are based on results of research programs by the treaty nations.

In this chapter we review the available information on reproductive and survival rates of Alaska fur seals in the late 1950s, when the population was thought to be in equilibrium, and present a life table for the population for this period. Noting that the population was apparently increasing at about 8%/year when it was at low abundance (see Chapter 10), we calculate the nature and magnitude of changes in the life table implied in the general density-dependent hypothesis.

This examination of the implied changes in the life table was suggested by our review of the hypothesis that vital rates are changing in response to population density (Smith and Polacheck, 1981). We found that in most cases the available data do not support the general hypothesis of density-dependent changes. Furthermore, even if the changes suggested in earlier interpretations of the data are real, the magnitude of these changes is insufficient to account for the high rate of increase in population size observed during the early part of this century. Thus, one objective of this chapter is to determine the magnitude of the changes in vital rates that we would expect given this high rate of increase. We then consider if the absence of evidence for such change is reasonable in view of the quantity of data that have been collected.

The information available for the Pribilof Island fur seal herd is most complete for St. Paul Island, where approximately 80% of pups are born each year. Because information from St. George Island is less complete, whenever possible we will consider only the St. Paul data, although parallel trends have been observed on both islands.

BACKGROUND

The data for this work come from the commercial harvest of 2 to 6-year-old males (Figure 1) and from the national scientific research programs. The

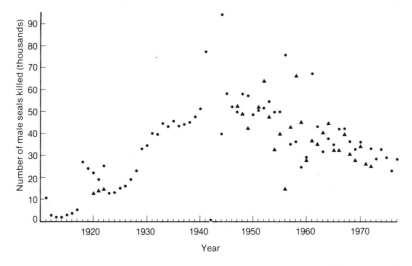

Figure 1 Number of male seals killed on St. Paul Island in the annual harvests, by year (●) and by year class (▲), from NPFSC (1961, 1969, 1971, 1975) and NMFS (1978).

research programs have fluctuated widely in scope and have encompassed many aspects of the biology of fur seals. The principal information from these programs relevant to this Chapter are (1) the counts of pups born from 1911 to

1924, (2) the counts of harem and idle bulls since 1911, (3) the determination of the age of the kill of males from tooth layers since 1947, (4) the pelagic samples of seals from 1958 to 1961, (5) the estimates of numbers of pups born since 1961, and (6) the numbers of pups found dead on the beaches each summer since 1911. These data are described in detail in Smith and Polacheck (1981), and in papers referenced therein. A brief description of the use we make of each of these sources of data follows.

Direct counts of pups on rookeries were made from 1911 to 1924, varying from counts on all rookeries to counts on a sample of the rookeries. Useful estimates or counts are available only for the years shown in Figure 2.

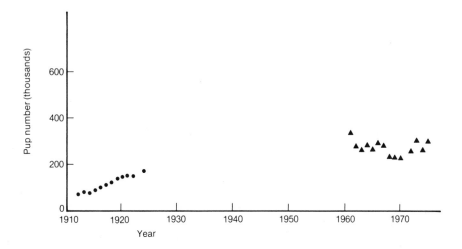

Figure 2   Counts (●) and estimates (▲) of the number of pups born on St. Paul Island, from Lander and Kajimura (1976).

The numbers of adult males on the rookeries, by reproductive class, have been estimated each year since 1912 (Figure 3). The most reliable data are available for the number of males actually holding harems, termed "harem masters." Fewer accurate data are also recorded for the number of adult males not holding harems, termed "idle bulls." The changes in abundance of harem and idle males provide the only measure of the numbers of males escaping the harvest, although uncertainties about survival rates make this difficult to interpret quantitatively.

The determination of age is accomplished by counting ridges on the surface of the teeth (Scheffer, 1955) and by counting layers inside the teeth (Anas, 1970). Age has been determined using these methods for a sample of the animals killed in the harvest since 1947. This permits estimation of the total numbers of males killed from each class, providing estimations of year class strength of males for ages 2 to 6. The estimates of total male kill by year class since 1947 are shown along with the annual kills of males in Figure 1.

From 1958 to 1974, age and reproductive condition were determined for females collected pelagically. Samples of reasonable size were obtained by U.S. researchers from 1958 to 1961 and are used here to estimate reproductive and survival rates.

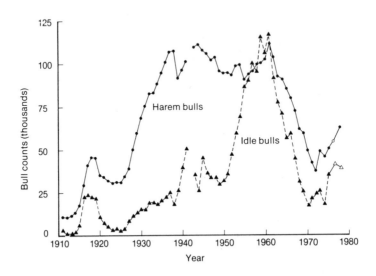

Figure 3   Counts of harem and idle bulls for St. Paul Island, from 1911 to 1977, from Lander and Kajimura (1976) and NMFS (1978).

Starting in the late 1940s pups were tagged in an attempt to estimate the number of pups born. Recovery information was based on the subsequent kill of males at ages 2 through 6. The estimates obtained from this procedure are not considered to be reliable (Chapman, 1964). Starting in 1961, estimates of the numbers of pups born were made based on marking pups by shearing a patch of fur, and subsequently recovering marked individuals in the same summer (Chapman and Johnson, 1968). These latter estimates are considered reliable and are shown in Figure 2 along with the earlier counts of pups born.

The number of pups found dead on the beaches prior to their leaving land in late summer had been estimated for a number of years since 1911. These counts are thought to represent nearly all the mortality occuring on land. They are useful in determining if the survival of pups on land has changed with increased population.

In analyzing these data we use standard statistical procedures for significance tests (Snedecor and Cochran, 1967) and follow Seber (1973) for estimating survival rates. In addition, we make use of discrete life-table analyses, following Mertz (1971) and Leslie (1945).

FEMALE SURVIVAL RATES

Previous Estimates

Survival rates of adult females have been estimated variously, based primarily on the age composition of pelagic samples taken in 1952 and from 1958 to 1961. Chapman (1961) estimated an average rate of 0.79 (rounded to 0.80) based on the age composition of the 1958-1960 U.S. pelagic samples. Nagasaki (1961) analyzed the pelagic samples from the United States for 1958, from the Japanese research program for 1952 and 1958, and the age structure of the females harvested on the Pribilof Islands in 1956 and 1957. He concluded that the sampling biases were such that". . . it is almost impossible, in the present state of study, to calculate accurate mortality rate by age based on the observed age-composition of catches". Chapman (1964) analyzed the U.S. pelagic samples for the years 1958 to 1961 combined, and estimated age-specific survival rates as the ratios of the "adjusted" numbers of each successive pair of ages. The "adjusted" numbers were obtained by fitting a Gompertz curve to the observed age distribution for seals at age 8 and older. This had the effect of smoothing the data and allowing extrapolation along this curve to ages less than 8. However, the fit of the smoothing curve to the observed numbers at each age is not very good with a systematic overestimation of the observed numbers sampled for ages 8 to 15.

Evidence for the Stationarity of the Age Structure

The age distribution of the female seals collected in the pelagic samples by U.S. research vessels is given in Table 1. In order to obtain unbiased estimates of survival rates, the age distribution of the female fur seal population must be stationary (i.e., constant size for the population and with constant proportions in each age class), and the sampling must be representative across the age classes for which survival estimates are made. General demographic theory states that if the population has been at a constant size for a certain period of time, the age structure will approach constant proportions from year to year.

It appears that the population probably reached a maximal, and perhaps constant, size in this century during the late 1940s and early 1950s. This is suggested by (1) the relatively constant annual kill of males between 1944 and 1955, (2) the declining rate of increase in the annual kills between 1930 and 1944, (3) the relatively constant counts of harem master bulls between about 1938 and 1961, and (4) the relatively constant counts of idle males between 1943 and 1951. These are all measures of the male segment of the population, but they suggest that the female segment was likewise relatively constant in size. The harvesting practices were relatively stable from the early 1920s until the early 1950s (Roppel and Davey, 1965).

The conclusion of an approximate constant size is brought into doubt by data for the kill from each year class, which are available since 1947 (Figure 1). The fluctuations in these numbers suggest that the number of seals at ages 2 to 5 in successive year classes varied considerably during this period; this might induce nonstationarity in the age distribution. The fluctuations in the kill by year class in the 1950s are also reflected in the annual kills of males (Figure 1). Similar variations in the annual kills are not seen prior to 1956 (disregarding the war years), which suggests that year class strength may not have affected age structure stationarity prior to the 1950s. Moreover, whereas the numbers killed by year class from 1947 to 1953 are consistent with the hypothesis of constant size, the data from 1954 and 1971 suggest a general but variable decline. It should be noted that the harvesting regime for males changed somewhat starting in the mid- to late 1950s as the annual catches began to decline (Roppel and Davey, 1965). However, these changes do not seem to explain the decline. This apparent decline, if real, should not have affected the age structure for the older animals between 1958 and 1961.

Table 1   Number of Female Seals by Age Class Collected by U.S. Researchers, by Year

Age	1958	1959	1960	1961	1958-1961
3	39	43	18	84	184
4	42	93	36	96	267
5	70	114	55	68	307
6	99	118	45	62	324
7	103	143	66	95	407
8	102	164	105	107	478
9	81	108	144	114	447
10	97	96	129	112	434
11	113	98	136	82	429
12	134	76	106	71	387
13	110	56	120	76	362
14	92	70	107	67	336
15	71	87	67	68	293
16	56	69	53	55	233
17	36	36	46	24	142
18	22	27	23	25	97
19	14	16	19	10	59
20-22	5	17	12	9	43

By 1958, when the first pelagic samples used here were taken, any irregularities in the age structure caused by the increase in size of the population since 1911 should have diminished to inconsequential levels if the population had approached a constant size. This was checked and confirmed by examination of the predicted age structures from a simulation model of population size. This model, described by Smith and Polacheck (1981), is structured around single-species density-dependent concepts. It assumes variously that fecundity, juvenile survival, and adult survival change with density such that the model mimics the observed growth of the population as reflected in the male harvest and, where available, the numbers of pups born. There are slight biases toward understimation in survival rates calculated from such simulated age structures for the late 1950s.

An additional complication in the representativeness of the pelagic samples is that substantial numbers of females of several age classes were killed beginning in 1956, in a management effort to reduce the numbers of pups born and thereby increase the survival of pups to age 3. The effect on the age structure of this harvesting was explored by further examination of the simulation model described above. The calculations were extended through 1961 with the age structure of the harvest of females from 1956 to 1961 incorporated into the model. The simulated age structures suggest that no effect of this harvest would have been observable for animals age 11 and older through 1961. Furthermore, the age structure probably did not show any effect of this harvesting for animals older than 8 in 1958, 9 in 1959, and 10 in 1960. The lack of effects for these ages is attributable to a combination of the magnitude and age structure of the female harvest.

New Estimates

Inspection of the data in Table 1 demonstrates that the samples are not representative of the animals in the younger ages as the proportions in each age class in the harvest show an increase with age up to the ages of 7 to 9 in all four years. There is considerable variability in the proportions observed in the age groups older than 8, however, which suggests the possible nonrepresentativeness of the samples between years. Following Seber (1973), the equality of the observed proportions between years within ages was tested with a chi-squared statistic. The test statistic for the data in Table 1 has a value of 122, with 36 degrees of freedom. As this is large ($p < 0.001$), there are likely some changes between years in the proportions within each age class. This could be the result of sampling biases between years or of nonstationarity of the age distribution.

The questions of the representativeness of the samples between years and the possibility of sample biases within years over the geographic areas sampled (Nagasaki, 1961) need to be investigated further. As the data are not readily available in a form that will allow such investigation at this time, and as these data have been used in the past (Chapman, 1961) to estimate survival rates, we have used them as presented, recognizing the need for more detailed investigation in order to better evaluate our results.

Owing to apparently changing survival rates with age, the most adequate estimator for survival for each age class is the ratio of numbers sampled at successive ages (Seber, 1973). Such estimates have not been presented elsewhere for the U.S. data, and are given here by year (Table 2) along with appropriate descriptive statistics (Seber, 1973). It is clear on inspection that the variability of the estimates increases markedly after approximately age 16, and that the samples are subject to high levels of variance in that they occasionally yield

Table 2 Survival (S) as Ratios of Observed Numbers of Females of Successive Ages in U.S. Pelagic Samples for Ages 8 to 20 by Year[a]

Age	$\hat{S}$	$\hat{\sigma}$	C	r
		1958		
8	0.79	0.092	−0.0117	−0.55
9	1.20	0.680	−0.0144	−0.50
10	1.17	0.161	−0.0122	−0.50
11	1.19	0.152	−0.0073	−0.45
12	0.82	0.106	−0.0062	−0.50
13	0.84	0.118	−0.0070	−0.49
14	0.77	0.122	−0.0086	−0.50
15	0.79	0.141	−0.0091	−0.47
16	0.64	0.137	−0.0091	−.048
17	0.61	0.165	−0.0177	−0.49
18	0.64	0.218	−0.0097	−0.33
19	0.21	0.136	−0.0238	−0.45
20	0.33	0.385	—	—
		1959		
8	0.66	0.067	−0.0054	−0.53
9	0.89	0.125	−0.0095	−0.52
10	1.02	0.147	−0.0081	−0.47
11	0.78	0.119	−0.0075	−0.49
12	0.74	0.130	−0.0164	−0.57
13	1.25	0.224	−0.0222	−0.50
14	1.24	0.200	−0.0113	−0.44
15	0.79	0.128	−0.0060	−0.44
16	0.52	0.107	−0.0109	−0.53
17	0.75	0.191	−0.0165	−0.46
18	0.59	0.187	−0.0116	−0.39
19	0.31	0.160	−0.0875	−0.67
20	1.40	0.820	—	—
		1960		
8	1.37	0.122	−0.0092	−0.45
9	0.90	0.109	−0.0064	−0.52
10	1.05	0.130	−0.0077	−0.46
11	0.88	0.101	−0.0131	−0.55
12	1.13	0.151	−0.0124	−0.47
13	0.89	0.119	−0.0134	−0.45
14	0.63	0.098	−0.0121	−0.52
15	0.79	0.145	−0.0064	−0.51
16	0.87	0.175	−0.0189	−0.42
17	0.50	0.128	−0.0167	−0.55
18	0.83	0.256	−0.0280	−0.36
19	0.32	0.148	−0.0286	−0.62
20	1.00	0.578	—	—

Table 2 (continued)

Age	$\hat{S}$	$\hat{\sigma}$	C	r
		1961		
8	1.07	0.107	−0.0085	−0.49
9	0.98	0.131	−0.0073	−0.46
10	0.73	0.106	−0.0060	−0.52
11	0.87	0.140	−0.0083	−0.53
12	1.07	0.177	−0.0084	−0.48
13	0.88	0.148	−0.0052	−0.52
14	1.02	0.175	−0.0074	−0.47
15	0.81	0.147	−0.0130	−0.41
16	0.44	0.107	−0.0094	−0.60
17	1.04	0.298	−0.0180	−0.37
18	0.40	0.150	−0.0137	−0.54
19	0.70	0.345	−0.0526	−0.36
20	0.29	0.229	—	—

[a]Also given are the estimated standard deviations ($\hat{\sigma}$) for early estimates, and covariances (C) and correlation coefficients (r) of successive pairs of estimates.

estimates greater than one. The estimates that are greater than one are not limited to those ages where the female kill might have had an effect.

The pairs of estimates of survival rates for successive age classes are negatively correlated, with the correlation coefficient depending only on the actual values of the survival rates for those ages. An expression for the correlation coefficient (r) in terms of the two annual survivals (S_x, S_{x+1} for ages x and $x + 1$) can be obtained by simplifying the expression for the correlation coefficient in terms of the covariance and the variances, as given in Seber (1973), obtaining

$$r(S_x, S_{x+1}) = -[(1 + S_x)(1 + S_{x+1}^{-1})]^{-1/2}$$

The magnitude of the second survival rate in a pair has more of an effect on the value of the correlation than does the magnitude of the first, and the correlation cannot exceed 0.7 in absolute value. Also for the survival rates in the neighborhood of those estimated for fur seals, the correlation coefficient will be around −0.5. The covariances and the corresponding correlations between pairs of estimates are given in Table 2.

The average across years of the estimated survival rates in Table 2, weighted inversely by their variances, are given in Table 3 (S), along with Chapman's (1964) estimates of survival rates S_c for these ages, for comparison. Note that the weighted average estimates are generally greater than Chapman's from ages 9 to 15, and lower from ages 16 to 19. These average estimates are shown in Figure 4.

These estimates of the survival rate are probably reliable for ages 12 and older. The estimates for ages 8 to 11 must be used with considerably less confidence owing to possible effects of the female harvest. We have little informa-

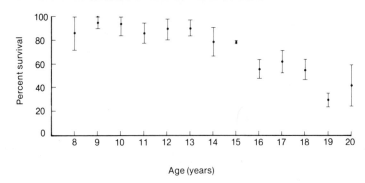

Figure 4 Estimates of annual survival rates by age with associated ranges of plus and minus one standard deviation, from Table 3.

tion on the survival rates for animals younger than 8 because of the apparent nonrepresentativeness of the sampling for these ages. Chapman's (1961) estimates of survival for ages 3 to 7 are very uncertain because of the lack of useful data for these ages and the poor fit of the Gompertz equation he used for extrapolation.

Table 3   Survival Rate Estimates for Female Alaska Fur Seals Pooled Across Years From Table 2 *(S)* and from Chapman $(1964)(S_c)$ [a]

Age	S	S_c
8	0.86 (0.144)	0.91
9	0.95 (0.058)	0.90
10	0.94 (0.099)	0.88
11	0.86 (0.085)	0.86
12	0.90 (0.090)	0.84
13	0.91 (0.066)	0.81
14	0.79 (0.118)	0.77
15	0.80 (0.004)	0.74
16	0.56 (0.080)	0.69
17	0.63 (0.091)	0.64
18	0.56 (0.085)	0.57
19	0.30 (0.063)	0.50
20	0.42 (0.177)	0.41

[a]Standard deviations in parentheses.

REPRODUCTIVE RATES

Previous Estimates

The pelagic samples of female seals (1958-1974) provide the basic data for estimating reproductive rates. The seals reproduce annually on the breeding islands, mate immediately after parturition, and have delayed implantation of the ovum. The females collected pelagically are classified as pregnant or non-pregnant on the basis of a detailed examination of the reproductive tract.

Chapman (1961) examined the proportion pregnant for the 1958, 1959, and 1960 samples. Utilizing his estimates of the age-specific mortality rates as discussed above, Chapman estimated the average pregnancy rate for seals 3 years of age and older at 0.60 (i.e., for a population with a stable age distribution, 60% of the females aged 3 and older are estimated to give birth each year). Nagasaki (1961) analyzed a mixture of pelagic and land-based samples for pregnancy rates. He concluded that there were differences in the pregnancy rate between the Japanese and U.S. pelagic samples for the younger ages, and that the land-based samples suffered consistent biases in the representativeness of reproductive condition. (Nagasaki's precise estimates are not tabulated, but are shown in his Figure 8.) Chapman (1964) further analyzed the pelagic samples from 1952, and 1958 to 1961 combined, for both U.S. and Japanese researchers. He also noted the difference between the U.S. and Japanese samples and concluded that his previous estimate of 0.60 for the Pribilof seals is the most reasonable.

Revised Reproductive Rate Estimates

The published analyses of pregnancy rates do not exhibit all the age-specific values (except perhaps Nagasaki's untabulated values). Also, the variability associated with these estimates is not given. We have calculated the values by year with their standard deviations (Table 4). The standard deviations are obtained by assuming binomial sampling within an age class (i.e., $\hat{\sigma}(P) = (p(1 - p)/n)^{1/2}$, where p denotes the estimated proportion pregnant and n denotes the number samples of that age). Combined age-specific pregnancy rates over the four years are also given as the average of the annual estimates weighted inversely with their variances. The averaged pregnancy rates from Table 4 are shown in Figure 5. Useful samples for pregnancy rates are available for years beyond 1961. These are not included in order to be consistent with the data from which survival rates were estimated above, but they do not affect the general conclusions of this chapter. These samples are discussed further in Smith and Polacheck (1981).

The problems of bias apparent in estimating survival from the pelagic samples are not as important in estimating pregnancy rates. For unbiased estimates in this regard females of any particular age must be equally likely to be in the sample, regardless of reproductive condition. The estimates of pregnancy rate by age are consistent over the four years shown in Table 4 and are similarly consistent for the data from 1962 to 1970 for females older than 5 years of age. There is some evidence of a declining pregnancy rate for ages 3, 4, and 5, between 1962 and 1970, but the changes are neither large nor consistent and may represent sampling variability and sampling area differences (A. Johnson, personal communication).

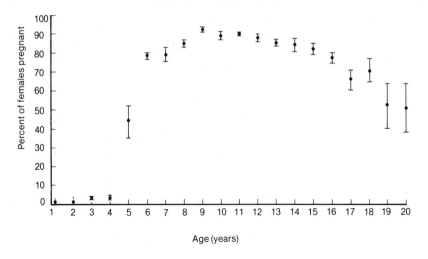

Figure 5 Estimates of annual reproductive rates as percent pregnant, by age, with associated intervals of plus and minus one standard deviation from Table 4.

SURVIVAL DURING THE FIRST SUMMER

The annual counts of dead pups can be compared to the estimates of numbers of pups born to get estimates of survival to the time the pups leave the islands at the end of summer. These data, up to 1951, have been analyzed by Kenyon et al. (1954), where it is noted that the estimated mortality rates had increased substantially just prior to 1951 and that this increase was associated with an apparent large increase in population size. Nagasaki (1961) presented an analysis of much the same data and also concludes that a density-dependent relationship was involved. However, the estimates of the number of pups born since 1924 used in both papers have subsequently been considered unreliable (see above and Chapman, 1964).

The same type of analysis has been extended up to 1976 using estimates of numbers of pups born from shearing and recapture estimates (Smith and Polacheck, 1981). In that analysis, the survival of pups to the end of the summer exhibits a decline with increased population size. This decline is apparent in the estimates of first summer survival in Table 5, which are calculated from data in Smith and Polacheck (1981, Table 3 of Appendix I in that reference).

Table 4   Proportion of Pregnant Female Seals by Age and Year of Sample $(\hat{p})$ and Weighted Average Proportion of Pregnant Over All Years $(\bar{p})^a$

Age	1958		1959		1960		1961		Average	
	$\hat{p}$	$\hat{o}(\hat{p})$	$\hat{p}$	$\hat{o}(\hat{p})$	$\hat{p}$	$\hat{o}(\hat{p})$	$\hat{p}$	$\hat{o}(\hat{p})$	$\bar{p}$	$\hat{o}(\bar{p})$
3	0.03	0.025	0	0	0	0	0	0	0.03	0.008
4	0.02	0.024	0.06	0.025	0.03	0.027	0.01	0.010	0.03	0.009
5	0.46	0.060	0.56	0.046	0.49	0.067	0.21	0.049	0.44	0.084
6	0.81	0.040	0.77	0.039	0.80	0.060	0.76	0.054	0.79	0.012
7	0.89	0.030	0.76	0.036	0.79	0.050	0.76	0.044	0.79	0.035
8	0.89	0.031	0.87	0.027	0.86	0.034	0.79	0.039	0.85	0.019
9	0.96	0.021	0.89	0.030	0.92	0.022	0.94	0.023	0.92	0.014
10	0.88	0.033	0.85	0.036	0.92	0.025	0.94	0.023	0.89	0.019
11	0.92	0.026	0.90	0.031	0.91	0.024	0.89	0.035	0.90	0.006
12	0.82	0.033	0.88	0.037	0.91	0.028	0.43	0.030	0.88	0.024
13	0.83	0.036	0.89	0.041	0.88	0.030	0.83	0.043	0.86	0.015
14	0.82	0.040	0.84	0.043	0.80	0.038	0.93	0.032	0.84	0.032
15	0.79	0.048	0.89	0.034	0.84	0.045	0.79	0.049	0.82	0.025
16	0.79	0.055	0.75	0.052	0.72	0.062	0.86	0.047	0.78	0.031
17	0.57	0.083	0.81	0.066	0.67	0.069	0.63	0.099	0.66	0.053
18	0.59	0.105	0.85	0.068	0.83	0.079	0.64	0.096	0.71	0.062
19	0.29	0.121	0.81	0.097	0.58	0.113	0.50	0.158	0.53	0.115
20	0.40	0.219	0.59	0.119	0.33	0.136	0.89	0.105	0.51	0.131

$^a\hat{o}(\bar{p})$ indicates standard deviations.

SURVIVAL TO AGE 3

Several estimates of the survival of younger seals are given in the literature (Chapman, 1961, 1964, 1973). This has been an area of great interest and research, since the primary hypothesis for the natural regulation of fur seal populations has been that the survival of younger animals varies with population size. Unfortunately, most of the data available to estimate juvenile survival rates are not easy to interpret. Certain key observations are missing in some critical years, and vital rates that would permit estimation of the missing information from other information are poorly known.

The basic technique used to estimate the survival rate for this period is to divide the number of male seals born into the sum of the male seals harvested and escaping harvest. The number escaping has not been reliably estimated. These estimates are discussed in Smith and Polacheck (1981), where it is concluded that reliable estimates of survival rates for these ages are not available.

In the absence of estimates of the actual survival rates, the ratio of the male kills from each year class and the estimates of pups born provide a lower bound to the actual survival rates. The available data are shown in Table 5, with the corresponding estimated lower bounds. Values are not presented for the 1950s because of the problems associated with the estimates of numbers of pups born during this period. Also included in Table 5 are estimated lower bounds for their survival from 1920 to 1922. These estimates are also based on the ratio of the actual kill of males from each year class to the estimated number of pups born. However, in this case, the actual kill from each year class was estimated from the size-frequency distribution of the male kill and may contain larger biases.

The average of these estimates of lower bounds for the survival rates for the years 1920 to 1922 (0.21) is significantly different from the average for the years 1961 to 1970 (0.26), but the difference is not in the expected direction. However, it is not clear that these lower bounds are comparable, since the number of males surviving the harvest at age 4, which is the primary factor causing these numbers to be less than the true survival rates, has presumable changed significantly with changing harvesting procedures (Roppel and Davey, 1965).

These lower bounds for the survival from birth to age 3 can be adjusted to the period from the end of the first summer to age 3 by dividing by the estimated first summer's survival rates indicated in Table 5. This adjustment serves to increase the difference between the two periods, although the latter period (when populations were higher), continues to have higher estimated lower bounds of survival rates.

We have explored in detail the general problem of estimating the survival rate of male seals from the annual kills and the annual counts of harem and idle bulls in Smith and Polacheck (1981). Calculations in that report suggest that the available information is consistent with a wide range of values of juvenile survival, but that a maximum value may be about 72%. This may be considered a minimal upper bound for the juvenile survival rate.

EQUILIBRIUM RATES

If available estimates of survival and pregnancy rates are asssumed to be for a period of time when the population was not markedly changing in size, the net rate of change should have been approximately zero. Making this assumption and ignoring any possible bias that would result from nonrandom sampling, we have estimates of pregnancy rates for all age classes, of survival rates for females ages 12 and older, and of survival rates for both sexes during the first summer of life. In addition, we have some estimates of lower bounds for survival of males from birth to age 3. We have no direct estimates of survival from ages 3 to 8, and some possibly biased estimates of survival from ages 8 to 12. The existing estimates can be used to calculate estimates for those values of survival rates for which we have little or no information and which are consistent with a population growth rate of zero.

Using standard life-table calculations, we computed the rate of survival of females between birth and age 3 that would result in a zero net rate of increase. This was done for a range of values of survival rates from ages 3 to 23 and for ranges of estimates of pregnancy rates and survival rates of older females consistent with the variances of the above estimates. The results are shown in Table 6.

Table 5 Estimated Number of Pups Born and Subsequent Harvest of Males from that Year Class on St. Paul Island, and Estimated Survival over the First Summer

	Pups Born[a] × 1000	Kill of Males[b]	Survival to Age 3[c]	First Summer's Survival Rate[d]
1920	143	14,751	0.206	0.972
1921	150	15,375	0.205	0.973
1922	159	17,050	0.215	0.983
1961	337	36,882	0.219	0.820
1962	278	34,991	0.252	0.829
1963	264	40,126	0.304	0.870
1964	285	44,882	0.315	0.920
1965	267	32,202	0.241	0.846
1966	296	32,285	0.219	0.924
1967	284	39,504	0.278	0.948
1968	235	30,266	0.258	0.887
1969	234	27,778	0.237	0.940
1970	230	34,188	0.297	0.906

[a]1920 to 1922 from Chapman (1961, his Table 2); 1961-1970 from Lander and Kajimura (1976, their Table 2).

[b]1920 to 1922 from Chapman (1961, his Table 3) as the sum of commercial and native kills; 1961-1970 from Lander and Kajimura (1976, their Table 6); 1970 from Marine Mammal Division (1977).

[c]Lower bound, assuming all kill was at age 3.

[d]Calculated from Smith and Polacheck (1981, Table 3 of Appendix I), for those years where the kill of males is known.

Table 6 The Calculated Survival Rates from Birth to Age 3 that Result in a Net Rate of Increase of Zero, Using Combinations of Ranges of Other Survival Rates and Pregnancy Rates

Survival Rates for Older Animals	$S_{3,4}$[a]	Pregnancy Rates		
		Low	Central	High
Low	0.80	0.90	0.85	0.81
	0.85	0.71	0.68	0.65
	0.90	0.57	0.55	0.52
	0.95	0.46	0.44	0.42
Central	0.80	0.71	0.68	0.65
	0.85	0.56	0.54	0.51
	0.90	0.45	0.43	0.41
	0.95	0.36	0.34	0.33
High	0.80	0.54	0.51	0.49
	0.85	0.42	0.40	0.39
	0.90	0.33	0.32	0.31
	0.95	0.27	0.25	0.24

[a]$S_{3,4}$ is the survival from ages 3 to 4.

The survival rates used between ages 3 and 12 were constructed by linear interpolation between the estimated value at age 12 and a range of values at age 3, as shown in Figure 6. The ranges of values for the pregnancy rate and the survival rate of older animals used in the computations in Table 6 were constructed as the weighted average estimates given in Tables 3 and 4 and these estimates plus (high) and minus (low) one standard error. These ranges of survival and pregnancy rates are not meant to represent confidence intervals, as discussed above, but rather to serve as a basis for considering the ranges of values of survival from birth to age 3 consistent with a zero rate of increase of the population.

It can be seen from Table 6 that within the range of uncertainty of the parameter estimates, allowable values of survival from birth to age 3 are generally higher than the lower bounds for this rate given in Table 5. To get values as low as the average estimated lower bounds from 1961 to 1970 (0.26) one must have rather high survival rates for all ages. Moreover, the lowest estimates of the adult survival rates tend to give biologically unreasonable estimates of the survival rates from birth to age 3. It can also be seen that within the sampling variability of the estimates of pregnancy and survival rates the calculations are rather more sensitive to changes in survival rates than to changes in pregnancy rates.

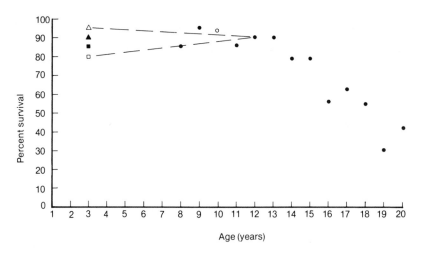

Figure 6 Survival estimates used in life-table calculations, and the method of interpolating values (dashed lines) between age 3 and 12. See text for details.

MAXIMUM POPULATION GROWTH RATES

While it is apparent that there are many uncertainties in our understanding of the reproductive and survival rates for Alaska fur seals when the population was at equilibrium, it is still possible to usefully address the question of what these rates would have had to have been when the population was increasing. The counts of pups made from 1912 to 1924 provide our best estimate of how fast the population is capable of increasing. This was about 8%/year (see Eberhardt, Chapter 10). This rate is also supported by the increases in the annual kills of males from around 1924 to 1931 (Figure 1).

The dominant feature of the survival and reproductive rates as estimated for fur seals for a rate of increase of zero is that they are rather close to biologically reasonable or even logically possible maximum values. Thus, there is little scope for changes in these rates in a density-dependent response to changes in population size that would result in a rate of increase of 8%. To explore the implications of this limited range on the density-dependent mechanisms that are possible or reasonable for this population, we have calculated the percentage changes in each of several rates that would be necessary were an 8% rate of increase to prevail. We have considered the changes in each rate as a percent of the difference between the maximum value and the rate corresponding to a zero rate of increase for the population, as illustrated in Figure 7.

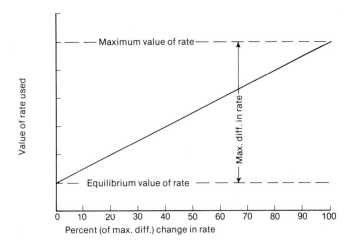

Figure 7 Diagram of the method of specifying changes in vital rates, as a percent of the difference between the biologically maximal value and the value at equilibrium. Values on the ordinate depend on the variable involved.

For survival the absolute maximum is obviously unity. We have little biological information to specify any values less than unity for this maximum. Estimated rates for ages 12 and 13 are around 0.94, suggesting that very high rates of survival are possible. Maximum values or survival rates may vary with age, as is observed for the estimated values. In particular, survival from birth through the first year probably has a lower maximum value than for any other age class, except perhaps for very old animals. Again, little information for a value below unity is available. However, it is not reasonable that the animals are capable of experiencing no natural mortality over any significant portion of their life span. In the present calculations maximum values in the range 0.96 to 1.0 were used for all ages above 3.

Maximum rates of survival from birth to age 3 are difficult to determine, as the animals are weaning, learning to feed in a new environment, and experiencing intensive growth during this period. Unlike pregnancy rates and adult survival rates, however, there is some information from the 1920s when the survival should have been higher if dependent on density. Chapman (1961, 1973) calculates the juvenile survival for this period as either approximately 31 or 36%, on the basis of estimates of the number of pups born, the number harvested from a year class (Table 5) and estimates of the possible number escaping kill. These values may represent maximal levels for this survival rate, but Chapman's calculations involve a number of assumptions that are not completely documented. We have further explored estimates of juvenile survival for this period with results which suggest that a rate of 54% may be a more reasonable upper bound (Smith and Polacheck, 1981).

Maximum values for reproductive rates are probably one pup per year for sexually mature females. Very little twinning has ever been observed (even in the period when the population was rapidly expanding). The sex ratio at birth is apparently even. This results in a maximum production of one-half daughter per sexually mature female per year. The proportion of the females that are sexually mature increases from age 3 up to age 6. The proportions may themselves be subject to changes as population size changes. For instance, it is observed that females collected in the western Pacific, primarily from Asian rookeries, show reproductive activity as much as one year earlier than those from the eastern Pacific. Thus for the younger ages the same maximum of one-half seems appropriate. The maximum reproductive rate for ages 1 and 2 is apparently zero as none have been observed to be sexually mature.

Using these maximum values for survival and reproductive rates, the life-table calculations suggest that if only reproductive rates are changed it is not possible to obtain a net rate of increase of 8% for any of the combinations rates considered in Table 6. The maximum rates of increase that can be obtained are at most a few percent. This is due to the estimated reproductive rates which are already high. The maximum rate of increase that can be obtained, for example, for the combination of rates labeled in Table 6 as "high" for survival rates and "central" for reproductive rates is 4.6 to 5.4%, depending on the value of survival from ages 3 to 4 used in determining the estimates for survival between age 3 and 12.

If only adult survival rates are changed, it is possible to obtain an 8% rate of increase for some of the combinations considered in Table 6. In particular, when the adult survival rates are low and when pregnancy rates are high, and given that the maximum survival rate possible is very close to 1.0, rates of increase of 8 to 10% are possible. It is necessary, in these cases, that the difference between the survival rates for a zero rate of increase and the maximum survival rates be large. This also requires that the pregnancy rates and the survival rates from birth to age 3 be high. Even so, the survival rates that are required to obtain 8% rates of increase are unreasonably high. Further, the percentage changes in adult survival rates that result in 8% rates of increase are on the order of 5 to 10%, changes that would be almost impossible to detect given the sampling variability apparent in Table 2.

If only juvenile survival rates are allowed to change, it is possible to obtain 8% rates of increase for many of the combinations in Table 6 with higher adult survival rates and pregnancy rates. These combinations result in smaller values of survival from birth to age 3 at equilibrium, and hence more range for possible changes in this rate. For example, for the combination of rates labeled "high" for survival rate and "central" for pregnancy rates in Table 6, the survival rate from birth to age 3 required to give an 8% rate of increase range from 61 to 79% for a corresponding range of survival from age 3 to 4 of 0.80 to 0.95. This range of three year rates correspond to unreasonably large annual rates of 85 to 93% if the rates were constant over the three years, and suggest that increases in juvenile survival of approximately 130% are required to obtain an 8% rate of increase of the population. Changes of this magnitude should have been relatively easy to detect, and are not suggested by the differences in the estimates of the survival rates in the 1920s and the 1960s, under any current interpretation of the data.

The general conclusion from considering the effect of changing one factor at a time is that the changes that must have occurred in the vital rates to account for the history of this population are not likely to involve only one of the three types of rates considered.

It is possible to consider changes in pairs of factors that could result in an 8% rate of increase. However, the number of possibilities and combinations becomes overwhelming quite quickly. Examination of some of these combinations gives an indication of the tradeoffs between changes in the different factors being considered can be obtained. Generally when two factors are being considered, only moderate changes in each are needed to obtain an 8% rate of increase. This is particularly worrisome, as the possibilities for detecting any such changes are related to the magnitudes of the changes. Thus as the numbers of factors changing increases, the possibilities for detecting the changes and hence determining which vital rates are involved, decrease.

DISCUSSION AND CONCLUSIONS

The Alaskan fur seal population has been considered as the one marine mammal population for which we have a good understanding of a population that has recovered from very reduced numbers while supporting a considerable harvest. In spite of this, the data and analyses presented here suggest that there are large gaps in our understanding of the dynamics of this population, both in the vital rates when the population was thought to have a rate of increase near zero and in the changes in the vital rates regulating the size of this population.

The major gaps in our understanding of the vital rates during a period of apparent zero population growth include estimates for the survival rates from ages 3 to 12, and for the survival rate from birth to age 3. While several analyses have been conducted to estimate values for some of these rates, such analyses are based on unsupported assumptions about several parameter values, and the sensitivities of the resulting estimates to these assumptions have generally not been considered.

Our analyses suggest that the survival of female seals from birth to age 3 must be substantially higher when the population is at equilibrium than the estimated lower bounds for this rate for juvenile males. Moreover, past estimates of survival rates for juvenile males (e.g., Chapman, 1961, 1964) when applied to females would require either or both adult survival and pregnancy rates near the upper range of the estimates presented in this chapter (e.g., Table 6) in order to be consistent with a net rate of increase that is nonnegative. This inconsistency between the estimates of the survival rates of juvenile males and the estimates of fecundity and survival rates of adult females has been reconciled in earlier papers by assuming that survival of juvenile females exceeds that of males by a constant factor (see Chapman, 1961, 1964, 1973).

However, given the lack of evidence for a differential rate of survival between males and females and the large uncertainties in all the estimated survival rates, there appears little reason to favor the assumption of differential rates of juvenile survival between the sexes in order to reconcile inconsistencies in estimates of the vital rates near equilibrium. Thus, within the ranges of the adult female survival and pregnancy rates presented here and what might be considered a reasonable range for estimates of the survival rate of juveniles near equilibrium based on estimates derived from the male kill, there are a large number of combinations of estimates that are consistent with a non-negative rate of growth without involving a differential survival rate.

The analyses of the changes in rates needed to obtain the observed high rates of increase in the 1920s indicate that there is not sufficient range in any one of the three major components of the life table (i.e., reproductive rates, survival rates from birth to age 3, and adult survival rates) for changes in one of these components to account for the observed dynamics of this population. These analyses strongly suggest that this population must be regulated by changes in at least two and possibly all three of these major components of its life table. If two or more of these components do vary with density, our calculations indicate that only small to moderate changes are needed in any single component to achieve a rate of growth of 8%. This fact may be responsible for the failure of the available data to provide convincing evidence for density-dependent changes and implies that an understanding of the mechanism regulating this population may be difficult to obtain.

While changes in more than a single major component of the life tables are biologically reasonable, the implications of this conclusion for managing this population are basically unexplored. The management for this population has been based on two concepts. The first concept, elaborated by Parker (1918), is that there are large numbers of male seals born that are not required to maintain the population. This has been shown to be true, at least in time frames of decades, as the populations continued to increase in the face of removal of large numbers of males. The second concept, introduced initially by Scheffer (1955), elaborated by Chapman (1961), and further developed (Chapman, 1964, 1973; Nagasaki, 1961; Bulgakova, 1971), is that changes in survival rates of young animals are responsible for the regulation of the population. The implication that the total yield from the population would be higher if both males and females were harvested has been drawn from this concept (Chapman, 1961, 1964). This deduction was tested by changing management policies from 1956 to 1966 to include a harvest of females. An increase in total yield did not result under this experimental harvesting regime. From the analysis put forward here it appears that this second concept on which management has been based, even if true, is insufficient in itself to account adequately for the dynamics of the population, and suggests the need to consider more complex models.

While our conclusions apply directly only to the Alaskan fur seal population breeding on the Pribilof Islands, the hypothesis that population regulation involves more than a single component of the life table should be considered for other large mammals which appear to have similar life tables at equilibrium with vital rates near their biological maximum (see Chapter 7). Mechanisms of regulation will likely be difficult to determine, since only small to moderate changes in vital rates may be occurring.

LITERATURE CITED

Anas, R. E. 1970. Accuracy in assigning ages to fur seals. J. Wildl. Manage. 34:844-852.

Bulgakova, T. Y. 1971. The estimation of the optimum sustainable kill from the Robben Island fur seal herd. VNIRO. Unpublished manuscript.

Chapman, D. G. 1961. Population dynamics of the Alaska fur seal herd. Trans. N. Am. Wildl. Nat. Resour. Conf. 26:356-369.

Chapman, D. G. 1964. A critical study of Pribilof fur seal population estimates. U.S. Fish Wildl. Serv. Fish. Bull. 63:657-669.

Chapman, D. G. 1973. Spawner-recruit models and estimation of the level of maximum sustainable catch. Rapp. P. V. Reun. Cons. Int. Explor. Mer 164:325-332.

Chapman, D. G., and A. M. Johnson. 1968. Estimation of fur seal pup populations by randomized sampling. Trans. Am. Fish. Soc. 97:264-270.

Ichihara, T. 1972. Maximum sustainable yield from the Robbin Island fur seal herd. Bull. Far Seas Fish. Res. Lab. 6:77-94.

Kenyon, K. W., V. B. Scheffer and D. G. Chapman. 1954. A population study of the Alaska fur seal herd. U.S. Fish Wildl. Serv. Spec. Sci. Rep. 12, 77 pp.

Lander, R. H., and H. Kajimura. 1976. Status of northern fur seals. Scientific Consultation on Marine Mammals, FAO, Bergen, Norway, 50 pp.

Leslie, P. H. 1945. On the use of matrices in certain population mathematics. Biometrika 33:183-212.

Marine Mammal Division. 1977. Fur Seal Investigations, 1976. U.S. Dept. Commer., Natl. Mar. Fish. Serv., Northwest Fish Center, Seattle, Wash., 92 pp. (Processed).

Mertz, D. B. 1971. Life history phenomena in increasing and decreasing populations. Pp. 361-399 *in:* G. P. Patil, E. C. Pielov and W. E. Waters (eds.). Statistical Ecology Vol. 2: Sampling and Modeling Biological Populations and Population Dynamics. Pennsylvania State University Press, University Park, Pennsylvania.

Nagasaki, F. 1961. Population study on the fur seal herd. Spec. Publ. Tokai Fish. Lab. No. 365, 60 pp.

Parker, G. H. 1918. The growth of the Alaskan fur seal herd between 1912 and 1917. Proc. Natl. Acad. Sci. 4:168-174.

Roppel, A. Y., and S. P. Davey. 1965. Evolution of fur seal management of the Pribilof Islands. J. Wildl. Manage. 20:448-463.

Scheffer, V. B. 1955. Body size with relation to population density in mammals. J. Mammal. 36:493-515.

Seber, G. A. F. 1973. The Estimation of Animal Abundance and Related Parameters. Griffin, London.

Smith, T., and T. Polacheck. 1980. The population dynamics of the Alaska fur seal: What do we really know? Report to the National Marine Mammal Laboratory, Seattle, Washington.

Snedecor, G. W., and W. G. Cochran. 1967. Statistical Methods. Iowa State University Press, Ames.

SIX

A Model of Lion Population Dynamics as a Function of Social Behavior

A. M. STARFIELD

P. R. FURNISS

G. L. SMUTS

INTRODUCTION

Since lions *(Panthera leo)* have few mammalian enemies, except their own kind, it is clear that their numbers are regulated both by food and social behavior (Bertram, 1973). This chapter describes an attempt to link these factors in the form of a comprehensive lion population model. Our objectives in constructing the model are outlined below.

Building a model of a biological system is the one sure way of confirming that one really understands the workings of that system, or, alternatively, of pinpointing areas of ignorance. More is known about the population biology and behavior of lions (Schenkel, 1966; Schaller, 1972; Rudnai, 1973a,b; Bertram, 1973, 1975; Smuts, 1976, 1978a; Smuts et al., 1978) than about most other large mammals, especially the carnivores. The first objective of the model was to see whether this knowledge could be incorporated into a realistic, self-consistent model.

Lion-cropping trails undertaken in the 5560 km² Central District of the Kruger National Park, South Africa, since 1974, show that lion numbers are soon restored by infiltration from surrounding areas and by increased reproduction and cub survival (Smuts, 1978a,b). Since the wildebeest *(Connochaetes taurinus)* population in this area is rapidly declining (Smuts, 1978b; Starfield et al., 1976), consideration is being given to more drastic lion cropping. A second objective of the model was to analyze this resilience of the lion population to numerical reduction and to explore different methods of control.

There is a continuing effort in the Central District of the Kruger Park to collect data on the dynamics of the lion population. The third objective of the model was to develop some insight into the information needed and to look for ways of interpreting the data and relating it to the conceptual structure of the model.

The approach taken throughout this chapter is epitomized by Hamming's (1962) dictum, "The purpose of computing is insight, not numbers."

STRUCTURE OF THE MODEL

The basic unit of our model, as of the real lion population as we view it, is the pride. A pride in the model is divided into six age and sex classes:

Adult males	5 years and older and all of the same age
Adult females	4 years and over
Subadult females	3 to 4 years
Subadult females	2 to 3 years
Large cubs	1 to 2 years
Small cubs	0 to 1 year

At each annual iteration of the model, mortality is calculated and the appropriate promotions from one age class to the next are performed.

The proportion of each class surviving from one year to the next is derived from a food factor having one of three values: "poor," "moderate," or "good." These values are largely defined in terms of survivals they produce rather than being related to a simple variable such as prey abundance. They are intended to be correlated with the ease of obtaining food at the most critical time of the year (i.e., when prey animals are most difficult to catch). The survivals, given a particular food factor, are the same for all pride animals over 2 years of age, with two different and lower survivals for the two cub classes. The adult survivals are further modified by age. The basic survival rates from Table 1 are used until age 8 years; thereafter survival declines steadily until certain death at age 16 years. The survival rates in Table 1 are based on rough estimates modified by experience with the model. The ages of males are specified precisely in the model, but for females a mean age is used. This mean age is adjusted whenever recruits, whose age is known exactly, join the pride or when a female dies. It is assumed that dying females are 2 years older than the mean.

Table 1   Basic Survival Rates as Used in the Model

Category of Lions	Food		
	Poor	Moderate	Good
Pride adults	0.90	0.95	0.97
Cubs aged 1-2	0.50	0.70	0.75
Cubs aged 0-1	0.20	0.60	0.75
Lodgers	0.50	0.70	0.75

Only adult females in a pride produce cubs. Field data (Schaller, 1972; Rudnai, 1973a; Bertram, 1975; Smuts et al., 1978) indicate that a female produces a litter approximately two years after her previous litter, provided that some of the cubs survive. If they all die in their first year, the female is likely to have a new litter after only one year. In the model the maximum fertility, which is the mean litter size of 2.9 cubs (Smuts et al., 1978), reduced to take account of the 6.7% older nonbreeding females, is expressed only if there are no surviving cubs from the previous year. If there are any cubs aged 1 to 2 years the maximum fertility is reduced by the number of newborn cubs per female that would, after natural mortality, give the existing number of cubs aged 1 to 2 years:

$$\text{cubs born} = (\text{max. fertility} \times \text{adult females}) - \frac{\text{cubs aged 1 to 2}}{\text{survival rate of cubs aged 0 to 1}}$$

The model contains a specified number of prides, each assigned to a territory which that pride is considered to hold. The territory is defined to be adjacent to a number of others, as shown schematically in Figure 1. The food values are independent for each territory and may be changed from year to year. In reality, lion movements and hence territories overlap (Schenkel, 1966; Schaller, 1972; Smuts, personal observation), but the prides remain distinct.

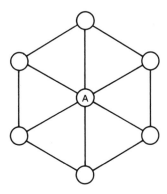

Figure 1 Schematic arrangement of territories for a model with seven territories. The central territory A is adjacent to six other territories.

There is no sex differentiation of the cubs until they are promoted to subadults (aged 2 to 3 years), at which stage sexes are assigned on the basis of a 1-1 ratio. Some of the female subadults may be recruited immediately into the pride as females aged 2 to 3 years. Any remaining females and all subadult males are expelled from the pride between their second and third years and form a "lodger group" of nomadic animals. The lodger groups thus formed are each assigned to a territory (initially the territory of origin), but are not part of the pride. The members of each lodger group are all of similar age and all originated from the same pride. This is not quite realistic since different lodger groups or individual lions may form associations which could last for any length of time (Schaller, 1972; Smuts et al., 1978). Generally, however, companionships are between animals of the same sex and age, and solitary animals are more likely to be accepted by other lone animals than by a group in which social ties already exist (Schaller, 1972).

Lodger groups within the same territory do not fight and kill one another. This is a safe assumption, since Schaller (1972) has also found that nomads tolerate one another (see also Smuts, 1978a) and further states, "This behavior seemed to be directly related to the kind of land tenure system: lions which defend no land generally accept strangers."

Lodger survival, like that of the prides, is affected by the food conditions of whichever territory they are in. Survival rates are somewhat lower than for adult pride animals (see Table 1) and under poor food conditions are considerably lower (Schaller, 1972). Lodger females do not breed. This is not quite true but is a fair assumption, since most cubs from these females either die or are killed (Schaller, 1972; Bertram, 1973).

The lodger groups may move from one territory to a neighboring territory. Such movement is dependent on food conditions in the territory they are in. If food is poor, a lodger group will certainly move, and if the new territory also has poor food conditions the group may move again up to a total of five times in a single iteration of the model (one year). If food is moderate or good a group will consider moving out of the territory only twice per year, with 0.5 probability of moving each time. Distinct from the food-induced movement, at the end of each iteration a check is made to see whether any of the territories are either vacant or contain prides without males. If so, a neighboring lodger group containing males will automatically be moved into that territory. This is based on the fact that the lodgers will notice the absence of pride males by the absence of roaring and marking (Schenkel, 1966; Smuts 1978a).

If the lodger group does move to a new territory the females will attempt to join the pride in that territory. Each territory has a defined "female capacity," indicating how many females older than 2 years will be tolerated by the pride. Young females will be recruited from the maturing cubs if there are vacancies (Bertram, 1973, 1975). If there are still vacancies these will be filled by female lodgers on a first-come first-served basis. If there are no vacancies the lodger females stay in their groups. The female capacity of each territory is affected by the food status, being raised by one in good years and reduced by one in poor years. Females are not expelled when conditions deteriorate, but dying females will not be replaced until numbers fall below the current capacity. In the real world, pride size (adult females) does not change markedly in the short term although there may be a small numerical increase in response to a series of good years (Bertram, 1973). In Kruger Park it was found that the total number of adult females remained constant over a period of 2½ years for six neighboring prides (Smuts, 1978c).

Lodger males join prides according to a different system. On entering a territory lodger males "challenge" the pride males. The strength of a challenge is defined as the number of challengers divided by the number of males in the pride. The challenge ratio is defined as the threshold strength (determined on the basis of the ages of the challengers and the pride lions, from Table 2) divided by the strength of the challenge. The threshold strengths are chosen so that a

Table 2   Threshold Strengths for Challenges as Used in the Model

Age of Pride Males (years)	Age of Challengers (years)			
	5	6	7	8
5	1.50	1.00	0.90	1.00
6	2.00	1.40	1.50	1.80
7	2.00	1.40	1.40	1.80
8	1.30	1.20	1.25	1.40
9	1.25	1.00	1.05	1.35
10	0.90	0.30	0.30	1.00

challenge ratio of 1 implies that the challenge has a 0.5 probability of succeeding. The chance of a challenge succeeding is defined in terms of the challenge ratio by Figure 2. If this ratio is greater than 1.2 (hopeless) no mortality occurs during the challenge, as it is assumed that the lodger males do not press their attempt. If a challenge is attempted but fails, both lodger and pride males suffer mortality with much greater risk to the challengers.

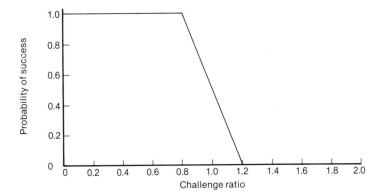

Figure 2   Probability of a lodger male lion's challenge succeeding as a function of the challenge ratio. (See text for details.)

If the challenge succeeds the old pride males are liquidated and the challengers become the new pride males. Bertram (1973) mentions that adult males ousted from a pride generally die within a short time. The successful challengers immediately kill a proportion of cubs in the pride, dependent on the number of challengers (mortality increases by 0.24 for each challenger). Since cub mortality rises after new males take over a pride (Bertram, 1973), we have assumed that with more males in the new group competition for females and consequently adverse adult male contact with cubs will be higher.

The challenge threshold strengths (Table 2) are rough estimates that were obtained by asking such questions as "are two 6-year-old pride males and three 6-year-old lodger males evenly matched?"

In a lodger group of mixed sex the females go off on their own if the males take over a pride. Although unimportant to the model, mixed groups usually break up before the males begin challenging pride males (Smuts, personal observations).

Finally, various cropping options are incorporated into the model. It is possible to crop from either the pride or the lodgers, or both, in any of the territories, and age and sex classes may be cropped selectively, either partially or completely.

The model was originally conceived and built as a stochastic model with probability distributions for births, deaths, recruitment, movement, and challenges. However, it was found virtually impossible to conduct a sensitivity analysis of the stochastic model, or to relate it in any way to reality (short of performing a phenomenal number of computer tests). A purely deterministic model was built, but was found to be unrealistic. For example, if four cubs aged 1 to 2 survive to become subadults, then in a deterministic model two of those subadults will be female and two will be male. In a stochastic model, it is possible for all four to be males. Clearly a lodger group consisting of four males will have a considerably greater chance of successfully challenging for leadership of a pride, and through infanticide this enhanced chance of success will alter the population dynamics of the model. The final model, is, therefore, semistochastic, that is, wherever possible, processes are deterministic, but where the variance affects the dynamics of the model, they are stochastic. Thus all survivals and births are deterministic, but as the numbers of lions in each class must be whole numbers, rounding is stochastic (e.g., 0.5 survival of three animals will give one or two survivors with equal probability). Recruitment of females is fully deterministic. The outcome of a challenge is stochastic, as are the sexing of subadults and the choice of neighboring territory when a lodger group moves.

The outcome is a model that gives results that are, in essence, repeatable, but that preserves important stochastic effects.

RESULTS

Two series of tests were used to investigate the model. The first series, based on a seven-territory configuration, was used to test the behavior of the model.

The seven-territory configuration is shown in Figure 1. All runs are compared against a standard simulation over a 15 year time span. In this standard simulation, five of the seven territories had "good" food conditions for all 15 years, while the remaining two has "moderate" food conditions. The values of all other parameters were those described in the previous section of this chapter. Realistic initial populations, for both prides and lodger groups, were assigned to each territory.

Figure 3 shows the pattern of changes in one pride of the seven (the pride in territory A of Figure 1) during the standard run. Similar diagrams can be drawn for any of the prides. Table 3 attempts to summarize the results of 11 runs of the seven-territory model with various alterations in parameter values. All runs were over a 15 year time span, and the entries in Table 3 are the means, per territory, for the last five years of each run. This was done to even out fluctuations from one year to the next and also to remove the influence of the population assigned to each territory at the start of the run.

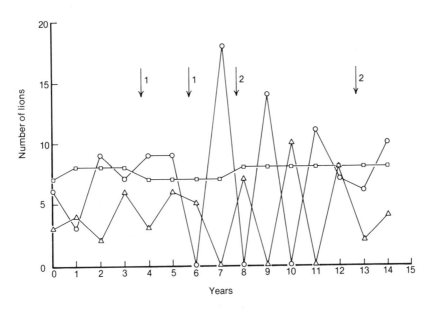

Figure 3 Fluctuations in the size of the pride in territory A of Figure 1, for 14 years of a standard simulation, in terms of cubs less than 1 year old (O), cubs between the ages of 1 and 2 (Δ), and pride animals older than 2 (□). Arrows indicate the takeover of the pride by new males; the number of new males is shown next to the arrow.

The results of one additional run of the seven-territory model are shown in Figure 4. Food conditions in all territories were fixed at "poor" for 15 years, and the lion population shows the ensuing decline.

One of the objectives of the model was to investigate the effects of different cropping strategies. A major consideration in lion control in Kruger National Park is the desire to protect local populations of rare antelope or concentrated

Table 3 Summary of the Results of a Number of Different Combinations of Assumed Parameter Values Used in the Lion Population Models

Case No.	Description	Number of Pride Animals Older than 2 years	Number of Cubs	Number of Adult Pride Males	Age of Pride Males (years)	Number of Lodgers per Territory	Number of Lodger Groups per Territory	Ratio of Lodgers to Adult Pride Lions	Successful Challenge Interval (years)
	Standard run	7.6	10.4	1.9	6.1	5.9	2.8	0.78	3.8
1	Standard deviation (estimated from 10 replicates)	(0.20)	(0.40)	(0.17)	(0.26)	(0.40)	(0.28)	—	(0.19)
2	Food one level lower for 4 years in 15	6.5	9.7	1.3	6.6	4.3	2.0	0.66	3.5
3	Lodger survival reduced	7.2	11.6	1.4	6.4	4.7	2.1	0.65	3.9
4	Cub survival reduced	7.4	10.1	1.7	6.0	4.8	2.5	0.65	3.6
5	Fertility increased by 20%	7.6	12.4	1.9	6.1	6.9	3.4	0.91	4.0
6	Fertility decreased by 20%	7.3	8.6	1.6	6.0	4.1	2.2	0.56	3.5
7	Successful challenges made easier	7.6	9.7	1.9	5.7	6.1	2.8	0.80	3.0
8	Food one level lower for all years	5.9	7.3	1.3	6.1	2.5	1.4	0.42	4.2
9	Food one level lower and fertility up 20%	5.7	8.7	1.1	5.9	2.9	1.6	0.51	4.2
10	Food one level lower and fertility down 20%	5.0	4.7	1.1	5.9	1.6	1.1	0.32	4.0
11	Food one level lower and successful challenges made easier	5.1	7.6	1.4	5.9	2.1	0.8	0.41	4.2

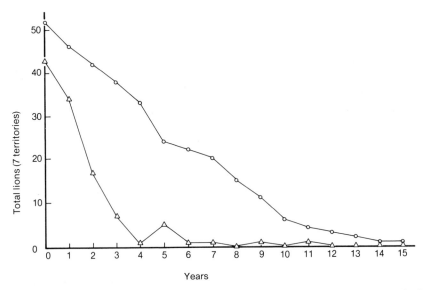

Figure 4 Decline in lion population size under poor food conditions, separately for pride lions older than 2 years (○) and lodger lions (Δ).

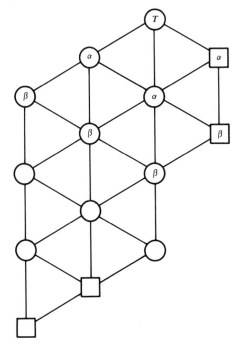

Figure 5 Arrangement of territories with good (○) and moderate (□) food conditions used in simulations of the effect of cropping policies A, B, C, and D (see text) on lion pride size. T denotes the target territory, α the adjacent territories, and β territories that are separated from the target by one territory.

129

breeding herds of wildebeest by reducing the lion population in the area. The seven-territory model was too small for this purpose, and so a series of model experiments were carried out using the 14-territory configuration shown in Figure 5. Food was kept constant at "good" for all territories except the four indicated, in which food was "moderate."

Four policies were tested. In all four the pride lions and all lodgers in the target territory (T) were eliminated in year 1. Policy A involved no other cropping. Policy B included removal of the lodgers from the three territories adjacent to the target territory. Policy C was similar to B but lodgers in the second ring of four territories, separated from the target territory by one territory, were removed as well. Policy D was as B, but the three prides adjacent to the target territory were reduced to half. In all cases cropping was confined to a single year.

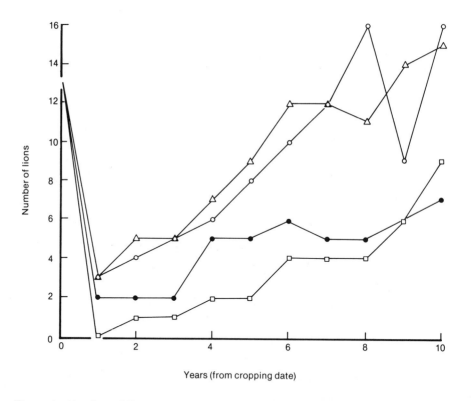

Years (from cropping date)

Figure 6   Number of lions older than 2 years in the target territory for the 10 years following cropping policy A ($\bigcirc$), B ($\triangle$), C ($\square$), and D ($\bullet$) (see text).

Figure 6 shows the number of lions older than 2 (i.e., the hunting animals) in the target territory over the 10 years after cropping. Finally Figure 7 compares the results from cropping policy C with those obtained using a revised model that allows for freer movement of lodger groups. These results, together with the other results described above, are criticized and discussed in the following section.

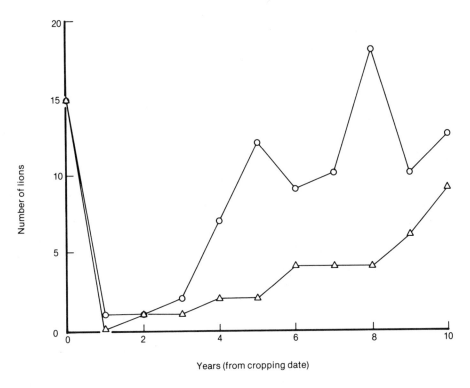

Figure 7   Number of lions older than 2 years in the target territory for the 10 years following cropping policy C (Δ) (see text) and a revised cropping model allowing freer movement of lodger groups (○).

DISCUSSION AND CONCLUSIONS

The following points arise out of the sampling of results presented in the previous section and also out of the large number of other studies we have conducted using the model.

In general, it was found that the number of pride animals varies very little according to our model, while the number of lodgers is far more variable, even in cases where parameter changes do not directly affect lodgers as such. Figure 3 shows that the number of mature lions in the simulated pride remains quite constant. The changes that do occur are largely changes in the number of males. Under good food conditions any vacancies among the females in the simulated pride are immediately filled. These results are in agreement with field observations (Bertram, 1973).

Oscillations in the number of cubs, as in Figure 3, were also a general feature of all simulated prides. These are caused by the reduction of female fertility due to the survival of the previous year's cubs, in conjunction with infanticide by successful challengers. If a large group of males takes over the pride, they will kill all cubs, the full fertility of the females will be expressed in the following year, and the cubs produced then will probably suppress all births the year after that. If a small group of males takes over the pride, they will kill only a portion of the cubs, which may accentuate or diminish the amplitude of the oscillations in cub numbers.

It may be worth considering whether the observed synchronization of litters in prides (Schaller, 1972; Rudnai, 1973a; Bertram 1975; Smuts et al., 1978) is a consequence of male infanticide, since the model, in which physiological synchronization is absent, exhibits synchronization.

From Figure 3 and also from the entries in Table 3 it can be seen that successful challenges by males occur at intervals of three to four years, which is realistic (Bertram, 1973). The results in Table 3 are also consistent with the observed mean age of 6.5 years for pride males in Kruger Park (Smuts et al., 1978).

Comparing run 2 in Table 3 with the standard run, run 1, it can be seen that occasional years of poorer food conditions can act as a significant control on overall lion numbers. In particular, lodger numbers are very sensitive to changes in food availability. This is highlighted by run 8 of Table 3 and also in Figure 4, where it can be seen that during uninterrupted severe conditions, the pride population drops by 50% in about five years, while the lodger population drops by about 50% in only two years. The influence of the territorial structure of the model is highlighted by the observation that none of the seven territories was unoccupied in the simulation for Figure 4 until year 10, by which time the total population had dropped to 6% of the initial population.

In recent years food conditions in the Central District of Kruger Park have been excellent, and there is, therefore, a lack of quantitative information on the lion population under less favorable conditions. These studies begin to provide some insight into aspects of lion population ecology that may merit study under such conditions. They also indicate that the ratio of lodgers to adult pride lions may be a useful indicator of food conditions and population trends.

The runs with different food values can also be interpreted (Table 3) as showing the effects of a general change in the survival rates used in the model. Specifically it was found that reducing the cub or lodger survival rates (runs 3 and 4 in Table 3) had little effect on pride animals, except for a slight reduction in the mean number of pride males. Both reductions, however, result in a clear decrease in the number of lodgers.

Looking at runs 1, 5, 6, and 7 in Table 3, it is apparent that under conditions of good food, altering the ease of takeover has virtually no effect on the population in comparison with the effect of increasing or decreasing the fertility. On the other hand, comparing runs 8, 9, 10, and 11 in Table 3, altering ease of takeover under moderate food conditions is seen to have the same order of effect as changes of 20% in the fertility. The relative importance of behavioral and physiological responses under different food conditions is an interesting insight that emerges from the model.

Referring to the simulated cropping experiments of Figure 6, it can be seen that with policies A and B the target population did not fully recover for about seven years. With policy C recovery was slower still, while policy D gave an intermediate result. In all cases the population remained depressed for some years after a single season's cropping.

Although it is difficult to make direct comparisons with cropping experiments in Kruger Park, the above results appear to be unrealistic. As mentioned in the introduction, cropping trials in Kruger Park showed a rapid restoration of numbers in the control areas. For example, following a 63% removal of lions in six territories, recolonization was 90% complete after only 17 months (Smuts, 1978a). While the proportions of the total population removed in the model simulation and the cropping trial are not strictly comparable, they are sufficiently similar to conclude that recolonization in the model is too slow. It is clear that the problem lies in the algorithm for lodger movement. The lodgers move slowly when food is good or moderate, as it was throughout the simulation runs. However, speeding up the movement, by making all lodger groups move five times per year, regardless of food, while accelerating recovery (as shown in Figure 7) still did not produce the rapid recovery met in real life.*

This is of particular significance in relation to one of the objectives of the model, which was to find out if one really understood the lion population dynamics. The vital role of lodger movement did not emerge until cropping trials were simulated (increased movement did not significantly affect the results of the seven-territory model).

Three possible modifications to the movement subroutine emerged:

1   A local vacuum caused by cropping may prompt a pride to split and the splinter group to move in and colonize a vacant territory; this is more likely to happen when food is difficult to obtain.

2   Lodgers may pass right through a territory without attempting to join the pride or even to meet it.

3   Lodgers are highly mobile and may investigate wide areas rapidly without remaining in an area. But if they discover an area with no or few lions they are likely to stay there.

None of these possible modifications has, as yet, been implemented in our model.

Another weakness of the model that has not as yet been investigated is that it does not allow individual lions or small groups of lodgers to meet and combine.

*Subsequent work using a 31-territory model with the algorithm for freer lodger movement produced results comparable with cropping experiments in Kruger Park. The above results may therefore not be "wrong", but could rather by representative of a smaller, self contained lion population.

Especially with regard to males, the fact that individuals may combine will have a twofold effect both in making it easier for a challenge to succeed and in greater infanticide when it does succeed. It is noteworthy that while other modifications, of no lesser biological significance, were incorporated in the model, this effect was ignored because it was difficult to model, and field observations do not, as yet, give an obvious lead on how to incorporate it into a model.

LITERATURE CITED

Bertram, B. C. R. 1973. Lion population regulations. E. Afr. Wildl. J. 11:215-225.

Bertram, B. C. R. 1975. Social factors influencing reproduction in wild lions. J. Zool. 177:463-482.

Hamming, R. W. 1962. Numberical Methods for Scientists and Engineers. McGraw-Hill, New York.

Rudnai, J. 1973a. Reproductive biology of lions (Panthera leo massaica Neumann) in Nairobi National Park. E. Afr. Wildl. J. 11:241-253.

Rudnai, J. 1973b. The Social Life of the Lion. Garden City Press, Hertfordshire, England.

Schaller, G. B. 1972. The Serengeti Lion. University of Chicago Press, Chicago.

Schenkel, R. 1966. Play, exploration and territory in the wild lion. Symp. Zool. Soc. London 18:11-22.

Smuts, G. L. 1976. Population characteristics and recent history of lions in two parts of the Kruger National Park. Koedoe 19:153-164.

Smuts, G. L. 1978a. Effects of population reduction on the travels and reproduction of lions (Panthera leo) in Kruger National Park, South Africa. Carnivore 1:61-72.

Smuts, G. L. 1978b. Interrelations between predators, prey and their environment. Bio. Sci. 28:316-320.

Smuts, G. L. 1978c. Predator control and related research in the Central District of the Kruger National Park (December 1976 to January 1978). Memorandum. National Parks Board of Trustees, Pretoria. 40 pp.

Smuts, G. L., J. Hanks, and I. J. Whyte. 1978. Reproduction and social organization of lions from the Kruger National Park. Carnivore 1:17-28.

Starfield, A. M., G. L. Smuts, and J. D. Shiell. 1976. A simple wildebeest population model and its application. S. Afr. J. Wildl. Res. 6:95-98.

SEVEN

Density-Dependent Processes and Management Strategy for the Northwest Atlantic Harp Seal Population*

P. F. LETT

R. K. MOHN

D. F. GRAY

INTRODUCTION

To study density-dependent mechanisms in wild mammal populations, it is helpful to be working with species that have fluctuated widely in abundance during periods of intense biological sampling. The Northwest Atlantic harp seal, *Pagophilus groenlandicus,* is one such species (Sergeant, 1976a; Lett and Benjaminsen, 1977). Although catch statistics are available for this population as early as the eighteenth century (Chafe et al., 1923), biological sampling did not begin until the early 1950s (Fisher, 1952; Sergeant, 1959; Sergeant and Fisher, 1960). This was a particularly opportune time to begin sampling, as the population had increased during and after World War II to a high level, estimated at 2.3 million seals (Lett and Benjaminsen, 1977). The sampling initiated at that time has continued and was intensified after 1960. In addition to catch-at-age information from the various fisheries, much data have been collected on maturity and fertility (Fisher, 1952, Sergeant, 1966, 1969, 1976a; Øritsland, 1971), on migration patterns (Sergeant, 1976a), on feeding habits (Sergeant, 1973; Kapel, 1975), on population census (Sergeant, 1975; Lavigne et al., 1977), and on mortality and sex ratios (Ricker, 1971; Ulltang, 1971; Benjaminsen and Øritsland, 1975; Lett and Benjaminsen, 1977).

Historically, harvesting levels on this stock have been particularly high. For example, the offshore harvest alone exceeded 450,000 seals during the period from 1830 to 1850, reaching a peak level of 687,000 animals in 1844. The average harvest was much lower (about 150,000) during the early 1900s. Recovery of the stock occurred during World Wars I and II. Hunting again became intense after 1945, especially after Norwegian vessels joined the hunt and took large numbers of adult females, which caused a rapid decline in the

---

*An expanded version of this chapter (by same title) was published as ICNAF Res. Doc. 78/XI/84 in the International Commission for the Northwest Atlantic Fisheries Selected Papers No. 5, 1979.

stock (Lett and Benjaminsen, 1977). Quotas were initiated in 1972, and the stock has been increasing since then (Benjaminsen and Øritsland, 1975; Lett and Benjaminsen, 1977). The more important observation, however, is that this stock has maintained an average annual yield of 275,000 animals over the past 150 years.

How has this population been able to survive for so long under such intense harvesting? The response to that question is the theme of this paper. Density-dependent mechanisms, possibly operating at behavioral and physiological levels, have given the stock resilience and stability in the face of widely varying annual exploitation. Indeed, it is upon the knowledge of these density-dependent mechanisms that much of our management advice hinges.

BIOLOGICAL SAMPLING OF THE STOCK

The Stock

The harp seal, as a species, reproduces in three widely separated populations located on pack ice around Newfoundland, Jan Mayen Island, and in the White Sea. Studies of skull and body dimensions (Khuzin, 1967) have shown that the Newfoundland, or western Atlantic, population is more distinct from the two eastern Atlantic populations than the two eastern populations are from each other. Sergeant (1976b) indicates that there is limited intermixing between the eastern and western populations.

The Newfoundland population is divided into two subpopulations — one reproducing on the southward drifting pack ice off southern Labrador, and the other in the Gulf of St. Lawrence. The herd that reproduces off Labrador in the spring (known as the Front herd) can be further divided into a northern and southern contingent. The position of the whelping patches depends on the food supply and the formation of rather loosely packed ice interspersed with sufficient leads of water. In early March (about the 8th or 9th), the younger females, which have less control over parturition than do older animals, are found on the ice when they give birth. The older females remain in the area of abundant food for approximately two more days, after which they too begin to whelp. Further details concerning the biology of this population are to be found in Lett et al. (1979) and the literature they review.

Age Composition Data

Data concerning catches taken from each age class form the basis of the assessment of many animal populations. Such data for the harp seal fishery in the Northwest Atlantic during 1974-1977 are given in Lett et al. (1979). Data for 1952-1974 are listed in Lett and Benjaminsen (1977). The determination of the actual age structure of the catch is a formidable problem for this fishery because of its diversity. The problem consists of producing weighted estimates of the numbers caught from each age for seals aged 1 and older by amalgamating the age frequencies from the various fisheries in their proper proportions.

Samples of seals from Notre Dame Bay, Newfoundland, consist primarily of animals that have not fully developed their adult characteristics, namely bedlamers. The net fisheries at La Tabatiere, Quebec, and southern Labrador yield samples of pregnant females and mature males as they migrate southward into the whelping areas. However, the samples from these areas in January show a preponderance of seals that will whelp for the first or second time; these animals arrive later than the older ones. Samples taken at St. Anthony, Newfoundland, seem to better represent the population's age structure, but separate catch statistics are not always available for these areas on the same scale as are the other sampling data. However, when such data were available, they indicate that, on the average, each of these fisheries tends to be roughly equivalent. Therefore, the catches were summed by age class without weighting to produce a description of the age composition for the yield by landsmen without serious error (Lett et al., 1979; Lett and Benjaminsen, 1977).

The catches by large vessels from the breeding areas and molting patches have a considerably different age structure from the catches taken by landsmen, and must therefore be treated separately (Lett et al., 1979; Lett and Benjaminsen, 1977). There is usually a high representation of seals aged 1 and 2 in these samples, because younger animals are segregated around the periphery of the adult herd and are more accessible to hunting. The overall age composition of the landsmen catch must be combined with that for animals aged 1 and older exploited by the larger vessels, weighting each frequency in accordance with their respective catches. If this procedure is not followed, a serious consistent bias would result, especially in the data for the last 15 years, owing to the increased interest of Newfoundland in the fishery (Lett and Benjaminsen, 1977).

Between 1952 and 1960, jaws of seals were collected for age determination on a regular basis from the landsmen's catch. This sector of the hunt may be considered as being well sampled (Sergeant and Fisher, 1960). During 1952-1954 and 1957-1959, samples of jaws were collected from the catches of large vessels. However, these samples were inadequate in 1957 and 1958, and it is unlikely that they represent the catch accurately. For this reason average catches by age class were used to represent the composition of the catch taken by large vessels. The attendant errors are possibly serious, since the catch of animals older than one year of age by large vessels represented between 87 and 96% of the total catch during this period. Sampling steadily improved after 1961, the catches from the molting patches by large vessels and the catches by landsmen being well represented in the samples. Annual data for catch by age class for 1961 and onward were obtained from various sources (Sergeant, personal communication; Øritsland, 1971; Benjaminsen and Øritsland, 1975). Sampling of both the landsmen and the large vessel catches was inadequate in 1972, but the available data are included in this analysis.

Samples of the West Greenland and Canadian Arctic catches, which represent an average of about 8% of the total catch, are excluded from this analysis, since no consistent sampling and catch records are available.

Samples of males from the molting patches are used to estimate natural mortality. However, as noted above, young seals (age 1) are usually segregated from the remaining age groups and are not consistently sampled in the molting patches (Benjaminsen and Øritsland, 1975). Furthermore, the closing date of the hunt affects this sample; the earlier the hunt in the molting patch is terminated, the lower is the proportion of females in the catch (Sergeant, 1965; Øritsland, 1971). Thus the age composition of adult seals varies, depending on the duration of the hunt, with the samples consisting mainly of male animals. At age 2 the sex ratio is about 50-50, but by age 10 about 80% of the molting animals sampled are male (Figure 1). This is primarily because many of the adult females remain in the water and are feeding to regain the energy lost during whelping and suckling.

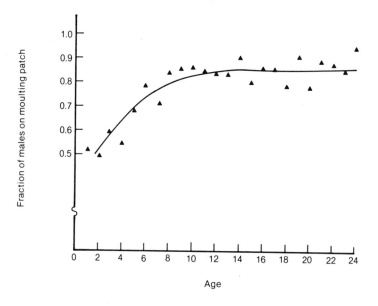

Figure 1   The fraction of males in the Norwegian catch of molting Northwest Atlantic harp seals from 1969 to 1974 (from Benjaminsen and Øritsland, 1975).

DETERMINATION OF BIOLOGICAL RELATIONSHIPS AND ESTIMATION OF PARAMETERS

The Instantaneous Rate of Natural Mortality (M)

The most elusive vital rate in population dynamics is usually natural mortality. Since the rate of exploitation of seals 1 year of age and older is very low, currently about 0.015 (Lett et al., 1977), the annual decline in numbers of these age groups is primarily due to natural mortality.

Natural mortality has been estimated using pup production determined by the survivorship method (Benjaminsen and Øritsland, 1975, 1976; Sergeant, 1975), using information on maturity, sex ratio, pregnancy rate, and the age composition of seals in the molting patches. The analysis assumes that there is no difference between the natural mortality rates of male and female harp seals. The validity of this assumption is based on the observation that males and females have similar growth rates and achieve equivalent maximum ages (Sergeant, 1973a). Since Lavigne et al. (1976) indicate that the metabolic rate and body size of seals are well correlated and Simms et al. (1959) indicate that mortality and metabolic rates of animals are related, it is unlikely that male and female seals have different natural mortality rates.

The age structure of the population was determined from samples of the catches by large vessels in the molting patches, as given in Lett and Benjaminsen (1977) and Sergeant (1977). These frequencies of each age class in the catch were multiplied by the proportion of males at different ages (Figure 1) to give an estimate of the population's age structure for animals 2 years of age and older. However, since partial recruitment may vary somewhat from year to year, only the data for animals from 5 to 22 years of age were used to calculate natural mortality.

With a knowledge of the maturity, sex ratio, pregnancy rate, and population structure, pup production could be estimated for the appropriate age groups. It was assumed, as in Lett and Benjaminsen (1977), that 6% of the breeding females are more than 25 years old. The natural mortality (M) for each age group was therefore calculated by the formula

$$M = \ln\left(\frac{_aN_t - {_aC_5}}{_{a+1}N_{t+1}}\right)$$

where $_aN_t$ is the number of animals of age a in the year t, and $_aC_t$ is the number of animals of age a in the catch in year t. For the period 1966-1977, M for animals from 5 to 22 years of age was estimated to be 0.10 with a standard error of 0.03. The maturity was estimated by linearly interpolating between estimates for 1966 and 1977.

Using the method devised by Ricker (1971), Benjaminsen and Øritsland (1975) calculated natural mortality to be 0.102 with a standard error of 0.011. Further consideration of these data resulted in a revised estimate of $M = 0.106$ (ICNAF, 1978). Lett and Benjaminsen (1977) estimated natural mortality to be 0.114 (using what may be an unreliable method), and Winters (1976) calculated M to be 0.115. Earlier estimates were somewhat lower; both Ricker (1971) and Ulltang (1971) estimated natural mortality at about 0.08 per year, while Sergeant and Fisher (1960) estimated the total mortality to be as low as 0.079 for the 1952-1954 period. The available evidence indicates, therefore, that the natural mortality for harp seals is about 0.10, the value used in the analysis described in this chapter.

No evidence for age-dependent natural mortality is available at this date (1978). This is not surprising considering the uncertainties in the data and the delicate changes possibly exhibited by this parameter (Lavigne et al., 1976). As more detailed data of this kind exist for the harp seal population than perhaps for any other stock of large mammals, a constant mortality for seals of age two and older is assumed. But to management, mortality in the first year is important as is its variation in relationship to some density-dependent mechanism, as suggested by Lett and Benjaminsen (1977).

Sequential Population Analysis

Sequential population analysis is a method of estimating the number of animals in a population, by age, for a period of years. It uses data concerning the numbers caught, by age (Fry, 1949; Murphy, 1964; Jones, 1964; Pope, 1972). The analysis of a year class starts with an estimate of the number alive in the last year for which catch data are available for that year class (terminal year-class size), and proceeds to estimate the numbers in previous years by adding estimated annual losses due to hunting and natural mortality. Inputs into the model are, therefore, the data concerning the numbers caught, their ages, an estimate of the natural mortality coefficient and estimates of the terminal year-class sizes. Further details concerning the application of the technique of sequential population analysis are to be found in Lett et al. 1979, especially as they relate to this study.

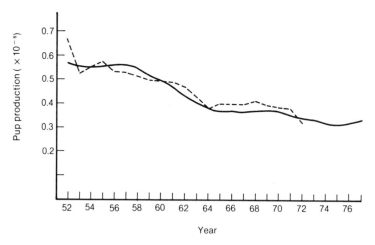

Figure 2   Pup production from sequential population analysis compared to pup production from Winters (unpublished data using regression techniques).

The estimates of the production of pups during 1952-1977 as produced by this method are illustrated in Figure 2. These estimates are similar to, but less variable than, estimates made by Winter (personal communication), which were based on regressions of indices of survival and pup catches. Final estimates of the numbers for each age as resulting from the sequential population analysis and exploitation rates are displayed in Lett et al. 1979.

A word of caution should go with these sequential population estimates. This type of analysis uses all the input data; there are no degrees of freedom to estimate error. Confidence limits on the estimates of pup production used to adjust the exploitation rates are indicated in Lett et al., 1979. The sample on which the population's estimated age structure is based was comprised of fewer than 1500 animals and most of these were less than 6 years old. Exploitation rates on animals one year of age and older are very low, and, even though the catch data are probably quite good, the resulting estimates of the numbers in each age class in the population are likely to have sizeable confidence limits. However, there is an overall consistency in the results, giving confidence to the estimates of the sizes of the total population and its trends.

Density-Dependent Mortality in the First Year

The estimation of pup production by cohort analysis by Lett and Benjaminsen (1977) gave more erratic results than would be expected for a marine mammal population. They postulated that the variation may be attributable to a fluctuating natural mortality rate of pups, which in some manner responds to exploitation.

The existence of density-dependent mortality was investigated, using the numbers of one year old animals estimated from the sequential population analysis in combination with estimates of the number of pups from information on maturity, pregnancy rates, and population estimates. Escapement was determined by subtracting the catch from the estimates of pup abundance for the period 1950-1976. The natural mortality of pups (M_0) was then obtained from the equation

$$M_0 = \left(\ln \frac{E}{N_1} \right)$$

where E is the escapement, and N_1 is the abundance at age 1 from the sequential population analysis. Owing to high year-to-year variability, the estimates were averaged over five-year periods as follows:

Period	Escapement $\pm$ 1 S. E.	Mortality $\pm$ 1 S. E.
1952-1956	219.2 $\pm$ 30.0	0.48 $\pm$ 0.11
1957-1961	344.2 $\pm$ 25.3	0.50 $\pm$ 0.14
1962-1966	150.6 $\pm$ 20.5	0.07 $\pm$ 0.22
1967-1971	137.8 $\pm$ 19.9	0.09 $\pm$ 0.09
1972-1976	197.6 $\pm$ 10.6	-0.08 ± 0.05

It is clear from this analysis that there has been a significant decrease in natural mortality that seems to be related to escapement (Figure 3). However, because of the way in which the points are spaced, it is impossible to determine the exact nature of the relationship. One hypothesis may be a continuous response, while another may imply a kind of step function (see also Chapter 23). The response presented in Figure 3 is a compromise between the two. Furthermore, it is interesting to note that the average of the natural mortality values in the above table ($M = 0.2$) is twice that of seals 1 year of age and older. This is the exact level proposed by Lavigne et al. (1976) for animals less than 1 year of age.

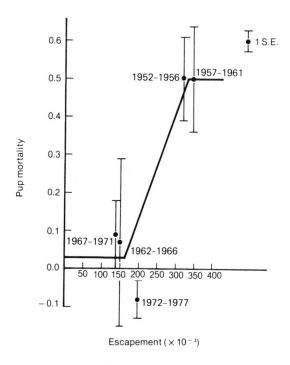

Figure 3 Estimated pup mortality as a function of estimated escapement.

Are these estimates of natural mortality real or simply anomalies of the manner in which the data were analyzed? In our opinion, this is a sound relationship. A continuous relationship exists only for values prior to 1962. The 1962 year class was 15 years old in 1977, and there may be a problem with the initial estimation of numbers at age. This error could be the result of a severely biased sample of molting animals, as it is unlikely that errors in estimates of the sex ratio or schedule of maturation would have much effect on the results. Furthermore, it is difficult to see how a consistent bias in sampling could result in such an abrupt change between the periods 1957-1961 and 1962-1966.

Density-dependent mortality in the first year of life has been shown for a number of species of large mammals (Chapter 23). Among seals, it has been demonstrated in the gray seal, *Halichoerus grypus,* breeding at the Farne Islands (Bonner, 1975) and for the northern fur seal, *Callorhinus ursinus* (Lander and Kajimuira, 1976). The most important cause of death in the young gray seal is starvation on shore. This occurs when the bond between the mother and her young is broken and the juveniles become separated from their food supply (Bonner, 1975). This mechanism cannot be hypothesized for harp seals, since the escapement occurs after the hunt for beaters, which is long after the animals have weaned. Mortality in the northern fur seal has been related to their high density on land, resulting in increased incidence of hookworm. Apparently, some critically high level of density is necessary before the parasitic infection reaches epidemic levels (Lander and Kajimuira, 1976). The infection causes internal hemorrhaging, weakening, and eventual death of the host. Further-more, there is some speculation that weakened animals are less able to survive oceanic storms (see also Chapters 5 and 10).

With virtually no knowledge of parasitic infection in harp seals, it is difficult to hypothesize the existence of a similar mechanism. The only reasonable hypothesis involves the food supply and the rigors of the first northerly migra-tion. Seals less than 1 year of age feed mainly on euphausiids in the surface layer during their stay off northeastern Newfoundland and their subsequent migra-tion to Greenland (Sergeant, 1973a; H. Fisher, personal communication). Before the northward migration begins, the seals form feeding aggregations. It is possible that the competition at this time for the limited food resources results in some seals being unfit to make the long migration northward. Sergeant (1973a) noted that very small juvenile seals do not attempt to migrate and therefore become separated from the major supplies of food. Lavigne et al. (1976) observe that lean seals have a higher metabolic rate than fat seals. Thus, a density-dependent mechanism that results in a lean seal would have an exaggerated effect on the animal's physical state, since more of its energy reserve would be required to maintain a constant body temperature and less could be used for gathering of food, eventually resulting in death. Intense competition for food at the high levels of escapement prior to 1962 may have been the reason for the much higher mortality rates estimated for the early periods. Given the levels of knowledge on this subject, there may be several equally plausible theories. In any case, a density-dependent relationship seems to exist and more research is required for its validation.

Density-Dependent Age of Whelping

For female harp seals, the current stage of maturity can be judged by examina-tion of ovaries for the presence or absence of a new corpus luteum. This struc-ture is obvious well before the time of implantation of the embryo, an event that is delayed several months in seals (Sergeant, 1973b). While the presence of a cor-pus luteum does not necessarily indicate a successful pregnancy, it indicates that maturation of a follicle has taken place. Øritsland (1971) used the backcalculation of corpus albicantia as an indication of the age of sexual maturity. This technique can be misleading in that the small scars persist only for a few years, perhaps three to four years for harp seals (Sergeant, 1973b).

It is possible that the mean age of maturity and its degree of concentration with certain age classes (standard deviation) has varied over the years. Variation in the standard deviation was checked by standardizing all the maturation curves to a mean age of zero and looking for inconsistencies in the rate of accumulation of mature animals within the population. Since all these curves seemed to be superimposed on one another, it was concluded that the standard deviation showed no variation.

Cumulative normal distributions and arcsine transformations were used to linearize the data for maturation in different years. It was found, in accordance with Lett and Benjaminsen (1977), that the arcsine transformation, using a range between 0° and 90°, gave the best fit to the data:

$$_t E_a = \sin(31.34 + 19.91\,a)$$

with $r^2 = 0.9$ and $F = 133.4$, where $_t E_a$ is the fraction whelping at age a in year t, r is the correlation coefficient, and F is the variance ratio.

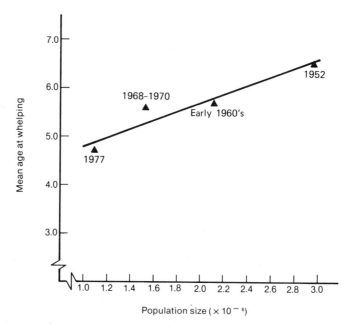

Figure 4   Mean age at whelping as a function of the population of animals 1 year of age and older from five years before.

The point of 50% maturity was interpolated from the various curves in different years and regressed against the population size of animals of age one and older (P_{t-n}) with different time lags (n) to see which gave the best fit. The lag producing the highest correlation was five years (Figure 4) and thus the functional regression developed to describe shifts in the mean age of whelping *(A)* yielded

$$A = 3.875 + 9.126 \times 10^{-7} P_{t-5}$$

with $r^2 = 0.90$ and $F = 20.4$. Consequently, the proportion whelping at age a in year t was obtained from the relationship

$$_t E_a = \sin[31.34 + 19.91\,a - (3.875 + 9.125 \times 10^{-7} P_{t-5})]$$

Sergeant (1966, 1973b) first proposed that the mean age of maturity was density dependent. Indeed, this phenomenon is known to occur in other marine and terrestrial mammals (Gambell, 1973; Markgren, 1969; Chapter 23). Lett and Benjaminsen (1977) developed a mathematical relationship for harp seals. Laws (1956, 1959) noted that, in Phocidae, sexual maturity is attained at a constant proportion of the final or asymptotic body weight (about 87%) and that it is attained at an earlier age when growth is accelerated. If the growth rate of juvenile harp seals is density dependent in a manner similar to the mortality rate, this should provide the mechanism for density-dependent maturity. The five year lag is certainly consistent with Laws' observations.

Density-Dependent Fertility Rate

Density-dependent fertility in harp seals was first noted by Lett and Benjaminsen (1977). All of the available data on the fertility rate found in the literature are presented in the following:

Year	Estimate of Fertility Rate	Reference
	Front	
1953	89.0	Sergeant (1966)
1964-1967	92.1	Øritsland (1971)
1968	95.5	Sergeant (1969)
1968-1970	97.8	Sergeant (1976a)
1976	95.4	Sergeant (1976a)
	Gulf	
1951-1954	86.0	Sergeant (1966)
1952	80.0	Fisher (1952)
1964	84.0	Sergeant (1966)
1965	90.0	Sergeant (1966)

These data are plotted against the population size of animals 2 years of age and older in Figure 5, indicating a fairly clear density-dependent relationship. The data were plotted against the population of older animals because the younger animals remain segregated from the herd and may not compete for the available food resources. Data for the same year for both areas were combined and the following equation was derived using functional regression:

$$F_t = 102.297 - 7.3734 \times 10^{-6} \times {}_2P_{t-1}$$

with $r^2 = 0.69$ and $F = 6.58$, where F_t is the pregnancy rate (%) in year t, t is the year in which the pups were produced, and ${}_2P_{t-1}$ is the population of animals 2 years of age and older in the preceding year (i.e., the year in which they became pregnant).

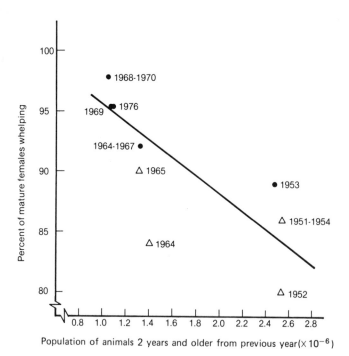

Figure 5   Percentage of mature females that whelp as a function of the population of animals 2 years of age from the previous year. Gulf population, Δ;   Front population, ●.

Variable fertility rates are well known in mammals (Chapter 23) and have been observed in at least three populations of whales (Gambell, 1973). Also, the unexploited population of Antarctic crabeater seals, *Lobodon carcinophagus;* which, like harp seals, have an unlimited extent of ice on which to whelp, has a low pregnancy rate of 0.76 (Øritsland, 1970). Markgren (1969) found that the ovulation rate in moose, *Alces alces,* was related to such factors as age, body size, nutrition and population density. Indeed, it is well known that fertility varies markedly in managed, utilized deer herds, *Odocoileus virginianus* (see Chapter 23). Nazarenko (1975) presents data on the White Sea stock of harp seals, indicating a fertility rate of only about 69%. The reason for the discrepancy between the White Sea and the Northwest Atlantic populations of harp seals is unclear.

SIMULATION

To synthesize the relationships described above into a useful tool, a simulation model was constructed. The model provides for the pregnancy rate, the maturation and pup mortality to operate in a density-dependent manner or be set at constant values. In order to demonstrate the effects of these density-dependent factors, analyses were conducted with one, two, or all three of the factors operating (Figures 6-9). Further details concerning the model, including a detailed description from which a computer program can be constructed, are found in Lett et al. (1979).

Population sizes as a function of time were projected with the hunt limited to the high Arctic and Greenland components. Related variables were set at the mean values presented in Lett and Benjaminsen (1977) and were retained in all projections, since they are presumed to be outside the control of management options but still must be taken into consideration.

Sustainable yields were estimated by projecting the 1977 population ahead for 50 years while varying hunting mortality. Such estimates of catch are not strictly sustainable yields because the projected population and catch after 50 years may be still slowly changing, whereas sustainable yields imply equilibria. The discrepancies should be small, however, and should decrease as the projected population at 50 years approaches the initial population size. For example, the maximum of the curve (Figure 8) is not far removed from the starting population of 1.39 million, and the area of concern for management is therefore not affected by this bias.

Each data point in Figures 7-9 is based on one run with the stochastic variables fixed at their mean values, unless error bars are present. The points with error bars represent the means of 10 runs with the stochastic variables drawn from their respective distributions as described in Lett et al. 1979. The exploitation rates used in the projections of sustainable yield were required to yield catches of pups and animals 1 year of age and older in the ratio of 80-20. For practical consideration, if the proportion of pups in the catch fell in the range of 0.195 to 0.205, the simulation was considered to be adequate. This level of performance generally required less than three trials for a given level of exploitation, once the user became familiar with the model.

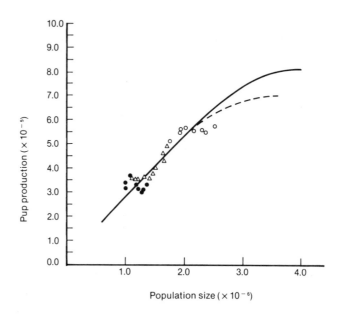

Figure 6 Pup production as a function of the size of the population of animals 1 year of age and older; historical data (points) and simulated output (lines). 1950's, ○; 1960's, Δ; 1970's, ●.

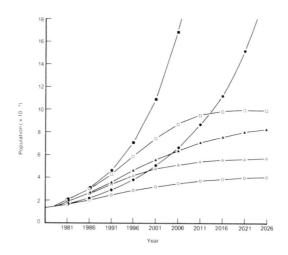

Figure 7 Projected size of the population of animals 1 year old and older as a function of time (with only high Arctic and Greenland catches) for various combinations of density-dependent controls. No density-dependent processes, ■; density-dependent mortality only, ●; density-dependent pregnancy rate only, □; density dependent mean age of first whelping only, ▲; density-dependent pregnancy rate and age at first whelping, Δ; all density-dependent processes, ○.

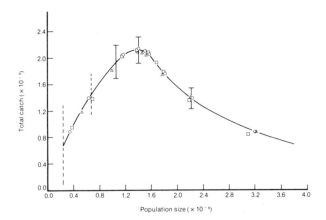

Figure 8 Sustainable yield for harp seals as estimates resulting from the use of three density-dependent controls. The "best" estimates of parameters in the equations for mean age of whelping, pregnancy, and pup mortality, ○; steeper pup mortality, Δ; and pregnancy rate relationships, □. Two standard deviations are indicated by vertical bars and are shown by dashed lines at points where simulated populations sometimes collapsed.

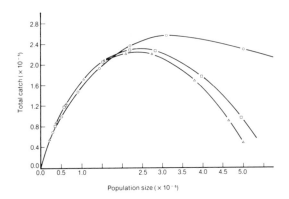

Figure 9 Sustainable yield for harp seals estimated with the use of one or two density-dependent controls expressed as a function of the size of the population of animals 1 year of age and older. Cases in which the mean age at whelping was incorporated as being density dependent, ○; the pregnancy rate in this case is constant at 0.92. Cases in which both density-dependent maturation and pregnancy rates were used, Δ and □. Situations in which the relationship for pregnancy rates have a 20% greater slop, Δ. Pup mortality is 0.2 in each case.

RESULTS AND DISCUSSION

Sequential Population Analysis

The sequential population analysis presented in this chapter shows that the population of seals which were 1 year old or older declined from 2.5 million in 1952 to 1.0 million in 1972 and increased to 1.3 million in 1977 (see Lett et al. 1979). Estimates by Lett and Benjaminsen (1977) indicate that the stock declined from 2.3 million in 1952 to a low of 1.0 million in 1968 and increased to 1.2 million in 1975. The analysis in Lett and Benjaminsen (1977) is a cohort analysis utilizing data for the years 1952 to 1975. Starting exploitation rates for 25-year-olds in 1975 were derived by averaging rates for ages 10 to 20 in 1973 and 1974, and pup mortality was assumed to be the same as the adult mortality at $M = 0.114$. Despite these differences from the present analysis, it is interesting that the population estimates from the cohort analysis agree so well with the present results.

Pup production, estimated from the sequential population analysis, declined from 570,000 in 1952 to 310,000 in 1975 and increased to 330,000 in 1977 (Figure 2). These estimates are very close to those of Winters (unpublished data) for 1953 to 1972. The cohort analysis of Lett and Benjaminsen (1977) gave estimates of pup production that varied considerably from those shown in Figure 2, due to the manner in which the starting values for hunting mortality were derived. Overall, however, these do show the same general trend of a decline from high values in the 1950s to a low of 310,000 in 1973. Reestimation of pup production from the cohort analysis and other data gave values of 290,000 in 1972, 310,000 in 1975, and 320,000 in 1977. Although the minimum occurs a little earlier in the cohort analysis than in the present analysis, the same trend is apparent with an increase in the most recent years.

Various methods have been used to estimate pup production in the Gulf of St. Lawrence, Sergeant (1975), using survivorship indices, estimated production at 120,000 in 1958 and 85,000 in 1967. Using catch and effort data, Lett et al. (1977) estimated that production averaged about 98,000 pups during 1964-1971, a value which seems reasonable, since the total kill of pups in the Gulf was 72,000 in 1971. Sergeant (1975) indicated that this figure was probably close to the total production, as the ships could not complete their quotas that year and seals could not be found after the hunt for branding. However, there was some escapement, because beaters were taken by landsmen of western Newfoundland during their northward migration through the Strait of Belle Isle. According to Lett et al. (1977), the total production in the Gulf in 1971 was about 90,000 pups. They also estimated production at 89,000 animals in 1975, in contrast to Lavigne's (1975) minimum estimate of 46,000, based on aerial photography. Lett et al. (1977) estimated production in the Gulf at 93,000 pups for 1977. Adding this figure to the 1977 estimate of 210,000 pups at the Front (Lavigne et al., 1977) gives a minimum estimate for overall production in 1977 of 303,000 pups, a figure only slightly below the estimates derived by cohort analysis and sequential population analysis.

Comparison of the Simulation with Other Models

The simulation model can be operated with up to three density-dependent controls. This compares with the model described by Capstick et al. (1976), which has only density-dependent maturity, and with the model of Lett and Benjaminsen (1977), which has density-dependent maturity and pregnancy. The effect of increasing the number of feedback controls is generally to stabilize the population in the model. This stability is reflected in the ability of the modeled system to respond to changes either in the environment or in fishing effort. Also, the incorporation of several feedbacks reduces the effect of an error in the estimation of a particular parameter. The sensitivity of the system is a measure of the effect that a change in an input parameter would have on output results, for example, population size or sustainable yield. The sensitivity of individual input parameters tends to be reduced in models with several feedback controls. Mohn (1977) demonstrated this in comparing the sensitivity of natural mortality in models with one and two feedback controls.

Qualitatively, it is desirable to consider the effect of including all three feedbacks in our model, because its sensitivity for a given parameter is generally reduced as more parameters are involved. For example, the response to increasing the pup catch would be a change in the mean age of whelping, an increase in the pregnancy rate and a decrease in pup mortality, thus making the final projected population size less sensitive to the change in pup kill than would be the case in a simpler model without density-dependent controls. Therefore, if the available data indicate the presence of density-dependent factors, it is generally beneficial to include all such factors in the simulation. However, because the total uncertainty in the results projected by the model is a function of the uncertainties associated with the individual parameters, density-dependent controls should not be included when the data do not warrant it.

Relationship Between Pup Production and Population Size

Catches of young harp seals have been higher than 600,000, and pup production in excess of the 1952 level was necessary to sustain the earlier levels of catch (Chafe et al., 1923). Therefore, recruitment curves were simulated under a number of assumptions to investigate the validity of certain density-dependent mechanisms (Figure 6). When all three factors (age of whelping, pregnancy rate, and mortality rate) are related to population density, the whelping rate decreases to 0.16 at a population size of 4.3 million. The natural mortality of pups is 0.5 at this population size, which gives a recruitment rate of

$$R = 0.16 \times \exp(-0.5) - 0.10$$

which balances the natural mortality among those 1 year of age and older.

When the fertility rate is held at 0.92 (a long-term average value), the natural mortality of pups is held at 0.2 and the maturity ogive is allowed to vary, the resultant recruitment curve diverges from that described above at a population of 2.2 million (Figure 6) and pup production increases to more than 800,000 at population of 3.75 million. At this level for stock size, the whelping rate is 0.20 compared with a value of 0.18 for the simulation with all three density- dependent factors operating. Thus, a mechanism that regulates the mortality of pups may have a significant effect on the rate of recruitment to the stock and on the maximum size that the population can achieve. This study indicates that pups may represent between 0.16 and 0.22 of the total population compared with Sergeant's (1975) statement that 0.20 to 0.25 of the total population are pups.

It is important to note that the stock-recruitment curves (Figure 6) do not have descending right limbs. Lett and Benjaminsen (1977) state that a recruitment curve for a stock that shows as little change as harp seals cannot have a descending right limb. Thus recruitment does not appear to follow either the Beverton and Holt (1957) or the Ricker (1954) recruitment curve (see Chapter 23), although Allen (1975) points out that fitting either curve would not greatly affect management decisions over a wide range of stock sizes. Allen (1973) shows a similar recruitment pattern for fin whales *(Balaenoptera physalus)*, although it is not clear from his raw data whether the shape is due to a changing population structure or a change in maturity. In general, a truncated sinusoidal function seems to fit the data from the sequential population analysis quite well (Figure 6), the autocorrelative nature of the points being a result of a fluctuating population structure. However, for the series of biological samples (1952-1977), the population has not been large enough to indicate which of the two recruitment curves is superior. Further discussion concerning the shape of such curves is found in Chapters 9-11, 14, 20, and 23.

Maximum Population Size

Sergeant (1975) has indicated that the population in 1952 may have been near the maximum population size on account of the poor condition of the molting adults. The present study indicates that, no matter what set of assumptions are used, the population of seals 1 year of age or older achieves a maximum of at least 4 million, compared with 2.5 million in 1952 from the sequential population analysis. However, the assumptions about the factors controlling the dynamics of the population severely affect estimates of the virgin stock size, as indicated by the 50-year projections (Figure 7). When no density-dependent factors are operating, the population increases continuously at a rate of 9%/year. Also, density-dependent pup mortality alone does not seem to constrain the population which continues to grow at a rate of 4%/year.

The growth of the population becomes asymptotic when the density-dependent pregnancy rate or the density-dependent age of whelping operate independently (Figure 7). Under the assumption of density-dependent pregnancy rate, the maximum population size is about 10 million, which is somewhat higher than the estimate of Lett and Benjaminsen (1977), but it depends also on assumptions concerning mean age of whelping and mortality in the first year.

Density-dependent age of whelping alone constrained the population to about 8 million animals. It can be concluded therefore that, of the three density-dependent relationships used in the simulation, that influencing the age of whelping has the greatest control over fluctuations in population size at equilibrium.

When both density-dependent pregnancy rate and age of whelping operate together, the maximum population size is projected to be about 5.5 million (Figure 7). Lett and Benjaminsen (1977) found that, under a similar set of assumptions, the population size never exceeded 4.1 million. However, in their model the aboriginal hunt in the Arctic and the landsmen hunt were both allowed to continue (but with no large vessel hunting activity), whereas in the present model no hunting activity is allowed except for a small amount of hunting in the Arctic, which is considered by some researchers (Capstick et al. 1976) as a part of natural mortality. With all three density-dependent processes operating, the projected maximum population size is 4.2 million seals (Figure 7).

Production Curves and Density Dependence

The fitting of general production models to data for harp seals, whether it be catch as a function of average exploitation rate or as a function of total population size from sequential population analysis, gives poor statistical relationships. Indeed, our efforts resulted in a coefficient of determination of only 4%. Using the simulation model with density-dependent mechanisms operating, four sets of data were generated as an experiment, each consisting of 50 ordered pairs of the population size of animals one year of age and older along with the total catch. The ratio of pups to adults was maintained at 80-20. The maximum sustainable yield (MSY) was 196,000 seals at a population size of 1.71 million for the first 50 years of data. The addition of another 50 points yielded an MSY of 179,000 at a population size of 1.61 million animals. For this set of conditions in the complete model, the MSY is 215,000 for a corresponding population size of 1.4 million (Figure 9). This constitutes at least a 9% error in the estimation of the MSY and an 18% error in the corresponding population size.

When only the density-dependent mean age of whelping was included in the simulation, Gulland's (1961) model could not be fitted to the first 50 years of data because the slope of the autocorrelative function of catch per unit population against population size was positive. With the addition of another 50 points, the MSY was determined to be 250,000 for a population size of 3.4 million. The corresponding "best" values were 255,000 and 3.1 million (Figure 9). These results, although reliant on the initial assumptions, indicate that fitting production models can lead to serious errors, especially in the stock size corresponding to the MSY.

The degree of density-dependence has a profound effect on the shape of the curves and the determination of MSY, as indicated by the use of two density-dependent factors (mean age of whelping and fertility rate) in the simulation (Figure 9) in contrast to allowing only the mean age of whelping to vary. The use of a fertility rate, which was 20% greater than the "best" estimate but still within the range of possibilities (Figure 5), did not have much effect on the MSY, which only decreased from 230,000 to 225,000. Higher fertility rates than those occurring on the average (0.92) would tend to increase the yield.

Lett and Benjaminsen (1977) indicated that the MSY is about 240,000 animals for a population size of about 1.6 million, while a recent update of the model (Lett et al., 1977) indicated that MSY is 220,000 for a population size of about 1.5 million. The basic difference between the two models is that a natural mortality coefficient of 0.114 was used in the former and 0.10 in the latter. Thus a 12% decrease in natural mortality rate leads to a decrease in the MSY estimate (from 240,000 to 220,000 animals) and in the associated population size. The results from the simulation model, with natural mortality of pups set at 0.2, indicate an MSY of 230,000 animals for a population size of 2.5 million (Figure 9). In comparing the results of the three models, it must be considered that, when an assumption changes, thus altering fundamental properties such as age structure, all of the other variables in the model must change accordingly. Thus some of the discrepancies in estimated MSY and population sizes are due to changes in the rates at which the mean age of whelping and the fertility respond to population size, or in general terms the degree of density dependence.

When all three density-dependent factors were allowed to vary, the MSY decreased slightly to 215,000 animals but the associated population size declined greatly to 1.4 million (Figure 9). The shape of the curve is primarily a function of the hypothesized relationship between escapement and pup mortality (Figure 3). At low stock sizes, the escapement is low and the mortality rate for pups is less than 0.1. At this low mortality rate, an increase in stock size results in a rapid rise in the production curve toward the MSY level. However, with the increase in natural mortality to its maximum level (0.5), the right limb of the curve declines rapidly at first, but the rate of decline decreases with increasing population once the escapement becomes greater than 300,000 animals.

In the general area of MSY (Figure 9), the populations size may vary from 1.0 million to 1.8 million while the sustainable yield varies only from 180,000 to 215,000. It is therefore difficult to specify a stock size that corresponds to an MSY as a variable management objective. The question then is how should the harp seal stock be managed in view of this kind of variability and uncertainty in the data.

The problems of environmental variability, in relation to general production models, in particular the Schaefer (1954) model, were discussed by Doubleday (1976) and Sissenwine (1977). Independently, Beddington and May (1977) also considered the problems of harvesting populations in a randomly fluctuating environment. Lett and Benjaminsen (1977) commented on the biological basis for an increase in the variability of the catch associated with a decrease in stock size. In the present model, a number of factors contribute to the overall variance, among them being uncertainties associated with natural mortality, with the uncontrolled aboriginal hunt in the Arctic and the hunt by landsmen of Newfoundland and Quebec. Although the incorporation of density-dependent mechanisms would be expected to stabilize a system, this may be offset to a degree by the uncertainty associated with the parameters used.

Density Dependence and Critical Stock Size

Doubleday (1976) and Beddington and May (1977) indicated that, if the size of the stock is below the level corresponding to the MSY, it is more likely to collapse than one above that level. Furthermore, they indicate that it is better to harvest at a constant exploitation rate than at a constant quota level, since the latter may cause the stock to fall below the sustainable yield level and not recover. For a harp seal population less than 800,000 animals, harvesting at a constant exploitation could lead to a collapse in the stock through natural fluctuations (Figure 9).

The critical stock size for harp seal is the level at which certain density-dependent factors cease to operate. Most density-dependent factors are the result of varying amounts of surplus energy affecting the physiology of the animals. If a population can gather, eat, and digest no more food than is now being utilized, the associated density-dependent factors can no longer have any effect on the population dynamics. For harp seals, the minimum age of maturity is about 4.5 years and a pregnancy rate greater than 0.98 has never been observed. Both biological limits are reached for a population size of about 800,000 animals, and any factors leading to lower population sizes (e.g., exploitation, disease, recruitment failure, inadequate food supply) cannot be compensated for. On the basis of the present analysis, this figure may be considered as the critical stock size.

LITERATURE CITED

Allen, K. R. 1973. Analysis of the stock-recruitment relation in Antarctic fin whales *(Balaenoptera physalus)*. Rapp. Cons. Explor. Mer 164:132-137.

Allen, R. L. 1975. A life table for harp seals in the Northwest Atlantic. Rapp. Cons. Explor. Mer 169:303-311.

Beddington, J. R., and R. M. May. 1977. Harvesting natural populations in the randomly fluctuating environment. Science 197:463-465.

Benjaminsen, T., and T. Øritsland. 1975. Adjusted estimates of year-class survival and production with estimates of mortality for northwestern Atlantic harp seals. Presented to Special Meeting of ICNAF Scientific Advisors to Panel A, Ottawa, Canada, 17-19 Nov. 1975 (unpublished manuscript).

Benjaminsen, T., and T. Øritsland. 1976. Age-group frequencies, mortality and production estimates for Northwest Atlantic harp seals updated from samples collected off Newfoundland-Labrador in 1976. Presented to Special Meeting of ICNAF Scientific Advisors to Panel A, Copenhagen, Denmark, 11-12 Oct. 1976 (unpublshed manuscript).

Beverton, R. J. H., and S. J. Holt. 1957. On the dynamics of exploited fish populations. Fish. Invest. London (2), 19:553p.

Bonner, W. N. 1975. Population increase of grey seals at the Farne Islands. Rapp. Cons. Explor. Mer 169:366-370.

Capstick, C. K., D. M. Lavigne, and K. Ronald. 1976. Population forecasts for Northwest Atlantic harp seals *(Pagophilus groenlandicus)*. ICNAF Res. Doc., No. 132, Serial No. 4018.

Chafe, L. G., W. A. Mann, and H. M. Mosdell. 1923. Chafe's sealing book: A history of the Newfoundland seal fishery from the earliest available records down to and including the voyage of 1923. Trade Printers and Publishing Ltd., St. John's, Newfoundland. 105 pp.

Doubleday, W. G. 1976. Environmental fluctuations and fisheries management. ICNAF Select. Pap. 1:141-150.

Fisher, H. D. 1952. Harp seals of the Northwest Atlantic. Fish. Res. Board Can., Atl. Biol. Sta. Circ., (Gen. Ser.) No. 20, 4 pp.

Fry, F. E. T. 1949. Statistics of a lake trout fishery. Biometrics 5:27-67.

Gambell, R. 1973. Some effects of exploitation on reproduction in whales. J. Reprod. Fertil. Suppl. 19:533-553.

Gulland, J. A. 1961. Fishing and the stocks of fish at Iceland. Fish. Invest. (Ser. 2), 23(4), 52p.

ICNAF. 1978. Report of Standing Committee on Research and Statistics. App. 1. Report of ad hoc Working Group on Seals. ICNAF Redbook 1978: 11-15.

Jones, R. 1964. Estimating population size from commercial statistics when fishing mortality varies with age. Rapp. Cons. Explor. Mer 5:210-214.

Kapel, F. O. 1975. Data on the catch of harp and hooded seals, 1954-74, and long-term fluctuations on sealing in Greenland (unpublished manuscript).

Khuzin, R. S. 1967. Variability of craniological features of the harp seal, *Pagophilus groenlandicus* (Erxlehen). Tr. Polyarn. Inst. Morsk. Rybn. Khoz. Okeanogr., 21:27-50. Transl. Ser. Fish. Res. Board Can. 1306.

Lander, H. R., and H. Kajimuira. 1976. Status of the northern fur seal. FAO Consultation on Marine Mammals, Bergen, Norway, ACMMR/MM/SC/34.

Lavigne, D. M. 1975. Harp seal, *Pagophilus groenlandicus*, production in the western Atlantic during March 1975. ICNAF Res. Doc., No. 150, Serial No. 3728.

Lavigne, D. M. 1976. Counting harp seals using ultraviolet photography. Polar Rec. 18:169-277.

Lavigne, D. M., W. Barchard, S. Innes, and N. A. Oritsland. 1976. Pinniped bioenergetics. FAO Consultation on Marine Mammals, Bergen, Norway, ACMRR/MM/SC/112.

Lavigne, D. M., S. Innes, and W. Barchard. 1977. The 1977 census of western Atlantic harp seal, *Pagophilus groenlandicus*. ICNAF Res. Doc. No. 62, Serial No. 5139.

Laws, R. M. 1956. Growth and sexual maturity in aquatic animals. Nature (London) 178:193-194.

Laws, R. M. 1959. Accelerated growth in seals with special reference to the Phocidae. Nor. Hvalfangsttid. 48:42-45.

Lett, P. F., and T. Benjaminsen. 1977. A stochastic model for the management of the northwestern Atlantic harp seal *(Pagophilus groenlandicus)* population. J. Fish. Res. Board Can. 34:1155-1187.

Lett. P. F., D. F. Gray, and R. K. Mohn. 1977. New estimates of harp seal production on the Front and in the Gulf of St. Lawrence and their impact on herd management. ICNAF Res. Doc., No. 68, Serial No. 5145.

Lett, P. F., D. F. Gray, and R. K. Mohn. 1979. Density-dependent processes and management strategy for the northwest Atlantic harp seal population. ICNAF Select. Pap. 5:61-80.

Markgren, G. 1969. Reproduction of moose in Sweden. Viltrevy (Stockholm), 6:125-299.

Mohn, R. K. 1977. Critical analysis of two harp seal population models. ICNAF Res. Doc., No. 64, Serial No. 5141.

Murphy, G. I. 1964. A solution of the catch equation. J. Fish. Res. Board Can. 22:191-201.

Nazarenko, Y. I. 1975. Sexual maturation, reproductive rate and missed pregnancy in female harp seals. Rapp. Cons. Explor. Mer 169:413-415.

Øritsland, T. 1970. Sealing and seal research in the southwest Atlantic pack ice. September-October 1964. In: M. W. Holdgate (ed.). Antarctic Ecology, Vol. 1. Academic Press Inc., London.

Øritsland, T. 1971. Catch and effort statistics for Norwegian sealing in the Front area. Newfoundland, 1971. ICNAF Res. Doc., No. 132, Serial No. 2638.

Pope, J. G. 1972. An investigation of the accuracy of virtual population analysis using cohort analysis. ICNAF Res. Bull. 9:65-74.

Ricker, W. E. 1954. Stock and recruitment. J. Fish. Res. Board Can. 11:559-623.

Ricker, W. E. 1971. Comments on the West Atlantic harp seal herd and proposals for the 1972 harvest. Presented to ICNAF Meeting of Panel A Experts, Charlottenlund, 23-24 Sept. 1971 (unpublished manuscript).

Schaefer, M. B. 1954. Some aspects of the dynamics of populations important to the management of the commercial marine fisheries. Bull. Inter-Am. Trop. Tuna Comm. 1:26-56.

Sergeant, D. E. 1959. Studies on the sustained catch of harp seals in the western North Atlantic. Fish. Res. Board Can., Arctic Unit Circ. No. 4, 38 pp.

Sergeant, D. E. 1965. Migrations of harp seals, *Pagophilus groenlandicus* Erxleben, in the Northwest Atlantic. J. Fish. Res. Board Can. 22:433-464.

Sergeant, D. E. 1966. Reproductive rates of harp seals *(Pagophilus groenlandicus* Erxleben). J. Fish. Res. Board Can. 23:757-766.

Sergeant, D. E. 1969. On the population dynamics and size of stocks of harp seals in the Northwest Atlantic. ICNAF Res. Doc., No. 31, Serial No. 2171.

Sergeant, D. E. 1973a. Feeding, growth, and productivity of Northwest Atlantic harp seals *(Pagophilus groenlandicus)*. J. Fish. Res. Board Can. 30:17-29.

Sergeant, D. E. 1973b. Environment and reproduction in seals. J. Reprod. Fertil. Suppl. 19:555-561.

Sergeant, D. E. 1975. Results of research on harp seals in 1975 with an estimate of production. ICNAF Res. Doc., No. 142, Serial No. 3715.

Sergeant, D. E. 1976a. Studies on harp seals of the western North Atlantic population in 1976, ICNAF Res. Doc., No. 124, Serial No. 4010.

Sergeant, D. E. 1976b. History and present status of populations of harp and hooded seals. Biol. Conserv. 10:95-116.

Sergeant, D. E. 1977. Studies on harp seals of the western Atlantic population. Int. Comm. Northwest Atl. Fish. Res. Doc. 77/XI/58.

Sergeant, D. E., and H. D. Fisher. 1960. Harp seal populations in the western North Atlantic from 1950 to 1960. Fish. Res. Board Can., Arctic Unit Circ. No. 5, 58 pp.

Simms, H. S., B. N. Berg, and D. F. Davies. 1959. Onset of disease and the longevity of rat and man. In: G. E. W. Wolstenholme and M. O'Connor (eds.). Colloquia on Ageing, Vol. 5. J. and A. Churchill Ltd., London.

Sissenwine, M. P. 1977. The effect of random fluctuations on a hypothetical fishery. ICNAF Select. Pap. 2:127-144.

Ulltang, O. 1971. Estimates of mortality and production of harp seals at Newfoundland. Presented to ICNAF Meeting of Panel A Experts. Charlottenlund, 23-24 Sept. 1971 (unpublished manuscript).

Winters, G. H. 1976. Estimation of mortality rates and surplus production of Northwest Atlantic harp seals. ICNAF Res. Doc., No. 127, Serial No. 4014.

EIGHT

Managing Gray Seal Populations
For Optimum Stability

JOHN HARWOOD _____

INTRODUCTION

The British population of the gray seal *(Halichoerus grypus)* provides an interesting case study for scientific management. The species breeds colonially on small islands, all within 70 km of the nearest mainland, and the numbers of pups born on some of the more accessible islands (particularly those in the Farnes group in Northumbria) have been counted annually for several decades (Coulson and Hickling, 1964; Bonner and Hickling, 1971; Hickling and Hawkey, 1975; Hickling et al., 1976, 1977). The numbers born on the more remote islands can be determined accurately by aerial photography (particularly if stereophotography is used), because the white-coated pups contrast well with the predominant background. The annual pup production of all British stocks over the last 13 years has been documented and undisturbed stocks have increased by 6 to 7% annually (Summers, 1978).

In this chapter I describe the way in which the data from this long history of field study have been used to derive life table parameters and density-dependent relationships for the British gray seal population. This information is then summarized in a population projection matrix with variable elements (see Chapters 5, 10, 15, 20-23) which is used to examine the effects of various management strategies.

POPULATION DYNAMICS

The relationship between pup production and total population size depends on age structure. No representative sample of all age classes exists for any British stock of gray seals, but more than 1000 females of breeding age were shot in control operations at the Farne Islands in 1972 and 1975 (Bonner and Hickling, 1974; Hickling et al., 1976). The age of a gray seal can be determined from the number of incremental layers in the cementum of a canine tooth (Hewer, 1964; Platt et al., 1974), and age structures have been determined for the Farne Islands samples. However, this material can also be used to calculate many life-table parameters for the female section of the population. If a population has attained a stable age distribution and adult survival does not vary with age, then the number of individuals in successive age classes should follow an exponential decay. The exponent of the function describing this curve will be the sum of the instantaneous adult mortality and the rate of increase of the population. The age-structures from the Farne Islands provide a good fit to an exponential decay and the rate of increase of the population over the appropriate time interval is known (Summers, 1978). Harwood and Prime (1978) found no significant difference between the mortality estimates from the 1972 and 1975 samples. The estimated mortality rate was 0.06 and the annual survival 0.94 with a standard error of 0.012. To allow for the fact that no gray seal older than 46 years has been recorded, they calculated an overall survival of 0.935, the weighted mean of a survival of 0.94 for animals between the age of five (the average age for first parturition) and 45 and a survival of zero for animals age 46.

Changes in the thickness of the cementum layers in the teeth provide an indication of the year in which each animal had its first pup (Laws, 1977). If fecundity is constant after an animal has its first pup, this information can be used to calculate age-specific fecundities in terms of the constant fecundity rate. The survival of pups from birth to weaning has been measured directly at the Farne Islands (Coulson and Hickling, 1964; Bonner, 1975; Hickling et al., 1977). Thus the only unknown parameters for this population are survival from weaning to first breeding (juvenile survival) and adult fecundity. There are an infinite number of pairs of values for these parameters which satisfy the familiar equation:

$$\sum_{x=0}^{x=46} l_x m_x e^{-rx} = 1$$

where l_x is the probability that a female survives from birth to age x, m_x is the mean number of female offspring born to a female of age x, and r is the intrinsic rate of increase for the population. However, only a few of these pairs of values are reasonable biologically. It is unlikely that the annual survival of juveniles will exceed adult survival and fecundity is unlikely to be higher than that recorded for other pinnipeds. Using these constraints, Harwood and Prime (1978) chose an adult fecundity of 0.9 and a juvenile survival of 0.49. Somewhat arbitrarily, they assigned the juvenile survival among the age classes so that survival in the first year was 0.66 and in subsequent years 0.93. The changes with time in the female segment of a gray seal population can be described by the following set of difference equations:

$$n_{0,t+1} = f_4 n_{4,t} + f_5 n_{5,t} + f_a n_{a,t} \tag{1}$$

$$n_{1,t+1} = S_0 S_1 n_{0,t} \tag{2}$$

$$n_{i+1,t+1} = S_j n_{i,t} \qquad i = 1,2,3 \tag{3}$$

$$n_{5,t+1} = S_a n_{4,t} \tag{4}$$

$$n_{a,t+1} = S_a(n_{6,t} + n_{a,t}) \tag{5}$$

where $n_{k,t}$ is the number of females of age k alive at time t, n_a refers to all animals over the age of five, f_k is the net fecundity for each age class, S_0 is survival from birth to weaning, S_1 is survival from weaning to age 1, S_j is annual juvenile survival, and S_a is adult survival. Estimates of these parameters for the Farne Islands gray seal population, which was increasing by 7% annually, are given in Table 1. The dynamics of the entire population are defined by equations (1) to (5). Gray seals are polygymous and, unless males are subject to heavy and selective hunting, there will probably always be a surplus of males. Information on the number of males is only necessary to estimate the total size of the population.

Table 1 Estimates of Population Parameters for the Female Segment of a Gray Seal Population with an Annual Rate of Increase of 7%[a]

Parameter	Symbol	Estimate
Net fecundity		
Age 4	f_4	0.08
Age 5	f_5	0.28
Over 5	f_a	0.42
Pup survival		
Birth to weaning	S_0	0.73
Weaning to age 1	S_1	0.90
Juvenile survival	S_j	0.93
Adult survival	S_a	0.935

[a] Values are derived from animals collected at the Farne Islands.

It is exceedingly unlikely that the population will show the 7% increase indicated by the values in Table 1 indefinitely. However, the only density-dependent relationship that has been described involves the survival of pups from birth to weaning (Coulson and Hickling, 1964; Bonner, 1975). The decrease in pup survival with an increase in the number of pups born for the Farne Islands stock can be described by a linear relationship, although this implies that survival will be negative at high densities. The decrease in survival appears to be caused by a disproportionate increase in the number of deaths due to desertion and injury at high densities. However, this effect is tempered by the fact that, provided suitable alternative islands are available, the number of pups born on a particular island tends to stabilize before pup survival decreases markedly. Harwood and Prime (1978) considered that the density-dependent process was described better by a function first suggested by Hassall (1975):

$$S_{0,t} = (1 + an_{0,t})^{-b} \tag{6}$$

where $S_{0,t}$ is the survival of pups from birth to weaning in year t, and a and b are constants. Using least-squares methods, equation (6) is a marginally better fit to the data than is a linear model. The best estimates of the constants are $a =$

0.0064 and $b = 0.09$. However, these values imply that the equilibrium population is several million, whereas current pup production is less than 2000! the residual sum of squares is increased by less than one percent if the mathematically more tractable form $a = 0.00016$, $b = 1$ is used, this also provides a more realistic estimate of the equilibrium population size. Equation (2) now becomes:

$$n_{1,t+1} = 0.9n_{0,t}(1 + 0.00016n_{0,t})^{-1} \qquad (7)$$

There is no evidence of a change in the age at first pregnancy for gray seals at the Farne Islands over the last 15 years, and changes in this variable have little effect on the rate of increase of the population (Harwood and Prime, 1978), a conclusion reached by Eberhardt and Siniff (1977) for more general models of marine mammal populations (see also Chapters 2, 22, 23). The rate of increase is susceptible to small changes in adult survival, but these are difficult to detect in practice. In the absence of other density-dependent processes, the size of the equilibrium population is inversely proportional to the value of a in equation (6). For the model described by equations (1) to (5), and using the values from Table 1, the rate of increase is zero when S_0 is reduced to 0.25. If pup survival is the major limiting factor, the current British population is well below its equilibrium level; recent estimates of pup survival range from 0.88 to 0.66 (Summers et al., 1975; Bonner, 1975; Anderson and Curry, 1976). Although it is unlikely that pup survival will ever decline to the level necessary to stabilize the population, the effects of high density are likely to extend into the postweaning period and survival during the first year of life could well be reduced to the necessary level. Additional density-dependent mechanisms, especially those acting on adult survival (see Chapter 23), are likely to play a role at higher population levels.

The value of the commercially exploited fish consumed by British gray seals has been put as high as $30 to $40 million annually (Parrish and Shearer, 1977). It is unlikely that natural factors will limit the present rate of increase of the population in the near future, and it can be argued that human intervention may be necessary to limit this increase or to actually reduce the size of the population. Attempts have already been made to control the animals breeding at the Farne Islands, which constitute the majority of the English population. A small industry has been developed in Scotland based on the harvesting of pups, but this has had little effect on the population size (Summers, 1978). Because life-table parameters and possible density-dependent relationships within the range of population sizes currently thought desirable are relatively well known, it should be possible to develop a biologically sound management policy.

Although some guidelines for the management of marine mammal populations do exist (Eberhardt and Siniff, 1977), these are based on the concept of "optimum sustainable population" and are designed for populations between the maximum sustained yield level (MSYL) and their unexploited equilibrium level. All the evidence indicates that the British gray seal population is well below MSYL and special criteria must be used for evaluating appropriate management strategies. These may have a more general application to other populations of large mammals.

POTENTIAL MANAGEMENT POLICIES

An enormous variety of management schemes can be devised, all intended to establish an equilibrium population. However, some of these are more immediately attractive than others. In particular, those based on the killing of pups alone would appear to have advantages. Pups can be killed in a relatively humane manner with little chance of wounding, there is a market for their skins and a small industry based on their exploitation already exists. If a proportion K of the annual pup production is killed, equation (7) becomes

$$n_{1,t+1} = 0.9n_{0,t}[(1 + an_{0,t})^{-1} - K] \tag{8}$$

If, on the other hand, a fixed quota of n_k pups is killed, the appropriate form is

$$n_{1,t+1} = 0.9n_{0,t}[(1 + an_{0,t})^{-1} - \frac{n_k}{n_{0,t}}] \tag{9}$$

Using the particular numerical values of equations (1) to (8), the unexploited equilibrium level occurs when $n_0^* = 3a^{-1}$, and the MSYL is $n_0^* = a^{-1}$ with $K = 0.25$.

Harwood (1978) considered the effects of management policies based on killing pups alone, and concluded that these had serious disadvantages. In particular, if a fixed quota for pups is used, it is impossible to establish a stable equilibrium below MSYL. In addition, the effects of a pup harvest are not reflected by changes in pup production for at least six years (the time from birth to first breeding). This time lag makes it particularly difficult to evaluate the effects of changes in management. Although it is possible to establish stable equilibria using a proportional pup quota, this is difficult to implement in practice. In addition, if the population is to be held below MSYL a large proportion of the pup production must be removed each year. Because changes in adult survival have a major effect on the dynamics of a gray seal population, killing a small number of adults each year can be an effective management policy. In particular, if the proportion of females taken each year is an increasing function of the previous year's pup production a population can be held in a stable equilibrium below MSYL with a fixed pup quota (Harwood, 1978). Because such an adult cull is density dependent, it reduces the destabilizing effects of the fixed pup quota. With such an adult cull, equation (5) becomes

$$n_{a,t+1} = (0.935 - bn_{0,t-1})(n_{5,t} + n_{a,t}) \tag{10}$$

where $bn_{0,t-1}$ is the proportion of females taken. Killing adult females also affects fecundity. Assuming a sex ratio of unity among the pups, equation (1) should be rewritten as

$$n_{0,t+1} = 0.08n_{4,t} + 0.28n_{5,t} + (0.42 - 0.5bn_{0,t-1})n_{a,t} \qquad (11)$$

This assumes that females are killed at the breeding colonies, this being the most practical procedure. Harwood (1978) considered the effects of management policies using equations (8)-(11) to describe the effects of adult and pup culls. Although these equations summarize the most obvious effects of these management policies they ignore other more complex biological effects. Taking pups alone seems to cause little disturbance, largely because pups are usually harvested at the end of the breeding season, but killing adults on the breeding beaches can cause widespread desertion (Summers and Harwood, 1978). In fact, the effects of disturbance can be as great as the direct effects of the cull. An estimate of the effect on fecundity can be incorporated into equation (11):

$$n_{0,t+1} = 0.08n_{4,t} + 0.28n_{5,t} + (0.42 - bn_{0,t-1})n_{a,t} \qquad (12)$$

where the disturbance causes desertion equivalent to the number of females killed.

In addition, the expression for pup survival [equation (8)] assumes that virtually no mortality occurs at low densities. However, there is undoubtedly some mortality, largely due to infection and animals being washed off breeding beaches (Anderson et al., 1979), at low densities (Coulson and Hickling, 1964; Bonner, 1975). Equation (8) can be modified to allow for this:

$$n_{1,t+1} = 0.9n_{0,t}[0.95(1 + an_{0,t})^{-1} - K] \qquad (13)$$

This equation provides as good a representation of the Farne Islands' data as any other monotonically decreasing function and incorporates a little more realism than some such functions. However, the qualitative conclusions of the analyses that follow are independent of the precise form of equation (13).

For any particular population size there is a set of values of bn_0 and K (or n_k) that will maintain a stable equilibrium. Suitable combinations for a population with a pup production of one-third MSYL are shown in Figure 1, both for the model used by Harwood (1978) and for the modified formulation using equations (12) and (13). The incorporation of desertion effects and the modification of the density-dependent function allow lower levels of harvesting to be used.

To choose an appropriate management policy, it is not sufficient to know that the equilibrium population is stable; among many other things, it is useful to consider the resilience of the equilibrium population to variations in the environment and in population parameters.

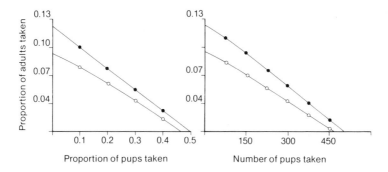

Figure 1 Combinations of pup and adult harvests that stabilize a gray seal population at one-third MSYL, for two different models under two harvest strategies: (*a*) proportion of the pup production harvested and (*b*) fixed number of pups harvested. For a given pup harvest the proportion of adults to be taken was determined by varying bn_0 in equations (11) and (12). (●) denotes calculations using the model described in Harwood (1978), and (○) denotes calculations incorporating the effects of desertion [equation (12)] and a modified pup survival function [equation (13)].

EVALUATION OF MANAGEMENT POLICIES

May (1973) pointed out that the local stability properties of simple population models could be described by considering the effects of small perturbations from equilibrium, and Beddington (1974) extended May's analysis to age-structured models. The stability properties of a particular Leslie matrix M, are described by the eigenvalues of a stability matrix M , where

$$M = M_{(n*)} + M'_{(n*)} H$$

$M_{(n*)}$ is the matrix M evaluated for the equilibrium values of the population vector n; $M'_{(n*)}$ is the matrix of derivatives of M with respect to the elements of n, evaluated at equilibrium; and H is a matrix each of whose columns is the equilibrium population vector $n*$. For the equilibrium to be stable the absolute value of the largest eigenvalue of M_δ must be less than one. In addition, the behavior around equilibrium is dominated by this eigenvalue (λ_1). The speed with which the population returns to equilibrium following a small perturbation is proportional to $\ln \lambda_1$, and the reciprocal of this value is the characteristic return time T_R (Beddington, 1978). This is analogous to the return time of simpler models (May et al., 1974). The difference equations describing the dynamics of a gray seal population with a fixed pup quota and an adult cull can be summarized in a Leslie matrix of the form:

$$\begin{bmatrix} 0 & 0 & 0 & 0 & 0.08 & 0.28 & 0.42 - bn_0^* \\ 0.9[0.95(1 + an_0^*)^{-1} - K] & 0 & 0 & 0 & 0 & 0 & 0 \\ 0 & 0.93 & 0 & 0 & 0 & 0 & 0 \\ 0 & 0 & 0.93 & 0 & 0 & 0 & 0 \\ 0 & 0 & 0 & 0.93 & 0 & 0 & 0 \\ 0 & 0 & 0 & 0 & 0.935 & 0 & 0 \\ 0 & 0 & 0 & 0 & 0 & 0.935 - bn_0^* & 0.935 - bn_0^* \end{bmatrix}$$

For this matrix the value of T_R depends on the size of K (or of n_k) and b (Harwood, 1978). The effect of a small disturbance on equilibrium populations with different return times is shown in Figure 2a-c; the time taken to return to equilibrium is proportional to the value of T_R.

More important than the response of the population to a single perturbation is its response to repeated environmental fluctuations. Beddington and May (1977) have shown that if random variation is introduced into the density-independent terms of a simple logistic model (i.e., variation in survival or fecundity) then the resulting variance in population size is proportional to T_R. However, Shepherd (1977) has pointed out that introducing random variation into a density- dependent term (such as the carrying capacity) results in an inverse relationship between population variance and T_R. The incorporation of an additive, random normal component (with mean zero and standard deviation 0.005) into the density-independent adult survival and fecundity terms in the age-structured gray seal model has the predicted effect on the coefficient of variation for the population (Table 2), and the population behavior following a perturbation remains unchanged (Figure 2d-f).

Table 2 The Relationship Between Population Variance and Characteristic Return Time for a Managed Population of Gray Seals when Random Variation Is Introduced into the Density- Independent Parameters of the model by adding a random normal component (mean 0, standard deviation 0.005) to adult survival[a]

Return Time	Coefficient of Variation (%)
11	0.90
22	1.34
35	1.50

[a]The values of other parameters used to achieve the appropriate return times are described in the caption to Figure 2.

The age-structured model does not contain parameters that act in the straightforward density-dependent manner exhibited by the carrying capacity of a simple logistic equation. The addition of a random element to the density-related parameters a and b in the model has an unpredictable effect on the variation of the population. However, variation in these parameters is unlikely to

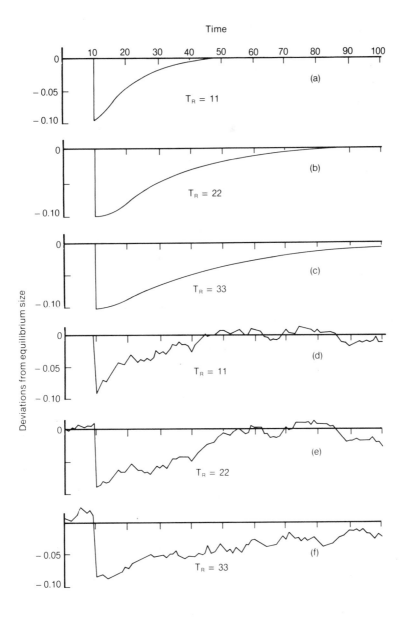

Figure 2 The effects of a 10% perturbation from equilibrium on simulation models of gray seal populations with different characteristic return times. The population size is expressed in terms of deviations from equilibrium size. In (a), (b), and (c), the models are completely deterministic; in (d), (e), and (f) a random normal component (mean 0, standard deviation 0.005) has been added to the adult survival. For (a) and (d) $T_R = 11$, $b/a = 0.1$, $K = 0.1$; for (b) and (e) $T_R = 22$, $b/a = 0$, $K = 0$; for (c) and (f) $T_R = 33$, $b/a = 0.010$, $K = 0.6$.

167

be of major importance in the dynamics of the population. The value of b will be set by the management policy. Although the proportion of adults taken (bn_0) will undoubtedly vary from year to year, the effect of this will be density-independent. Variations in a are equivalent to changes in the number of available breeding sites. If management policies are intended to maintain the population well below MSYL it is unlikely that variations in the number of potential breeding sites will have any discernible effect on the population provided existing sites remain available. It is therefore likely that most of the variability in a gray seal population maintained below MSYL will be the result of variations in mortality and fecundity rates. In this case the magnitude of the characteristic return time can be a realiable indication of the potential resilience of the population. This suggests a possible criterion for evaluating rival management policies: any policy that produces a return time greater than that for an unexploited population should be implemented with caution, since it may increase the potential variability of the population.

The return times of a population subject to proportional and fixed pup culls are shown in Figure 3 using both the formulation in Harwood (1978) and a model incorporating equations (12) and (13). The elaborated model results in an overall increase in T_R, because the stabilizing effect of the density-dependent function is reduced. However, the general conclusion that taking adults in a density-dependent fashion reduces T_R holds for all values of K and most values of n_k (Figure 4).

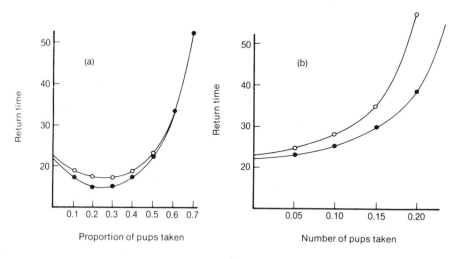

Figure 3   The relationship between characteristic return time and pup harvest for a gray seal population from which only pups are taken: (a) taking a proportion of the pup production, (b) taking a fixed number of pups each year. The number of pups is shown as a proportion of the pup production at MSYL. Results are for the model described in Harwood (1978) (●), and for the same model incorporating equation (13) as a modified pup survival function (○).

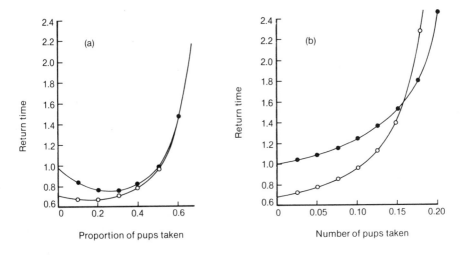

Figure 4   The relationship between characteristic return time (expressed as a proportion of the return time for an unexploited population) and pup harvest for a gray seal population from which adults are taken in a density-dependent way. The modified pup survival function [equation (13)] and the effects of desertion [equation (12)] have been incorporated. Panel (*a*) represents taking a proportion of the pup production, and (*b*) taking a fixed number of pups each year. The number of pups is shown as a proportion of the pup production at MSYL. Results are shown when no adults are taken (●) and when an adult kill equivalent to 1% of the pup production at MSYL is taken (○).

Although the skins of gray seals have a commercial value, the carcasses are difficult to transport and process. It is possible to make a crude estimate of the potential financial yield from different management policies in terms of the value of the skins produced, although such an analysis is very naive in economic terms. A policy which concentrates on taking pups will always give a higher financial return than one concentrating on adults, partly because pups are killed before there has been much loss from natural mortality (see Chapter 23) but also because there is little demand for the skins of adult seals. In the following calculations the arbitrary assumption that one pup skin is worth four adult skins has been made, the predictions would only be altered if adult skins were more valuable than pup skins. In general, for all population levels below MSYL, T_R rises rapidly with the value of the harvest if a fixed quota of pups is used, but much less rapidly if a proportional quota is used (Figure 5*a*). In fact, yields with a value up to 50% of that at MSYL can be obtained without T_R exceeding that for an unexploited population. If a level of three-quarters MSYL is acceptable, yields with a value up to 40% of that at MSYL can be obtained with a fixed quota, and much higher levels are possible with a proportional quota (Figure 5*b*).

DISCUSSION

The preceding analyses have been based entirely on theoretical considerations. To implement a management policy successfully more attention must be paid to practicality. Although the characteristic return time is a useful measure of the resilience of a managed population of gray seals, its calculated value depends entirely on the structural form of the model used to describe it, and in particular on the form of the relationship between pup survival and population size. In fact, any monotonically decreasing relationship between these two variables gives qualitatively similar predictions. Population sizes have been expressed with reference to the MSYL, but it is not necessary to calculate this level exactly in order to design a suitable management policy. Since, in this case, management is concerned with restricting the population to a specific range over which the form of the density-dependent relationship is known, appropriate harvesting levels can be set on the basis of existing information. However, a careful check must be kept on the behavior of the population since the effects of taking adults are by no means simple. Even greater care is necessary if the desired population size lies outside the range for which good biological data exist. The possibility of additional, undetected density-dependent effects must also be considered.

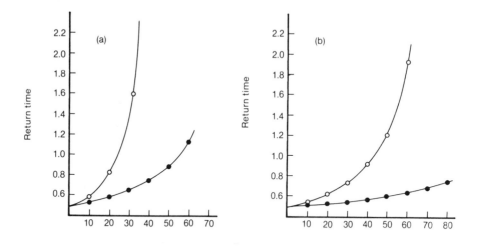

Value of harvest (as percentage of MSY)

Figure 5   The relationship between characteristic return time (expressed as a proportion of the return time for an unexploited population) and the value of the harvest (expressed as a proportion of the maximum sustainable yield) for a gray seal population from which both pups and adults are taken. (a) Population maintained at one-third MSYL, and (b) population maintained at three-quarters MSYL. Results are shown when a proportion of pup production is taken (●) and when a fixed number of pups is taken each year (○).

Although they may not be important over the specified range, it would be unwise to ignore their possible effects.

Taking a proportion of the annual pup production inevitably has less effect on the stability of the population than taking a fixed quota of pups (which tends to counteract the stabilizing effects of the density-dependent relationships), but there are considerable difficulties in administrating such a system. In particular, the quota must be set before the breeding season. Although the pup production can be predicted there will inevitably be errors. Since the harvesting will be divided among a number of individuals, there are problems of allocation if the quota is reset every year. It is impossible to regulate the effort devoted to capturing pups because they are particularly vulnerable on land, and, weather permitting, it is possible to take almost the entire production of some colonies. However, it is possible to monitor pup production for the entire population with considerable accuracy, and this monitoring is necessary to set the proportion of adults to be taken, so that the pup quota can be regularly adjusted for variations in pup production. This would bring the stability properties of this type of management closer to those of a proportional quota, without a great increase in administrative detail.

LITERATURE CITED

Anderson, S. S., J. Baker, J. H. Prime, and A. Baird. 1979. Mortality in grey seal pups: Incidence and causes. J. Zool. London. 189:407-417.

Anderson, S. S., and M. G. Curry. 1976. Grey seals at the Monach Isles Outer Hebrides, 1975. International Council for the Exploration of the Sea. CM 1976/N:9 mimeo.

Beddington, J. R. 1974. Age distribution and stability of simple discrete time population models. J. Theoret. Biol. 47:65-74.

Beddington, J. R. 1978. On the risks associated with different harvesting strategies in the harvesting of baleen whales. Rep. Int. Whaling Comm. 28:165-167.

Beddington, J. R., and R. M. May. 1977. Harvesting natural populations in a randomly fluctuating environment. Science 197:463-465.

Bonner, W. N. 1975. Population increase of grey seals at the Farne Islands. Rapp. P. V. Reun. Cons. Int. Explor. Mer 169:366-370.

Bonner, W. N., and G. Hickling. 1971. The Grey seals of the Farne Islands: Report for the period October 1969 to July 1971. Trans. Nat. Hist. Soc. Northumbria 17:141-162.

Bonner, W. N., and G. Hickling. 1974. The Grey seals of the Farne Islands 1971 to 1973. Trans. Nat. Hist. Soc. Northumbria 42:65-84.

Coulson, J. C., and G. Hickling. 1964. The breeding biology of the Grey seal, Halichoerus grypus (Fab.), on the Farne Islands, Northumberland. J. Anim. Ecol. 33:485-512.

Eberhardt, L. L., and D. B. Siniff. 1977. Population dynamics and marine mammal management policies. J. Fish. Res. Board Can. 34:183-190.

Harwood, J. 1978. The effects of management policies on the stability and resilience of British grey seal populations. J. Appl. Ecol. 15:413-421.

Harwood, J., J. H. Prime. 1978. Some factors affecting the size of British Grey seal populations. J. Appl. Ecol. 15:401-411.

Hassell, M. P. 1975. Density-dependence in single-species populations. J. Anim. Ecol. 44:283-295.

Hewer, H. R. 1964. The determination of age, sexual maturity and a life-table in the grey seal *(Halichoerus grypus)*. Proc. Zool. Soc. London 142(4):593-624.

Hickling, G., and P. Hawkey. 1975. The Grey seals of the Farne Islands: The 1974 breeding season. Trans. Nat. Hist. Soc. Northumbria 42:93-97.

Hickling, G., P. Hawkey, and L. H. Harwood. 1976. The Grey seal of the Farne Islands: The 1975 breeding season. Trans. Nat. Hist. Soc. Northumbria 42:107-114.

Laws, R. M. 1977. The significance of vertebrates in the Antarctic marine ecosystem. Pp. 411-138 *in*: G. A. Llano (ed.). Adaptation within Antarctic Ecosystems. Proceedings of 3rd Symposium on Antarctic Biology. Smithsonian Institute, Washington, D.C.

May, R. M. 1973. Stability and Complexity in Model Ecosystems. Princeton University Press, Princeton, New Jersey.

May, R. M., G. R. Conway, M. P. Hassell, and T. R. E. Southwood. 1974. Time delays, density dependence, and single species oscillations. J. Anim. Ecol. 43:747-770.

Parrish, B. B., and W. M. Shearer. 1977. Effects of seals on fisheries. International Council for the Exploration of the Sea. CM 1977/M:14 mimeo.

Platt, N. E., J. H. Prime, and S. R. Witthames. 1974. The age of the grey seal at the Farne Islands. International Council for the Exploration of the Sea. CM 1974/N:3 mimeo.

Shepherd, J. G. 1977. The sensitivity of exploited populations to environmental "noise," and the implications for management. International Council for the Exploration of the Sea. CM 1977/F:27 mimeo.

Summers, C. F. 1978. Trends in the size of British Grey seal populations. J. Appl. Ecol. 15:395-400.

Summers, C. F., R. W. Burton, and S. S. Anderson. 1975. Grey seal *(Halichoerus grypus)* pup production at North Rona: A study of birth and survival statistics collected in 1972. J. Zool. London 175:439-451.

Summers, C. F., and J. Harwood. 1978. Indirect effects of seal culls. International Council for the Exploration of the Sea. CM 1978/N:14 mimeo.

NINE

Population Dynamics of
the Yellowstone Grizzly Bear

DALE R. MCCULLOUGH

INTRODUCTION

The grizzly bear *(Ursus arctos horribilis)* range in the continguous United States
has shrunk greatly from the original distribution throughout most of the western
plains and mountains. Now they occur primarily in a few mountain wildernesses
in Wyoming, Montana, and Idaho. One of the most substantial remaining
populations occurs in Yellowstone National Park and the surrounding mountains, covering a total area of about 12,000 to 14,000 km² (Cole, 1974;
Craighead et al., 1974). From 1925 to 1958, crude estimates placed the grizzly
bear population between about 200 and 300 animals (Craighead and Craighead,
1967; Cole, 1971). In 1959 the Craighead research team began a study of the
grizzly bears that continued until 1970, when unreconcilable differences
developed between the Craigheads and the Park Service. These differences
centered on the method of closing open pit garbage dumps in the park, and the
impact it was having on the grizzly bear population. On the basis of a
mathematical model of the population the Craighead team became convinced
that the bear population was threatened with extinction (Craighead et al., 1973,
1974), and this information soon found its way into the public press (see, e.g.,
Seater, 1973; Craighead, 1973). The Park Service, as represented by Glenn
Cole, denied these claims (Cole, 1971, 1973, 1974). The controversy concerning
this issue (see, e.g., Johnson, 1973; Cowan et al., 1974; Gilbert, 1976; Cauble,
1977) is beyond the scope of this chapter. This chapter will present a new
analysis and synthesis of the available data and give an alternate interpretation
of the dynamics of this population.

TOTAL POPULATION SIZE

The number of grizzlies in the Yellowstone ecosystem (Yellowstone National
Park and surrounding wilderness) has been a point of contention. The bears accounted for at the garbage dumps by Craighead et al. (1974) ranged from 154 to
202 (Figure 1). Disagreement between the Craighead team and the Park Service
centered on whether a separate segment of the population lived in backcountry
areas and did not visit the dumps, and hence, was not included in the above
counts. Craighead and Craighead (1971) maintained that nearly all bears visited
the dumps, on the basis of their few observations of grizzlies in backcountry areas

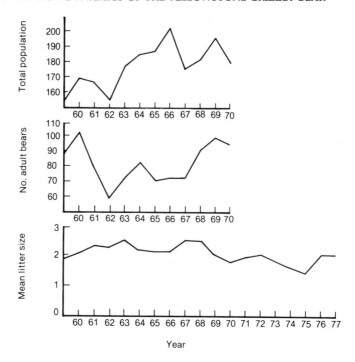

Figure 1   Observed changes in total and adult population sizes of Yellowstone grizzly bears for 1959 to 1970 (above and middle) and mean litter size for 1959 to 1970 (below). Data on mean litter size from 1971 to 1977 are from Cole (1973, 1976), Meagher (1977) and Knight et al. (1976).

in the summertime. Also, they observed their marked bears in these areas. Cowan et al. (1974) pointed out the fact that bears marked at the dumps were observed in the backcountry did not speak to the question of whether other bears did not come to the dumps. Cole (1973) maintained that a backcountry popula-tion of 50 to 100 grizzlies was present in the ecosystem, and that the total popula-tion was 250 to 300. Based on estimates of the ratio of marked bears in the park compared to the ratio of marked to unmarked bears killed outside the park (Cowan et al., 1974) concluded that the Craigheads' censuses at garbage dumps were accounting for 58% of the total population. Therefore, to the average of 177 bears visiting dumps, should be added 132 uncounted bears for a total of 309 grizzlies in the ecosystem. Subsequently, Craighead et al., (1974), using the same approach, concluded that their censuses were 77% efficient, and that 52 bears needed to be added for a total population of 229.

The results of both Cowan et al., (1974) and Craighead et al. (1974) on mark-ed to unmarked ratios are subject to the same criticism. They compare the ratio of marked to unmarked living bears at the dumps with the ratio of marked to unmarked bears killed outside the park. This approach is questionable because it compared living segments of the population inside the park with dead segments outside the park when the number of bears marked varied greatly over time because of varying marking efforts and unknown mortality of marked bears. Thus, it is impossible to determine how many bears were marked and alive at a given time, or what the average number of marked bears was over the period of study.

The comparison of the ratio of marked to unmarked bears dying (of all causes) within the park to that of bears dying outside the park circumvents these difficulties because the numbers are cumulative over time. Cowan et al. (1974) did not have access to these data, but they were published by Craighead et al., (1974). Using data from 1959 through 1967, prior to closure of the dumps, 45 dead bears in the park were marked and 29 unmarked, or 60.81% were marked. Outside the park, 23 were marked and 45 unmarked, or 33.82% of the total were marked. For the entire study period (1959 to 1970) there were 66 marked and 58 unmarked dead bears found inside the park (52.23% marked), while outside, 31 of 103 dead bears were marked (30.1%).

If it is assumed that all the bears in the ecosystem are completely mixed, as suggested by the Craighead team, the ratio of marked to unmarked bears dying outside the park should be the same as in the park. Therefore, the null hypothesis that there was no difference between the marked to unmarked ratios inside and outside the park can be tested. Rejection of the null hypothesis would indicate a significant number of bears in the ecosystem did not visit garbage dumps regularly.

A χ^2 test for two independent samples, using the data described above, gave highly significant differences (for 1959 to 1967, $p = 0.0013$; for 1959 to 1970, $p = 0.0005$). The null hypothesis was rejected, and it was concluded that a significant proportion of the bears in the ecosystem were not regular visitors to the garbage dumps, and hence, were not included in the censuses made by the Craighead team.

The difference in the ratios inside the park and outside the park can be used to estimate the number of uncounted bears. On the basis of a mean censused population for 1959 to 1967 of 174, the difference in ratios indicates 137 uncounted bears for a total population of 314 (see Cowan et al., 1974, for methods of calculation). Thus the efficiency of the Craighead team was 55.4%. For 1959 to 1970 there was an average of 177 bears, for a total population of 312 for the ecosystem. This yields a census efficiency of 56.73%. Since the latter period includes the entire study, and the greatest accumulation of bear mortalities, it is considered the best estimate of the total population for the Yellowstone ecosystem.

RECRUITMENT OF JUVENILE CLASSES

Craighead et al. (1974) concluded that reproductive rates were essentially constant from 1959 through 1967, but that they declined after 1967. They included Cole's (1973) data for 1971 to 1973, obtained after the end of their field studies.

Craighead et al. (1974) attributed this decline in reproductive rate to abrupt closing of garbage dumps. Cole (1974) agreed that the average litter size decreased, but he argued that in compensation for lowered population density, younger females were breeding and producing smaller litters thereby lowering average litter size. As evidence for this view, Cole (1974) pointed to the high ratio of yearlings to females.

Craighead et al. (1973) assumed no compensatory responses in their data collected from 1959 to 1970: "A depressed reproductive rate probably would not be compensated by an increase in yearling survival as long as the population recruitment was subjected to unusual population stresses." Furthermore, their assumption of a stable age distribution implies that the relationship of reproduction to adult age classes was not changing. The implicit assumption of no compensatory mechanisms is obvious from their model which is based on the assumptions of a stable age distribution and constant rates of reproduction and mortality.

The analysis to be developed here is based on the population censused at garbage dumps by Craighead et al. (1974). Recruitment of cubs is used instead of reproduction per se, since birth occurs during winter and the Craighead data were obtained in summer (June 1 to August 30). In the absence of embryo counts, there is no way of determining mortality of offspring between birth (during denning time) and midsummer. Thus, the recruitment of cubs is a measure of the number of cubs born and surviving to approximately 6 months of age. Likewise the recruitment of yearlings is a measure of survival to 18 months of age, and that of 2-year-olds the survival to 30 months of age. Ideally, recruitment should be measured as the addition to the adult segment at the population through recruitment of the youngest adult class which Craighead et al. (1974) regarded as the 5.5 years of age. But the sample upon which the number of 5-year-olds is based was too small (six animals) for analysis. The numbers of 3- and 4-year-olds were based on field observations, and because they could not be distinguished they were combined into one category. Precise analysis is difficult, therefore, since it is not possible to determine the specific breakdown of these two age classes by year.

If there were no compensatory mechanisms in the population, then the number of young recruited (R) to a given juvenile age and the recruitment rate (R/N) should be independent of the size of the adult population (N) producing the recruits. Thus, a regression of R/N on N should produce a more-or-less random scatter of points with a slope not significantly different from zero.

Least-squares regressions produced high coefficients of determination and negative slopes indicating that recruitment was not independent of adult population size (Table 1, Figure 2). In fact, recruitment of young is highly dependent on the size of the adult population. As the number of adults increases, the recruitment rate declines rapidly. For example, 71% of the variation in the specific recruitment rate R/N for cubs is explained by the number of adults. The slope is significantly different from zero at $p < 0.0005$. A similar regression of absolute number of cubs against number of adults gives a similar

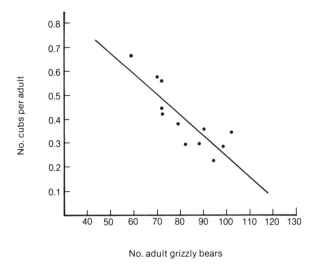

Figure 2   Relationship between recruitment rate (cubs per adult) and number of adult grizzly bears in the population. Fitted regression line is shown.

result, demonstrating that the negative slope is not a bias introduced by using N on both axes. Furthermore, to examine the combined impact of adults in the population on juveniles in the population, the total number of juveniles was regressed on the total number of adults. This regression was significant and the slope was negative $(r^2 = 0.3481, p = 0.0316)$, which confirms the earlier relationships. Since number (rather than a rate) was used on the y axis, this relationship cannot be an artifact of the R/N on N regression. Therefore it is concluded that the number of juveniles recruited was strongly compensatory, with high populations of adults recruiting few juveniles and low numbers of adults recruiting high numbers of juveniles.

Alone, these results are sufficient to invalidate the two assumptions of the model by Craighead et al. (1974). The negative correlation between recruitment of young and adult population size demonstrates beyond doubt that the Craighead et al. (1974) data indicate a very strong compensatory factor. Similarly, because birth and survivorship (as discussed later) of young animals declines as the number of adults increases and vice versa, a stable age distribution could not have been maintained over the range in adult numbers contained in the data from Craighead et al. (1974).

The next problem is to determine at what point in the life of juveniles the mortality was occurring. It could have been restricted only to the first months as cubs or it may have continued through to affect the yearling and 2-year-old classes. The first problem then was to determine if the influence of adults on survivorship of juvenile classes was lagged in time. The highest coefficients of determination were obtained between number of juveniles of the year-class cohort of the year in which the adult number was determined and those adult numbers.

Thus, the best relationships were between adults of a given year, with cubs of that year, yearlings in the following year, and 2-year-olds two years later (Table 1). Thus, this set of regressions were used as the best estimators of survivorship from cubs through 2 years of age.

In biological terms, the influence of adults on recruitment of juveniles was most strongly felt by the year class over time rather than all the age categories at that point in time. As an example, the regression of yearling on adults in the same year gave an r^2 of 0.226, while the yearlings of the following year it was 0.598. It is tempting to assume that this occurred because the influence of adults on the year class was felt mostly in the age class (cubs), and the high correlations with older ages in the year class was dependent on this fact. However, as will be shown below, this was not the case, and survivorship continued to be influenced. Therefore, it is apparent that this set of regressions gives reasonable predictions of survivorship of year classes over time that depend only on the initial adult population.

Table 1   Regressions of R/N on N, Where R is Recruitment of Various Juvenile Age Classes and N Is Number of Adults in the Population[a]

R/N	n	r^2	Significance Level
Cubs/Adults	12	0.714	0.0005
Yearlings/Adults	11	0.598	0.0053
2-year-olds/Adults	10	0.645	0.0051
3- to 4-year olds/Adults	8	0.246	0.2218
Cubs/Adult females	7	0.394	0.1315
Cubs/Adult males	7	0.751	0.0116

[a]See text for further explanation.

When the recruitment rate (R/N) for cubs through 2-year-olds was transformed to number of young recruited to a given age and plotted against N, a set of parabolic curves was obtained (Figure 3). Maximum recruitment of cubs was reached at an adult population of 60 to 70, yearlings at 60, and 2-year-olds at 50. These curves are presented as numbers of young produced and surviving depending upon initial adult population size in Figure 4. Some mortality between cub and yearling ages occurred even at low adult population densities. However, mortality between yearling and 2-year-olds was virtually absent up to an adult population of 60. Above this level mortality rates of all juvenile classes increased rapidly.

The correlation of recruitment of 3- to 4-year-olds three years after birth with adults was substantially poorer ($r^2 = 0.246$) than the other correlations (Table 1) and the slope of the regression was not significantly different from zero although the trend was negative. This was mainly due to differing survivorships of 3- to 4-year-olds from those of younger age classes (as considered further below) although smaller sample size ($n = 8$) and variable survival of each age class within the class contributed to the poorer correlation.

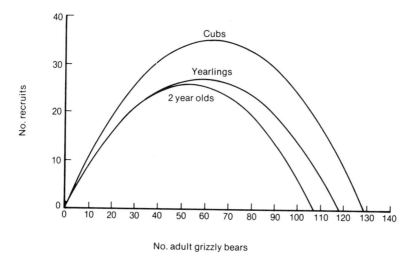

Figure 3 Prediction of number of grizzly bear recruits to various juvenile age classes dependent upon adult population size based on regressions in Table 1.

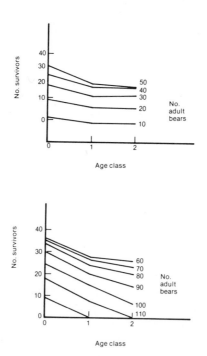

Figure 4 Survivorship of juveniles as dependent upon adult population size: low adult population size is represented in the top panel, high adult population size in the lower panel.

179

Survivorship of 3- to 4-year-olds was analyzed further by regressing their survival on the combined number of 2-year-olds of the two previous years, the age classes giving rise to the 3- to 4-year-olds (Figure 5). The regression was statistically significant with a negative slope ($r^2 = 0.4636$, $p = 0.0302$). It is apparent that low numbers of the combined 2-year-old category produced high rates of survival, and high numbers gave low rates of survival. For the lowest numbers, survivorship was approximately 100%. Some values exceed 100%, but with the exception of the 159.1% value (Figure 5) these are easily explained by sampling error. This one large discrepancy probably reflects a large error in field classification by the Craighead research team.

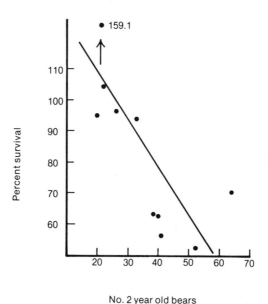

Figure 5   Relationship between survival to age 3 to 4 years for Yellowstone grizzly bears and the number of 2-year-old bears from the previous two years. The extreme value of 159.1% survival appears to be an error in field classification. Fitted regression line is shown.

Recognizing the regression of recruitment rate for cubs with the number of adults as the best estimator allows a more precise estimation of the reproductive rate. Therefore, we can reexamine the dispute between the Craigheads and Cole over reproductive rates. The mean number of adults for the period of 1959 to 1967 was 77.3, while that of 1968 to 1970 was 94. The number of cubs recruited for these populations, as estimated from the regressions was 34 and 27.5, respectively. The mean number of females in the population for the latter period was 45.3 (Table 3 in Craighead et al., 1974). The number of females in the population prior to 1964 is not available, but based on the regressions in Figure 7, the number of females can be estimated for 1959 to 1963. Combining these with the known number from 1964 to 1970 gives an average number of adult females of 43.8 for the period. Thus, the reproductive rate for the 1959 to 1967 period was $34/43.8 = 0.776$, while for the 1968 to 1970 period it was $27.5/45.3 = 0.607$. The rates estimated by Craighead et al. (1974) were 0.6516 and 0.6090, respectively. Their estimates were derived by dividing the mean annual litter size observed in the field each year by the average interval between litters (3.4 years) as determined from 30 known females. This procedure leads to an underestimation of the reproductive rate for 1959 to 1967 by Craighead et al. (1974), as can be seen by comparing them with the realized recruitment of cubs by the number of individual females identified in the population. From 1964 through 1967 there were 174 adult females which produced 126 cubs for a rate of 0.7241. Clearly, the estimating procedure of Craighead et al. (1974) gave results too low, and the realized rates and the rate estimated from the regressions mentioned above are in much closer accord. For the same period, Craighead et al. (1974) estimated an average reproductive rate of 0.6558. For 1968 through 1970 the realized rate was 0.5956, while the Craighead et al. (1974) estimated rate was 0.6090. These estimates plus the one obtained from the regression (0.607) are all in close agreement. For the combined data from 1964 to 1970 the realized rate was 0.673, while the average estimated reproductive rate of Craighead et al. (1974) was 0.636.

In conclusion, the rates derived from the recruitment rate regression agree most closely with the realized rates, and are considered more reliable than those derived by the methods of Craighead et al. (1974).

Nevertheless, their conclusion that the reproductive rate after 1967 was lower than for 1959 to 1967 appears correct. However, while they attribute the lower rate to stresses placed on the population by closure of open pit garbage dumps, in actuality, it can be accounted for entirely on the basis of the high number of adults in the population (as expressed in suppression of recruitment of cubs).

A final question on reproductive rates is how the rate was adjusted. It could have been due to lowering the average litter size, decreasing the proportion of females producing litters, or a combination of both variables. In Figure 6 the mean litter size per female (from field observations; see Figure 1) and the realized recruitment per females are regressed on the number of adults. The slopes of the regressions are negative and significant at the 0.90 level of confidence.

Comparing the predictions of the two regressions over a range of values for adult population size shows that the precentage of females with cubs varied from high at low adult population size to low at high population size. At the average adult population size of 82.57 the percentage of females with litters was 31.03, while for 60 adults it was 37.19 and for 100 adults, 25.66%. When adult population was run in multiple regression on litter size and percentage of females with litters, a statistically significant relationship was obtained ($R^2 = 0.7952$). Thus, both litter size and percentage of females with litters were involved in changes in reproductive rate, with litter size accounting for about 32% of the identified variation and the percentage of females with litters about 48%. Similarly, when

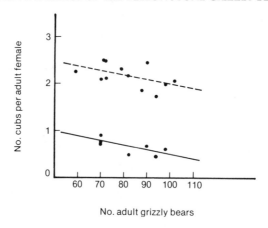

Figure 6   Relationship between Yellowstone grizzly bear litter size (dashed line) and cub recruitment (solid line), expressed as cubs per adult female, and the size of the adult population. Fitted regression lines are shown.

the number of cubs recruited was regressed in multiple stepwise regression on the percentage of females with litters and mean litter size ($R^2 = 0.9806$) the percentage of females with litters accounted for 77% of the known variance. Picton (1978) reported a negative relationship between severity of climate (hard winter) and mean litter size, but including his climate index as a third variable in the stepwise regression was not significant; indeed it lowered the level of significance obtained with the first two variables.

CAUSE OF SUPPRESSION OF RECRUITMENT

In the previous section, it was clearly demonstrated that recruitment of juvenile age classes was inversely related to the size of the adult population.

To test whether it was adults only or the total noncub population size that was responsible for suppression of cub recruitment a series of regressions were run. The suppostion was that if food resources were limiting, through effects produced by adults, the competition from 3- to 4-year-olds, 2-year-olds, and yearlings would have an additional suppressing effect when compared to that of adults alone. The regressions were run with first adults plus 3- to 4-year-olds, then with adults plus 3- to 4-year-olds plus 2-year-olds, and finally with adults through yearlings combined. With the inclusion of each successive additional juvenile class the correlation dropped substantially. It was apparent, therefore, that it was only the adults that were influencing cub survival.

Further analysis was conducted to determine the influence of sex of adults on cub recruitment. Data concerning sex were taken from Table 3 in Craighead et al. (1974). Sample size was 7, since data were available only from 1964 to 1970. The regression of cub recruitment on number of adult females (Table 1) was not significant ($r^2 = 0.394$). Examination of the data will show a range of females of only 40 to 48. Most of the variation in number of adults was in males. Thus, large variation in recruitment of cubs occurred despite only minor change in the number of adult females, suggesting the outcome of cub recruitment was largely independent of the number of females. Regression of recruitment rate on number of adult males gave a high correlation ($r^2 = 0.751$) which was highly significant (Table 1). A similar result was obtained by Stringham (1980) in an interpopulational comparison of five grizzly and brown bear populations, including the Yellowstone population. The conclusion that adult males were responsible for most of the suppression of recruitment of cubs seems inescapable. This is also the conclusion reached for bears in a more general sense by Bunnell et al. (Chapter 4).

Two hypotheses would seem to account for the effect of males on cubs: (1) that adult males directly killed the cubs, and (2) that being dominant, adult males excluded females with young from the choicest feeding sites in the dumps. Cole (1973) believed that cubs were killed directly, and Park Service personnel reported finding three dead cubs killed in one night at the Rabbit Creek dump. Craighead and Craighead (1967) also reported males killing cubs (see also discussion of J. Craighead, pp. 244-245 in Herrero, 1972), and Troyer and Hensel (1962) have reported males killing cubs in brown bears in Alaska. Stokes (1970) discussed this factor relative to concentration at garbage dumps. Herrero (1972) proposed that the high aggressiveness of grizzly bears evolved because of the need for females to protect cubs in open areas where there are not trees for cubs to climb for escape.

Rogers (1977) has discussed why adult males might be selected to kill cubs in black bears, and the same arguments can be made for grizzly bears. It is known that female grizzlies do not conceive until they wean their previous offspring, which may occur after the yearling or 2-year-old stage (Craighead et al., 1969; Mundy and Flook, 1973; Pearson, 1975; and others). Thus, females are receptive only every two to three years. In a polygymous species, such as the grizzly bear, dominant males can increase their individual reproductive success by killing cubs of females they have not copulated with, and hence offspring that are not their own. By killing the cubs, the female is likely to come into estrus one or two years sooner than if the cubs were reared, and the dominant male, because of his status, is likely to be the sire of the new litter. Females, because of the prior investment in the litter, should be selected to defend the cubs vigorously, but not to the extent of seriously imperilling their own life, which would preclude further reproductive effort. This interpretation would seem to agree with the discussion of bear behavior (pp. 243-254 in Herrero, 1972), and it parallels similar behavior by dominant males in langur monkeys *(Presbytis entellus)* in India (Hrdy, 1974) and lions *(Panthera leo)* in Africa (Schaller, 1972).

Even though the evidence is clear that adult males will kill cubs, it is not known if a sufficient number are killed to account for the negative relationship of cubs to number of adult males. Similarly, there is little direct evidence to indicate that competition is involved. Thus, while the influence of adult males on survivorship of young bears is clear, the mechanism of this suppression is speculative. Indeed, it may be a combination of direct killing and food competition.

SEX RATIOS

In a sample of 78 cubs obtained from 1959 to 1971, Craighead et al. (1974) found that 59% were males and 41% females. This difference was not statistically significant, and they concluded it was due to sampling procedures. For their model they used a 50-50 ratio because they "were unable to account for so large a sex imbalance on theoretical grounds and did not observe any factors which would have favored survival of male cubs during the first 4 months of life."

However, if one combines the data of Craighead et al. (1974) for juveniles from cubs through 4-year-olds, there is a significant difference from a 50-50 ratio (χ^2 test, $p = 0.0418$) with males predominating (60%). Furthermore, there is good theoretical reason for a possible male bias at conception. The widely accepted theory of sex ratio selection was presented by Fisher (1930). Fisher's principle, as it has become known, is that because each sex contributes 50% of the genes to each generation, natural selection should result in equal parental investment in each sex. If there is differential mortality of one sex or the other before the end of parental care under typical conditions, the total investment in the sex with higher mortality would be less than in the sex with greater survivorship. Therefore, natural selection should favor a shift in the sex ratio at conception towards the sex with the greater mortality. Because differential male mortality is commonly observed in polygynous species such as the grizzly bear, it is frequent that males predominate at conception.

It is important, now, to see if the data from Craighead et al. (1974) point to differential mortality by sex. The juvenile sex ratio is unbalanced in favor of males and the adult sex ratio is unbalanced in favor of females (53.7%). Thus, males must be suffering disproportionate mortality at some stage. Moreover, from 1964 to 1970, when sex of adults was available, the adult population varied from 70 to 98; however, the number of females varied only from 40 to 48. This suggests that most of the flux occurred in males, while female numbers remained relatively constant. The regressions of number of females and number of males on total number of adults are shown in Figure 7. There was a strong shift in sex ratio with males predominating for large populations and females predominated in small populations. The mean population size for the period, 82.4, gave the proportions of the sexes (53.7% female) derived by Craighead et al. (1974). These results can be expressed as a regression of the percent of the population that was male on population size wherein the percentage of males increases with population size ($Y = 5.935 + 0.4812x$; $r^2 = 0.8599$; p = 0.0026. Clearly, differential mortality of males had to be present to account for this shift.

Craighead et al. (1974) attributed this shift in sex ratio to male mortality in the adult stage, but their own life-table calculations (their Tables 9 and 12)

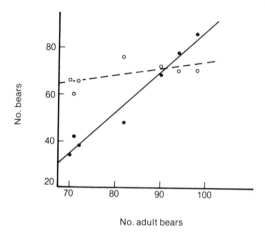

Figure 7   Relationship between number of adult male (solid line) and adult female (dashed line) grizzly bears and the total adult population size. Fitted regression lines are shown.

shown expectance of further life of males and females at 5.5 years of age to differ too little to account for the pronounced shift of sex ratios in adults with changing population size. The earlier demonstration in this paper of the inverse relationship of number of adults to number of juveniles in the population showed that age structure and, hence, survivorship, varied remarkably. Craighead et al. (1974) determined age structure in 1966, the peak population (202) year of their study. This age structure, which they treated as representative, was one of the most unstable combinations of adults and juveniles. In actual fact, there was no representative or stable age distribution at any time during the study.

If one considers sex ratio of juveniles, the total sample of living cubs (78) was too small to examine for each year. Therefore, to increase sample size, adult population sizes by year were grouped into three categories consisting of the four lowest years, four intermediate, and four highest years. The percentage of the cub recruits during those years for the three categories are shown in Figure 8. It can be seen that the percentage of cubs that were male declined with increasing population size. This was the relationship expected, and if it carried over to the other juvenile classes, a shift in sex ratio from male bias to female bias could be accomplished mainly by mortality of males in the juvenile classes. Furthermore, if one examines the data on known mortalities of juveniles according to whether adult population size increased (i.e., few juvenile mortalities) or decreased (more juvenile mortalities) during years of adult population increases, the sex of known mortalities of juveniles was 58.06% male (Table 2), or slightly less than their prevalence in the population (Figure 8). However, during years of stable or declining adult populations, 70.73% of known mortality of juveniles was male, which was much greater than their proportion in the population (Table 2). If sex of known mortality for adults is examined the same way, 59.38% of known mor-

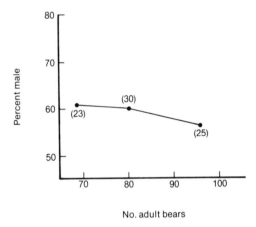

Figure 8   Relationship between the percentage of surviving cubs that were male and the size of the adult bear population with sample sizes in parentheses.

tality was male during population increases, and this exceeds the proportion of males in the population at low adult population sizes (Table 2, Figure 7). During stable or declining adult populations, known mortality was 54.55% male, or approximately the ratio in the population at high adult population sizes (Table 2, Figure 7).

These results are rather ambiguous because small sample sizes require grouping data and specific age and sex data are not readily available. Furthermore, known mortality was not correlated with population size of either total population or number of adults or population change, suggesting that the sample may not be entirely representative. And, many animals in the known mortality were of unknown sex and age. The results are further confused by time lag effects, as will be considered next. But overall, the results suggest that the shifts in sex ratio of adults with adult population size (Figure 7) are produced primarily by differential survival of the sexes of juvenile recruits to the adult population, with differential mortality of adult males amplifying the relationship somewhat. Thus, overall, a disproportionately great number of males were present at conception, and when adult populations were low, both sexes of cubs had high survivorships, and the proportion of males in the population of adults increased as the number of adults increased. At high adult populations cub production was low and survivorship of males declined drastically. Mortality of cubs reduced recruitment of new males to the adult population, and some differential mortality of adult males decreased the proportion of males in the population as total adult population declined.

TIME LAGS

The picture emerging from this analysis is an oscillatory pattern of adults increasing, with concomitant shift in sex ratio to males, with decrease in recruit-

Table 2   Relationship of Sex Ratio in the Living Population to Sex Ratio in the Known
Mortality

	Adults		Juveniles	
	In Population	In Mortality	In Population[a]	In Mortality
	Adult Population Increase			
% Male	42.6	59.4	60.3	58.1
% Female	57.4	40.6	39.7	41.9
	Adult Population Decline			
% Male	46.3	54.6	59.1	70.7
% Female	53.7	45.4	40.9	29.3

[a]Based on sex distribution in cubs, Figure 8.

ment of cubs. The oscillatory behavior is introduced by the delay in sexual
maturity until about 4.5 years of age, and the relatively large proportion of
juveniles in the population. Thus, there is a time lag in the compensatory relationship between adults and juveniles in the age structure.

The driving variable in the system is the number of adult grizzlies, which fairly
directly determines the number of cubs recruited. Furthermore, the survivorship
of these recruits to 2 years of age is also quite directly conditioned by adults at
the time of birth of the year class (Figure 4). It is the 3- to 4-year-olds that are the
variable age classes, being affected like juveniles under some circumstances, and
like adults in others. Unlike the younger juveniles, whose survivorship is most
dependent on the number of adults in the year of birth of their year class, 3- to 4-
year-olds are as closely correlated with the adults in the population at that age
($r^2 = 0.267$) as with the adults at the time of birth of their year class ($r^2 =
0.246$), and neither is statistically significant. Furthermore, while number of
cubs produced and survivorship through 2 years of age are closely related (Figure
4), the reverse is true of 3- to 4-year-olds. If the survival of 3- to 4-year-olds is
regressed on the number of 2-year-olds in the two previous years (i.e., the
antecedents of the 3- to 4-year-olds), a strongly negative relationship is obtained,
which is statistically significant (Figure 5). Thus, strength of the year class is inversely related to survivorship. With about 25 or fewer combined 2-year-olds,
survivorship is nearly 100%, but as the number of 2-year-olds increases, survivorship to 3 to 4 years of age declines rapidly (Figure 5).

This points to a relationship between number of adults and number of 3- to 4-
year-olds as shown schematically in Figure 9a. The actual data do not fit this
pattern precisely (Figure 9b) because the oscillatory tendency in adults is readily
interrupted by either exceptional survival or mortality of either 3- to 4-year-olds
or adults, or both if they are inverse (e.g., a high survivorship of 3- to 4-year olds
may be offset by a high mortality of adults in a given year). These stabilizing
events tend to occur when the rate of change of the adult population is near zero
at an increasing population, as is shown in Figure 9c. Because the rate of change
in the adult population includes both the recruitment of 4-year-olds and the

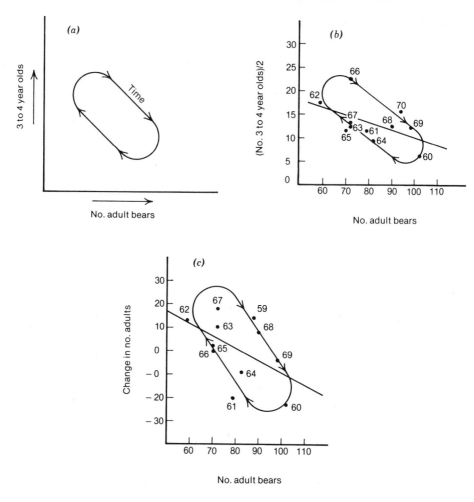

Figure 9 (*a*) Schematic relationship between number of 3- to 4-year-old grizzly bears and number of adults illustrating a time lag. (*b*) Observed relationship between 3- and 4-year olds (divided by 2) and the adult grizzly bear population size. Numbers of points indicate the year, and the solid line is a least square regression fit. The time lag curve is fitted by eye. (*c*) Observed change in number of adult grizzly bears in relation to adult population size. Numbers by points indicate the year, and the solid line is a least squares regression fit. The time lag curve is drawn by eye.

mortality of adults, the overall fit of the time sequence to the time lag curve is substantially better than for Figure 9*b*. The relative importance of recruitment to adult mortality is shown in Figure 10, where rate of adult population change is regressed on approximate number of 4-year-olds. If the number of 4-year-olds is about 10 to 12, mortality of adults is roughly offset, and population of adults shows little change. If greater numbers of 4-year-olds are present, the adult population increases, and recruitment must exceed adult mortality. Conversely, if 4-year-olds are fewer than about 10, adult mortality exceeds recruitment, and the number of adults declines.

Note the cluster of points in Figure 10. These are points where the adult population is arrested at zero population change on the ascending part of the time lag curve in Figure 9c. However, once a high recruitment in conjunction with a low adult mortality occurs, the peak and decline of the next oscillation will occur.

The periodicity of the oscillations, from peak to peak, tends to be about 10 years. The oscillation of the adult population from 1959 to 1970 can be discern-ed in Figure 1. And although small numbers of females result in the high variance in mean litter size, the general inverse relationship of litter size with adult population size can be seen in this figure. The generally low litter sizes following 1970, and increase in the last couple of years suggest another oscilla-tion in adult population size has taken place.

If there were no time lag in this oscillation, the relationship of adults to juveniles would suggest that total population should show little variation. For ex-ample, at an adult population of 60, there would be 114 juveniles for a total population of 174. At 100 adults there would be 79 juveniles for a total of 179, or only slightly greater than in the first case. But, because of time lags, the relation-ship of juveniles to adults varied considerably, as was reflected in the observed total grizzly population (Figure 1).

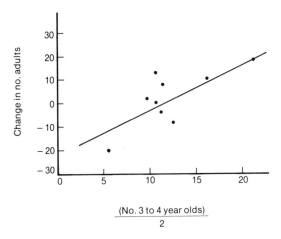

Figure 10   Relationship between the change in number of adults and the number of 3-to 4-year-olds (divided by 2). Fitted regression line is shown.

EXPECTED RESPONSE OF THE POPULATION TO EXPLOITATION

Yellowstone National Park is a protected area, and as such, hunting is excluded. Therefore, the mortality of grizzly bears will be largely natural, with a small number of accidental deaths and problem bears removed for human safety. Man-caused mortality under usual circumstances will be minor. A controlled harvesting program could introduce stability, and dampen out the oscillatory behavior of the age structure and could also eliminate time lags. This could be accomplished most readily by cropping adults, to stabilize their number at about 70 (park population, not the entire ecosystem) or below. Because the park population is protected and because considerable protection has been extended to the Yellowstone ecosystem, the oscillatory pattern that has been observed (Figure 1) can be expected to continue, with number of adults varying approximately inversely with mean litter size. Thus, a series of years with below average litter sizes is not sufficient cause for despair; nor are above average litter sizes justification for smugness. Unfortunately, declines in average litter size from this cause, which is self-correcting, could not be distinguished from declines from more serious reasons. It would be most reassuring to have a reliable index to the size of the adult population to be sure that complacency is justified.

Although exploitation is not anticipated for this population, it is still instructive to examine the response of the population to harvest to gain some estimate of the ability of the population to sustain removals. Because such a program of harvest, if applied to adults, would result in dampening the oscillatory behavior of the population, it is reasonable to derive equilibrium values without time lags. This is the same as assuming that the mortality of adults is fixed at a given number, and maintained over a considerable time. The equilibrium value of the age structure will come to balance so long as maximum sustainable yield (MSY) is not exceeded by the mortality.

Figure 11 gives a Ricker-like, stock-recruitment model (Ricker, 1954) of the park population of grizzly bears, with equilibrium age structure. Recruitment of cubs through 2-year-olds were determined from Figures 3 and 4, and recruitment of 3- and 4-year-olds was derived from Figures 5 and 10. Since data were not available to determine differences in survivorship of these two age classes, they were assumed to be the same. The carrying capacity K is approximately 81 adults, the point at which average recruitment of 4-year-olds is balanced by chronic mortality of adults (Figure 10). Chronic mortality (McCullough, 1979) is defined as mortality due to factors which debilitate gradually, such as malnutrition, nonvirulent diseases and parasites, and old age. This is opposed to those factors that kill by injury and trauma, such as shooting, predation, or virulent disease. The estimate of the equilibrium for recruitment of 4-year-olds and chronic mortality of adults may be slightly on the low side, because it is based on the probability of equal abundance of 3- and 4-year-olds (Figure 10). Perhaps there are slightly fewer 4-year-olds. Thus while 12 4-year-olds is the maximum, 8 would seem to be the likely minimum, and this would shift the estimate of K in Figure 10 to only 84 adults, a relatively minor amount. Recalculation from the survivorship data of Craighead et al. (1974) gives an equilibrium adult mortality of about 8 bears per year, and Avrin (1976) calculated an equilibrium mortality of 10 bears per year. Total population at K would be 188 at 81 adults or 181 at 84. The higher estimate of K of 84 adults is used since, in fact, some mortality in

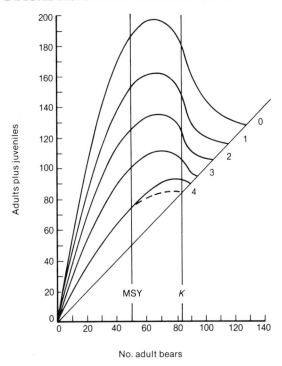

Figure 11 Stock-recruitment model of the Yellowstone grizzly bear population at equilibrium values (no time lags). See text for further explanation.

the population was human-induced in control of problem bears. Of course, if the population were actually allowed to grow to K, time lags in age structure would begin, and an equilibrium age structure would not be maintained.

MSY (calculated as maximum removal of adults that can be balanced by recruitment of 4-year-olds) occurs at an adult population of 50, where 24.5 4-year-olds are recruited. The total population is 186. Thus, MSY measured in terms of 4-year-olds is 13.2% of total population and 49.0% of the adult population. The estimate of Cowan (1970) that bears can sustain a kill of 5% of the total population would seem to be quite conservative. Introducing a 5% mortality with this model, assuming a total population starting at K would result in a population of 84 adults if human kills substitute for natural mortality. If the 5% mortality is in addition to the 5% natural mortality at carrying capacity, then the total mortality is 10 percent of the population, a much less conservative figure. For the white-tailed deer on the George Reserve, the substitution effect of hunting mortality for chronic mortality was a negative curvilinear line which passes from the number of adults at MSY through the number of adults at carrying capacity (McCullough, 1979). Thus, as human kill increased, chronic mortality decreased. A similar relationship is shown for the Yellowstone Park grizzly bear population in Figure 11 by the dashed line. It should be noted that the intersection with carrying capacity is axiomatic, since this is the point of balance between recruitment of 4-year-olds and the mortality of adults in this unhunted population. Similarly, the evidence points to negligible chronic mortality at MSY. The adult population size did not fall this low under the protection afforded (Figures 1 and 9c), and presumably increased man-included mortality would be required to hold the adult population to 50 at equilibrium. Furthermore, at an adult population of 50 almost no mortality would occur between the yearling and 2-year-old ages (Figure 4) and the number of 2-year-olds (about 25) is low enough that survivorship to 3 to 4 years of age would be virtually 100% (Figure 5). Thus, while the end points of the dashed line in Figure 11 are consistent with the data, the shape is arbitrary. Assuming the estimation to be reasonable, the relationship of hunting kill to chronic mortality would be as given in Table 3. A hunting mortality of about 5% would result in a total mortality of about 9% which is within the sustainable limit of 13.3%. Cowan (1970) also reported a kill of about 8% in Yugoslavia, and a kill in Sweden of about 7% with an additional 5 to 6% natural mortality which were sustainable. Thus, the estimates given here seem reasonable.

These results, which are based upon the population using the garbage dumps in the Craighead et al. (1974) study, can be equated to the total ecosystem population by assuming a census efficiency of 56.73%, as estimated earlier. These results (Table 4) suggest that the ecosystem population could sustain a total annual mortality of about 43 bears at MSY, and that under total protection, a chronic mortality of about 21 bears per year could be expected.

It may be well to caution against uncritical acceptance of these predictions. At best, they are approximations, since all models are deficient when measured against the real world. Moreover, the data on which this analysis is based were collected during a period when food at garbage dumps was available to the bears. Even though both the Craighead study team and the Park Service research biologists were in agreement that there is sufficient natural foods to sustain the bear population, no one has carefully evaluated the energetic cost of gathering such dispersed foods. It seems questionable to me that digging out rodent burrows, digging roots, catching trout, collecting berries, and searching out carrion could yield a net energy gain equal to feeding from a garbage pile. The very attractiveness of the dumps to the bears would seem to argue otherwise.

On the other hand, the recognition that adult males were the class of bears most influencing survival of young suggests that dispersal of bears away from the dumps may well improve the recruitment of young, even if net energy coming into the population is somewhat reduced. This would be particularly true if it could be shown that the mortality was due to direct killing of young bears by the adult males while concentrated at the garbage dumps.

MSY AND POPULATION STABILITY

Several other characteristics of the model in Figure 11 should be discussed. First, note the actual adult population oscillated from about 60 to 100 (Figures 1 and

Table 3   Estimated Relationship Between the Number of Bears Dying as Attributed to Hunting Mortality and Chronic Mortality for the Yellowstone Park Grizzly Population, as Derived from Figure 11

Total Population	Adult Population	Chronic Mortality	Hunting Mortality	Hunting Mortality (% of pop.)	Total Mortality	Total Mortality (% of pop.)
186	50 (MSY)	0.0	24.5	13.17	24.5	13.17
196	60	3.5	20.2	10.31	23.7	12.09
196	70	7.0	13.5	6.89	20.5	10.43
189	80	8.8	4.7	2.49	13.5	7.14
181	84 (K)	8.0	0.0	0.00	8.0	4.42

Table 4   Comparison of Population Estimates for Various Management Schemes for Yellowstone Park and the Total Ecosystem Based on a 56.73% Census Efficiency

	Total Population	Number Adults	Total Mort.[a]
Management Scheme		Yellowstone Park	
MSY	186	50	24.5
Maximum Population	197	65	22.5
Protection	187	81	12.0
		Yellowstone Ecosystem	
MSY	328	88	43.2
Maximum Population	347	115	39.7
Protection	330	143	21.2

[a]Includes chronic mortality plus harvest.

9c). From Figure 11 it can be seen that these fluctuations centered on K, oscillating above and below about equally. In the absence of time lags, lower total populations would be obtained at high adult numbers, and high total populations at low adult numbers. With time lags there should be little relationship between adult numbers and total numbers, and, indeed, this is the case with the actual data ($r^2 = 0.013$, $p = 0.728$).

Moreover, at equilibrium, little change in total population would occur whether total protection allowed the population to remain at K or a high harvest rate reduced the population to MSY. Total population would change but little with exploitation, although age structure and turnover would be shifted substantially. Thus, in this case, change in age structure is a far more sensitive parameter in determining what is happening with the population than is total population.

However, if exploitation exceeds MSY, total numbers drop precipitously, showing the great susceptibility of the population to overexploitation. This vulnerability is particularly apparent if most of the mortality is on adults, since there is a five-year delay between birth and adulthood. Thus, even very low rates of exploitation will push the population to extinction if the adult population is reduced below that at MSY. This explains the population history of grizzly bears which have been characterized as showing gradual declines followed by precipitous crashes (Storer and Tevis, 1955; Craighead et al., 1974). And, if the population survives, once overexplitation has occured, extreme protection is required to allow population recovery. Because of slow rate of maturation of juveniles, buildup takes an inordinate amount of time.

COMPARISON WITH OTHER YELLOWSTONE GRIZZLY BEAR MODELS

The compensatory model outlined here differs substantially from the model of Craighead et al. (1974). It considers shifts in reproductive rate, age structure, and sex ratio in a compensatory, dynamic relationship. It responds much more favorably to exploitation than the rather rigid model they propose.

It can also be compared to the model of Arvin (1976) who used a compensatory relationship between adults and recruitment of juveniles in combination with the fixed survivorship rates of adults as reported by Craighead et al. (1974). Simulations with his model showed the oscillatory behavior of the adult segment of the population, time lag in adult and juvenile age structure, and the stability effect of increased harvest on the oscillatory behavior of the population. Nevertheless, having only recruitment behave compensatorily rather than both recruitment and adult mortality resulted in (1) doubling the time period of the oscillation from about 10 to about 20 years; (2) greatly increasing the oscillatory behavior of the total population; (3) constraining the adult oscillation between 80 and 110 at low hunter kill, instead of the observed 60 to 100; and (4) an estimated MSY of 13 for the Yellowstone Park segment of the grizzly population, or just over one-half the 24.5 estimated here. These differences demonstrate that a partially compensatory model may be more realistic than a fixed model, but that predictions may still vary substantially from a more completely compensatory model. Even the model developed here may be deficient in total assessment of compensation. For example, Craighead et al. (1969) noted that age at first pregnancy varied substantially, and 3-year-old females were observed to copulate commonly, although they did not bear young. Hensel, et al. (1969) reported breeding by 3-year-olds in brown bear *(U. A. middendorfi)* in Alaska. If some females reached maturity and produced cubs at 4 years of age instead of 5, the resiliency of the population to exploitation would increase, a topic previously explored by Bunnell and Tait (1977).

In summary, this work goes well beyond that of Craighead et al. (1974), who presented a noncompensatory model for the grizzly bears of Yellowstone National Park. Reexamination of their data suggests several compensatory mechanisms in this population. The number of juveniles recruited is negatively correlated with the size of the adult population. Delay in sexual maturity until 5 years of age results in an oscillatory tendency with a periodicity of about 10 years in the relationship between adult and juvenile segments of the population. Compensation occurs both in recruitment of juveniles and mortality of adults. Moreover, sex ratio biased toward males at bith and differential survival of males through the oscillatory period results in pronounced sex ratio shifts in the adult population. Incorporation of compensation in models results in much more favorable predictions of the impact of increased mortality on the viability of the population. Nevertheless, the model demonstrates that long delays in sexual maturity result in high vulnerability of the population to overexploitation.

LITERATURE CITED

Arvin, D. E. 1976. A numerical model of the Yellowstone grizzly bear population and its management implications. Master's thesis, University of Michigan, Ann Arbor. 51 pp.

Bunnell, F. L., and D. E. N. Tait. 1978. Bears in models and reality-implications to management. Pp. 15-23 *in:* C. J. Martinka and K. L. McArthur (eds.). Bears—their biology and management. Bear Biol. Assoc. Series No. 4, 375 pp.

Cauble, C. 1977. The great grizzly grapple. Nat. Hist. 86:74-81.

Cole, G. F. 1971. Preservation and management of grizzly bears in Yellowstone National Park. Bioscience 21:858-863.

Cole, G. F. 1973. Management involving grizzly bears in Yellowstone National Park 1970-72. U.S. Dept. of Interior, Natl. Park Serv., Nat. Resour. Rep. No. 7. 10 pp.

Cole, G. F. 1974. Management involving grizzly bears and humans in Yellowstone National Park, 1970-73. Bioscience 24:335-338.

Cole, G. F. 1976. Management involving grizzly and black bears in Yellowstone National Park 1970-75. U.S. Dept. of Interior, Natl. Park Serv., Nat. Resour. Rep. No. 9, 26 pp.

Cowan, I. McT. 1970. The status and conservation of bears *(Ursidae)* of the world—1970. Pp. 343-367 *in:* S. Herrero (ed.) 1972. Bears: Their biology and management. IUCN Publ. New Ser. No. 23, Morges, Switzerland.

Cowan, I. McT., D. G. Chapman, R. S. Hoffmann, D. R. McCullough, G. A. Swanson, and R. B. Weeden. 1974. Report of the Committee on the Yellowstone grizzlies. National Academy of Sciences Report. 61 pp.

Craighead, F. C., Jr. 1973. They're killing Yellowstone's grizzlies (interview). Natl. Wildl. 11:4-9, 17.

Craighead, J. J., and F. C. Craighead, Jr. 1967. Management of bears in Yellowstone National Park. Environmental Research Institute and Montana Cooperative Wildlife Research Unit Report. 113 pp.

Craighead, J. J., and F. C. Craighead, Jr. 1971. Grizzly bear-man relationships in Yellowstone National Park. Bioscience 21:845-857.

Craighead, J. J., M. G. Hornacker, and F. C. Craighead, Jr. 1969. Reproductive biology of young female grizzly bears. J. Reprod. Fert. Suppl. 6:447-475.

Craighead, J. J., J. R. Varney, and F. C. Craighead, Jr. 1973. A computer analysis of the Yellowstone grizzly bear population. Montana Cooperative Wildlife Research Unit, University of Montana, Missoula. 81 + 61 pp. of appendix.

Craighead, J. J., J. R. Varney, and F. C. Craighead, Jr. 1974. A population analysis of the Yellowstone grizzly bears. Univ. Mont. For. Conserv. Exp. Sta. Bull. 40. 20 pp.

Fisher, R. 1930. The Genetical Theory of Natural Selection (reprint, 1958). Dover, New York.

Gilbert, B. 1976. The great grizzly controversy. Audubon Mag. 78:63-92.

Hensel, R. J., W. A. Troyer, and A. W. Erickson. 1969. Reproduction in the female brown bear. J. Wildl. Manage. 33:357-365.

Herrero, S. (ed.). 1972. Bears—Their biology and management. IUCN Publ. New Ser. No. 23, Morges, Switzerland. 371 pp.

Hrdy, S. B. 1974. Male-male competition and infanticide among the langurs *(Presbytis entellus)* of Abu, Rojasthan. Folia Primatol. 22:19-58.

Johnson, A. S. 1973. Yellowstone's grizzlies: Endangered or prospering? Defenders Wildl. News, October:557-568.

Knight, R., J. Basile, K. Greer, S. Judd, L. Oldenburg, and L. Roop. 1976. Yellowstone grizzly bear investigations: Annual report of the Interagency Study Team. U.S. Dept. Interior, Natl. Park Serv., Misc. Rep. No. 10. 75 pp.

McCullough, D. R. 1979. The George Reserve deer herd: Population ecology of a *K*-selected species. University of Michigan Press, Ann Arbor, Michigan.

Meagher, M. 1978. Evaluation of bear management in Yellowstone National Park, 1977. Natl. Park Serv., Yellowstone Natl. Park Res. Note No. 8. 18 pp.

Mundy, K. R. D., and D. R. Flook. 1973. Background for managing grizzly bears in the national parks of Canada. Can. Wildl. Serv. Rep. Series No. 22. 35 pp.

Pearson, A. M. 1975. The northern interior grizzly bear *Ursus arctos* L. Can. Wildl. Serv. Rep. Series No. 34. 86 pp.

Picton, H. D. 1978. Climate and reproduction of grizzly bears in Yellowstone National Park. Nature 274:888-889.

Ricker, W. E. 1954. Stock and recruitment. J. Fish. Res. Board Can. 11:559-623.

Rogers, L. L. 1977. Social relationships, movements, and population dynamics of black bears in northeastern Minnesota. Ph.D. Dissertation, University of Minnesota, Minneapolis. 194 pp.

Schaller, G. B. 1972. The Serengeti lion: A study of predator-prey relations. University of Chicago Press, Chicago.

Seater, S. 1973. Vanishing point: The grizzly bear. Environ. Qual. 4:28-34.

Stokes, A. W. 1970. An ethologist's views on managing grizzly bears. Bioscience 20:1154-1157.

Storer, T. I., and L. P. Tevis, Jr. 1955. California Grizzly. University of California Press, Berkeley.

Stringham, S. F. 1980. Possible impacts of hunting on the grizzly/brown bear, a threatened species. Pp. 337-349 *in:* C. J. Martinka and K. L. McArthur (eds.). Bears—their biology and management. Bear Biol. Assoc. Series No. 4, 375 pp.

Troyer, W. A., and R. J. Hensel. 1962. Cannibalism in brown bear. Anim. Behav. 10:231.

TEN

Population Dynamics of the Pribilof Fur Seals

L. L. EBERHARDT

INTRODUCTION

As seen in other chapters in this book, there is now a substantial quantity of literature that deals with optimal management strategies for wild animals. Most, if not all, such schemes assume that various kinds of information are available when, in fact, a great deal of such information is rarely available for natural populations. Usually it is assumed that the size of the population, its age and sex structure, as well as age-specific reproductive and mortality rates are known. Harvests are assumed to be accurately controlled, and the number of removals known exactly. In the actual practice of wildlife management, information of this kind is, at best, only partially available, and the estimates are usually biased to an unknown degree.

A considerable impetus toward obtaining better estimates of the "vital statistics" of certain populations comes from the Marine Mammal Protection Act of 1972 (Public Law 92-522). A key provision of the Act is the requirement of maintaining an "optimum sustainable population." Some abiguity of language permits a range of interpretations, but a working definition has evolved as including the range between the maximum sustainable yield level (maximum net productivity) and carrying capacity (maximum sustainable population level). A more detailed discussion of possible interpretations and policies appears in Eberhardt and Siniff (1977) and Eberhardt (1977a,b).

There are three classes of problems to be dealt with. One is the estimation of certain key rates and parameters. This is very difficult, but has been the subject of a good deal of research and practical study. The second problem is one of determining how some of these parameters vary with population density. The greatest uncertainties exist in this area, and overshadow most of the other difficulties, but progress is being made (Chapter 23). A third, but interrelated, issue is one of deducing an optimum strategy for management.

One of the most extensive sets of data on a population of marine mammals is that for the Pribilof fur seals. A partial assessment of that data is attempted in the present paper, with reference to population dynamics and the problems enumerated above. Another view is given by Smith and Polacheck (Chapter 5), who also present background on the history of the herd and methods of study.

ESTIMATES OF VITAL STATISTICS

In combination with population size, the main ingredients for an analysis of population dynamics are age-specific reproductive and survival rates. Reproductive rates are most readily measured, and seem to be much the same for a number of the pinniped species. Survival rates are much more difficult to obtain, and mainly constitute estimates of a single rate for fully adult animals of all ages. Most of the available estimates of survival over the first few years of life appear to be either educated guesses or inferred to balance out a life table. The most notable exception comes from studies of the northern fur seal *(Callorhinus ursinus)*. Extensive censuses of pups on rookeries, combined with tallies of young males made at harvesting times, make it possible to estimate survival over the first three years of life directly.

A detailed analysis of the extensive data collected on the Pribilof fur seals has not yet become available. Chapman (1973a), however, gave estimates of the number of pups born and of male recruitment to age 3 on St. Paul Island through 1965. He indicated that female recruitment exceeded that of males somewhat, and utilized a factor of 1.1 to 1.0 as the ratio of females to males at age 3. Using that factor, and assuming that half of the pups born are females, yields the survival estimates of Table 1. Chapman also estimated the annual survival rate of adult females to be 0.89. This is in general agreement with estimates given for several other species of pinnipeds, summarized by Eberhardt and Siniff (1977).

Table 1   Apparent Survival of Female Fur Seals on St. Paul Island

Year	Number Born (thousands)	Recruitment at Age 3 (thousands)	Survival to Age 3
1920	71.5	28.4	0.40
1921	75.0	28.8	0.38
1922	79.5	30.5	0.38
1950	225.5	88.1	0.39
1951	223.5	84.6	0.38
1952	219.0	99.3	0.45
1953	222.5	74.0	0.33
1954	225.0	48.1	0.21
1955	230.5	58.4	0.25
1956	226.5	21.9	0.10
1957	210.0	63.9	0.30
1958	193.5	89.4	0.46
1959	167.5	66.3	0.40
1960	160.0	47.1	0.29
1961	168.5	57.2	0.34
1962	139.0	46.1	0.33
1963	132.0	51.7	0.39
1964	142.5	57.0	0.40
1965	133.5	43.9	0.33

Extensive pelagic collections of fur seals, made from 1958 to 1974, provide an important source of data on age-specific pregnancy rates. Chapman (1961) gave the data for the first three years of the collections (1958-1960) and various reports of the North Pacific Fur Seal Commission include the more recent data. A tabulation of the 1958 to 1974 data appears in the 1975 Annual Report on fur seal investigations by the Marine Mammal Division of the Northwest Fisheries Center, National Marine Fisheries Service. Full details of the collections have not yet been published, although a cooperative study between Canadian and U.S. scientists has been underway since cessation of collections in 1974. Because the collections were made in different pelagic areas in various years, there may be difficulties in comparing annual pregnancy rates. Estimating adult female survival rates from age structure of the collections is hampered by the low survival to age 3 apparent in the mid-1950s (Table 1). This factor, plus the general decline in abundance of the Pribilof herd beginning about that time, means that the age structure will not have been constant in recent years.

If it is assumed that the herd was stationary for a considerable period prior to the early 1950s, age samples representing animals born in those years might be used to estimate adult survival by the methods studied by Chapman and Robson (1960). Using their "segment" method (described below), and data from the year classes born prior to 1952, an average rate of 0.91 for adult female survival was estimated. This is higher than Chapman's 0.89, but the standard errors for the individual years are large enough to make the difference of questionable significance. The data concerning age structure suggest decreasing survival beyond age 16. Consequently, the segment calculations were terminated at age 16. Survival beyond age 16 was calculated by pooling the 1958 to 1967 data and estimating survival from the ratios of successive pairs of age classes. Although there are few data on wild populations that adequately support such a conclusion, it nonetheless seems reasonable to suppose that large mammals are subject to the "U-shaped" mortality curve of demography.

The data for the Pribilof fur seals can thus be used to approximate the essential ingredients of population dynamics, as represented by the "Lotka" equations. As Cole (1954) demonstrated for species having a short annual period of births, these equations can be written as summations, rather than in the usual integral form. The equations are

$$1 = \sum_{x=a}^{x=\infty} \lambda^{-x} l_x m_x \tag{1}$$

$$\frac{1}{\beta} = \sum_{x=0}^{x=\infty} \lambda^{-x} l_x \tag{2}$$

$$c_x = \beta \lambda^{-x} l_x \tag{3}$$

where a = age of first reproduction

λ = e^r = finite rate of population increase

m_x = age-specific reproductive rate (female births per female)

l_x = survival from birth to age x

β = "birthrate per capita"

c_x = proportion of population aged x

As is well-known, the equations pertain to populations exhibiting a stable age structure, and changing at a rate, r, the intrinsic rate of increase.

Figure 1 shows the rates derived above for fur seals. In constructing the figure, a survival rate of 0.40 to age 3 has been assumed (Table 1), with survival in the third year assumed to be at the adult rate (0.91), while that of the second year was arbitrarily set at 0.80. It may be noted that only survival to age 3 is required in a solution of equation (1), so that the main effect of specific rates in the first three years is on the shape of the estimated age structure [equation (3)]. The lower curve of Figure 1 shows the products $l_x m_x$ which sum to R_0, the net rate of reproduction. The combination of an assumed constant survival rate for ages 4 to 16 and a flat-topped reproductive curve results in the $l_x m_x$ curve being essentially triangular in shape, with peak productivity achieved at age 6. It is, however, important to note that a given year class will make substantial contributions to productivity for more than 10 years.

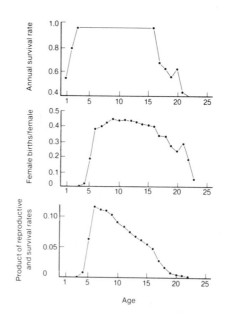

Figure 1   Survival and reproductive rates (m_x) and the product of survival to age x and the corresponding reproductive rate ($l_x m_x$) for Pribilof fur seals.

Using the data of Figure 1, equation (1) can be solved iteratively for an estimate of r, the intrinsic rate of increase. For the data given, the estimate is $r=0.006$, or virtually that of a stationary population. There are various uncertainties that need to be considered in connection with this result. One is that the drop in survival after age 16 may be unrealistic. However, it can be noted that the reproductive rates drop off rather sharply at about that age, which supports a notion of declining vitality. Furthermore, if pregnancy rates are declining sharply, then a somewhat higher survival in these older ages would not have much overall effect on productivity in any case.

A second question is whether survival to age 3 might actually be somewhat higher. The value selected is not the highest of Table 1, but does seem to be the maximum supportable by the bulk of the observations. It is true, however, that the data are actually based on male survival to age 3 (with an arbitrary adjustment factor of 1.1). Conceivably juvenile female survival may not always be proportional to that of males. Yet a third issue is the estimation of adult survival rates. The problems here are examined in the following section.

Problems in Estimating Survival From Age Structure

The method used here is the "segment" method of Chapman and Robson (1960). They utilize two statistics:

$$n = n_0 + n_1 + n_2 + \cdots + n_k$$

$$T = n_1 + 2n_2 + 3n_3 + \cdots + kn_k$$

where the n_i represent the number of individuals in particular age classes in a segment of the age structure. Thus n_0 may be the number of 7-year-olds, n_1 the 8-year-olds, and n_k the 16-year-olds. Survival is estimated by an iterative solution of

$$\frac{T}{n} \frac{s}{1-s} - (k+1) \frac{s^{k-1}}{1-s^{k+1}} \tag{4}$$

In the work reported here, Newton's method was used to iterate to solutions for s.

Methods of the type exemplified by equation (4) depend on "stationarity" of the population, that is, assume a constant population level and a stable age distribution. Since this assumption can generally be questioned for natural populations, various simulations were carried out to study the consequences of deviations from the assumed conditions. Obviously there is a very wide range of ways in which this underlying assumption can fail. The study reported here considers only those circumstances in which a stable age distribution has been induced at a given rate of change of population size and is followed by a shift to conditions approximated by Figure 1, that would ultimately yield a stationary population.

The shift from a population changing at a rate r to stationary conditions necessarily requires modification of either the l_x or the m_x series. In order to start the simulations with a stable age distribution, initial age vectors were computed from equation (3). In one series this was accomplished simply by introducing the desired value of r in equation (2) and maintaining the l_x values of Figure 1. The initial age vector is then calculated from equation (3), using the value of β obtained from equation (2). This is equivalent to increasing some (or all) of the m_x values in equation (1) and thus raising the birth rate. The second mode of change utilized manipulations of the l_x series until the desired rate of increase was achieved [calculated from equation (1)], without changing the m_x series. In both cases, these changes (birth rate or survival rate) served only to generate an initial age vector. The actual rates used in the simulation were those appropriate to a stationary population, and approximated those of Figure 1.

Since one of the issues involved in considering the effect of changing age structure is induced oscillations in population number, it seemed worthwhile to work with a smaller population and to utilize stochastic processes rather than solely the deterministic model of equations (1) to (3). Hence an initial population of 1000 individuals was used in each of the simulations, and 30 replicate simulations were averaged to trace the mean value of a population response under given initial conditions. Survival of each individual in each year was randomly determined with probabilities equal to the age-specific survival rates of Figure 1. Similarly, individuals aged three or older were subjected to a random draw to determine reproductive success, again according to the rates of Figure 1. The Chapman-Robson "segment" method, as described above was used to calculate survival for the group of individuals aged 6 to 16. These survival calculations were conducted in each year of the simulation. Each simulation run was either 30 or 50 years in length. Only females are considered in the simulation.

Behavior of the simulation model was checked initially by runs starting with an age vector calculated for stationary conditions, that is, with parameters essentially those of Figure 1. Since these initial runs indicated a small downward trend in population levels, adult survival was increased from about 0.91 as in Figure 1 to 0.92. This yielded the results of Figure 2, which shows a nearly constant population level (lower panel) and mean survival varying slightly about the expected value (0.92).

A second simulation run, extending over 50 years, was started with an initial age vector constructed according to equations (2) and (3), with $r = 0.08$ and survival rates as described above. The results (Figure 3) show an initial drop of about 10% (lower panel) in population size, followed by a "recovery" and at least one further oscillation of small magnitude. The pronounced initial drop can be ascribed to the fact that the initial age structure is that of a population growing rapidly (for marine mammals) and thus containing proportionately more younger animals than a stationary population. These younger animals do not begin to contribute importantly to reproduction until age 5, and make a maximal contribution at ages 6 to 8 (lower panel of Figure 1), when the rate of "recovery" (lower panel of Figure 3) is a maximum.

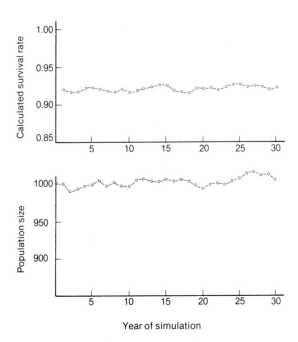

Figure 2 Calculated survival rate and population size of a simulated population, starting with an age vector constructed according to the parameters used in the simulation as described in the text.

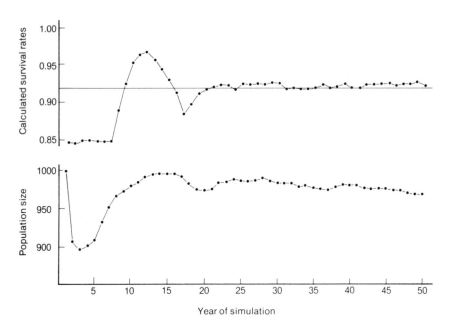

Figure 3 Calculated survival rate and population size of a simulated population, starting with an initial population vector derived from $r = 0.08$.

The apparent survival in this simulation (upper panel of Figure 3) is constant to age 7, reflecting the fact that individuals "born" under the stationary conditions do not enter the survival calculations until age 6. Since the data for the first "year" of the simulation are those of the calculated initial age vector, the actual sequence of stochastic events does not appear in the graphs until the second of the 50 plotted points. The apparent survival rate for the first years of the simulation (about 0.85) can be calculated from equation (3), which can be represented for the ages between 2 and 16 as

$$c_x = \frac{\beta P_0 P_1}{P^2} \left(\frac{P}{\lambda}\right)^x \tag{5}$$

where P_0 = survival from birth to age one, P_1 = survival from age 1 to 2, and P = annual survival from age 2 to 16. Since $P = 0.92$, and $r = 0.08$, the apparent survival over the range used in calculations (age 6 to 16) is

$$\frac{0.92}{e^{0.08}} = 0.849$$

As noted above, the initial age vector of a growing population contains proportionately more young animals than would exist under stationary conditions. When the stationary conditions prevail, as they do in all the simulations reported here, smaller numbers of young are "born" so that when these individuals reach the initial age (6) used in survival calculations, the effect is one of raising the apparent survival rate. This begins an oscillation in apparent survival which in this case persists importantly until about year 20, after which mean survival reflects the stationary rate (0.92).

The effect of an initial survival vector derived from a decreasing population (Figure 4) is to reverse that of a simulation started from a growing population (Figure 3). This simulation again used the rates of the previous efforts, but with $r = -0.05$. The apparent survival in the initial years is now

$$\frac{0.92}{e^{-0.05}} = 0.967$$

Since the initial population vector now has proportionately fewer young animals than under stationary conditions, the population is thus proportionately high in breeding-age individuals. This yields a very small initial surge of growth followed by a decline to an apparent steady-state level of about 950 individuals. Apparent survival rates again oscillate significantly out to about year 20.

Results reported thus far pertain to the use of initial age vectors constructed with the same survival rates as those used in the stationary population, and thus in effect, based on increased reproductive rates. A further simulation was conducted by constructing an initial age vector based on increased survival rates in equation (1) to yield $r = 0.08$. The rates thus arrived at were (1) survival to age 3 of 0.55, and (2) an adult survival of 0.97. The only change in juvenile survival

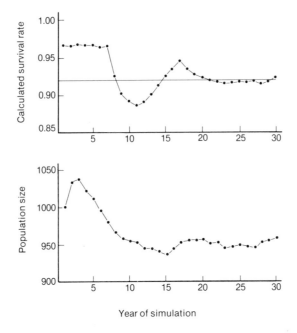

Figure 4   Calculated survival rate and population size of a simulated population, starting with an initial population vector derived from $r = -0.05$.

was in the first year of life, with survival in the second and third years maintained at the rates previously used. The rate of increase actually achieved by the stochastic model was checked by a simulation run conducted at these rates (30 replications were used here, as before). The observed rate of growth was $r = 0.066$.

Results of a simulation based on this new initial vector (Figure 5) are rather less variable than those described above. The initial apparent survival is now calculated as

$$\frac{0.97}{e^{0.08}} = 0.895$$

Fluctuations of both population size and apparent survival are not as marked in this example as in the previous cases. Survival estimates are nonetheless appreciably biased for about 15 years. One reason for the smaller fluctuation would seem to be the smaller apparent bias in the initial survival estimates. This is, of course, only a "canceling-out" effect, inasmuch as the survival rate under which the initial vector was constructed was 0.97. It is interesting that the population in this simulation reaches a steady state at a level about equal to the initial value, while those in the preceding examples dropped below the starting value. It seems likely that this is a consequence of the initial conditions.

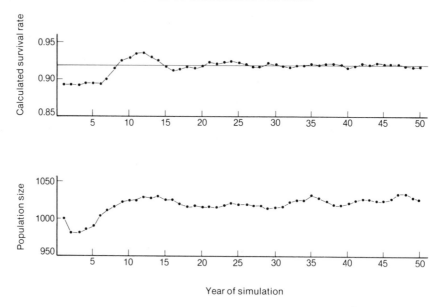

Figure 5 Calculated survival rate and population size of a simulated population, starting with an initial age vector based on increased survival rates in equation (1) to yield $r = 0.08$.

Figures 6 and 7 show the initial age vectors for the simulations described above. In each figure, the solid lines represent the stationary conditions, which each of the simulated populations presumably reached after 20 years or more. One can thus deduce something of the reasons for the behavior of the curves in Figures 2-5 from study of the relationship of the initial age vector to its ultimate form.

A feature not explored above is that of varying only pup (first-year) survival. It has been suggested (Eberhardt 1977a) that population self-regulation may depend on a sequence of changes in "vital statistics." The main such factor may be survival in the first year of life. Hence it would seem worthwhile to investigate this aspect in more detail. In general, it appears that the main factor in the bias is the initial displacement calculated from equation (5). This may then provide a useful index to the likely degree of bias.

The above simulations suggest the possibility of some substantial biases in survival estimates derived from age structures. It is, of course, not necessarily true that the apparent survival rates resulting from the simulations discussed here would actually be used in practice. Various tests (Chapman and Robson 1960) can be utilized to detect anomalies in the age structure, so that investigators may be aware of nonrandom fluctuations. A realization that a problem exists is, however, not very helpful unless an alternative scheme is available for estimating survival rates.

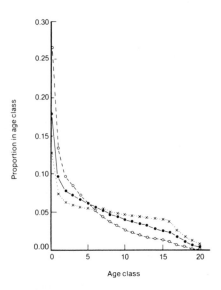

Figure 6 Proportions by age class for initial age vectors for the stationary population of Figure 2 (●), the population with $r = 0.08$ of Figure 3 (○), and the population with $r = -0.05$ of Figure 4 (✕).

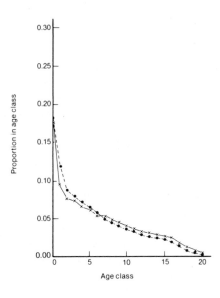

Figure 7 Proportions by age class for initial age vectors for the stationary population of Figure 2 (✕) and the population with $r = 0.08$ of Figure 5 (●).

207

MODELS FOR DENSITY DEPENDENCE

There are now a substantial number of theoretical models representing density dependence in the literature. The concern here, however, is mainly with applications of such models to actual populations. These usually take the form of stock-recruitment curves and, for marine mammals, most of the applications take the form of postulating a particular curve and somehow inferring parameters. The main exception lies in the work of D. G. Chapman (1961, 1973a), who has studied the stock-recruitment relation with actual data on the Pribilof fur seals.

In a study of harp seals *(Pagophilus groenlandicus)*, Allen (1975) has proposed that either the Beverton and Holt, or the Ricker stock-recruitment curve might be applied. However, the population growth rates achievable by marine mammals are sufficiently low that these two curves will in fact be virtually identical (cf. Eberhardt, 1977c). Allen (1975) actually applied the Beverton and Holt curve, and concluded that a sustainable harvest of 51% of the population might be achieved. Since even under rather generous assumptions as to parameters (Eberhardt and Siniff, 1977) marine mammal populations are unlikely to achieve growth rates exceeding 10% per year, this seems an unrealistic conclusion. An examination of Allen's (1975) work suggests that the calculations may have in fact been on a per generation basis.

Both Allen (1975) and Lett and Benjaminsen (1977) assume for harp seals that density dependence operates through reproductive rates. Allen assumes that reproductive rates are a linear function of the size of the population of pups, while Lett and Benjaminsen shift both the ages of initial reproduction and the overall pregnancy rate (as functions of numbers of older seals in the population). However Chapman's (1973) data for fur seals may be interpreted (Table 1) as showing mortality (to age 3) to be density dependent. The same seems true for gray seals (Chapter 8). Evidence from other species of large mammals, (see Chapter 23) leads me to believe that survival of very young individuals may be the proximate factor in population self-regulation (see chapters 2, 8, and 22), and that the stresses then operating may reduce growth rates and thus delay attainment of sexual maturity (cf. Eberhardt, 1977a).

The classical logistic model indicates that maximum sustainable yield (MSY) will be realized at a population level 50% of the maximal or asymptotic level. Some fisheries management models (Fox, 1970; Pella and Tomlinson, 1969) give maximum sustainable yield (MSY) levels well below the 50% point. Using information on reproduction and survival Fowler (see Chapter 23) has shown that large mammals in general have MSY levels well above the 50% point. Three sets of data on recruitment versus parental stocks also suggest that the MSY point for marine mammals may be above the 50% level supporting the general conclusions of Fowler (1981, Chapter 23). These include the fur seal data studied by Chapman (1973a), data on Antarctic blue whales *(Balaenoptera musculus)* given by Gulland (1971), and Antarctic fin whales *(Balaenoptera physalus)* from the study of Allen (1973) (see also Chapter 14). All three sets of data yield a slope of about 10% near the origin when recruit numbers are plotted against parental stock. This may then be a measure of the maximal rate of growth attained by the respective populations.

From a management point of view, the key issue is one of determining the stock level at which maximal recruitment takes place. Since the three sets of data suggest a nearly linear relationship between recruits and stock at the lower stock levels, there is also an implication of considerable theoretical interest. This is that the mechanisms of density dependence may not operate at lower stock levels (see Fowler, 1981). These possibilities will be explored in the following sections.

A Model for Density-Dependent Juvenile Survival

A simple model, used by Eberhardt (1977a), may serve to explore the data of Table 1. The model is

$$S = S_0 [1 - e^{-k(N_{max} - N)}] \tag{6}$$

where S is survival in the first year, N is the current pup population size, and S_0, k, and N_{max} are constants. As can be seen S_0 is a maximum survival rate, N_{max} denotes the maximum pup population size, while k controls the rate of departure from maximal survival rate as the population increases in size. Using the data of Table 1 with this model amounts to assuming constant survival from age 1 to 3.

In its present form, the model poses some difficulties for fitting. It can, however, be rescaled by using the difference between current and maximal population level as the independent variable, that is, let $x = N_{max} - N$, so that

$$y = S_0 [1 - e^{-kx}] \tag{7}$$

This model can readily be fit by nonlinear least squares. This does not, of course, provide a least-squares estimate for N_{max}. However, this is readily approximated by repeated fits, in which N_{max} is varied across a range of values to find a minimum in the sum-of-squares of the deviations. The program used in the present study is that of Watson et al. (1977).

Since a number of the pup population estimates have been revised rathea arbitrarily (following discovery of substantial overestimation from a tag-and-recovery method), the "best" fit to the data of Table 1 may not necessarily be the best representation of the actual situation. The purpose here is to provide a basis for further exploration. A variety of other models should also be considered. It should be noted here that there is independent evidence, in harvest and age structure data, of the low survival of the 1950s, so that the question above has to do with relative position of some points, and not with the sharp depression in survival rates itself.

Table 2 contains data from the nonlinear least-squares fit. Two of the fitted curves appear in Figure 8, one being that for the minimum variance given in

Table 2 ($N_{max} = 235$, $S_0 = 0.372$, $k = 0.161$) and the other having the lowest rate parameter of the set shown ($N_{max} = 290$, $S_0 = 0.383$, $k = 0.024$). From Table 2 and Figure 8 it seems clear that the fitting does not provide much information about the rate parameter k. Given the data in hand, one would infer from either curve that the maximum sustained yield point would be in the neighborhood of 200,000 female pups on St. Paul Island.

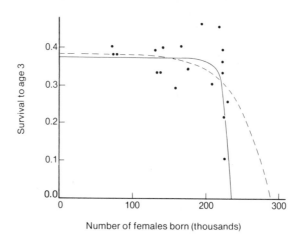

Figure 8 Two curves fitted to data on survival to age 3 of Pribilof fur seals on St. Paul Island, one (solid line) being that for the minimum variance given in Table 2 and the other (dashed line) having the lowest rate parameter of the set shown.

Table 2 Parameters from Nonlinear Least-Squares Fits to fur Seal Survival Data

N_{max}	Asymptote (S_0)	Rate (k)	Variance About Regression
230	0.377	0.238	0.00995
235	0.372	0.161	0.00554
240	0.374	0.104	0.00563
245	0.374	0.078	0.00580
250	0.374	0.063	0.00594
260	0.375	0.045	0.00614
275	0.378	0.032	0.00632
290	0.383	0.024	0.00644
300	Program failed to converge		

Two Stock-Recruitment Curves

Chapman (1973a) noted that there is as yet no satisfactory way to choose a particular curve for a given population. He suggested that an alternative may be a nonparametric approach, namely, to find a maximum sustained yield point without postulating a particular functional form for the stock-recruitment relationship. As Chapman noted, there are various difficulties to be overcome before such an approach can be wholly satisfactory. One of the problems is the choice of intervals for use in locating the maximum yield point. The choice of intervals may influence the location of the MSY point.

Chapman's (1973a) nonparametric approach to finding the MSY point for fur seals was based on a simple "catch" model:

$$C = R_1 + R_2 - \frac{a_1 P}{n} - a_2 P$$

where the catch C, is the sum of the 3-year-old male R_1 and female R_2 recruits less the number of replacements for adult males and females needed to sustain the herd. Thus an annual replacement rate a_2 is required from 3-year-old females to maintain the adult female population P. A similar rate a_1 is required for breeding adult males, whose numbers are calculated by applying a sex ratio n to the number of adult females. Chapman calculated the rate as

$$a_1 = \frac{0.36}{(1 - 0.2)^4} = 0.88$$

where the numerator is an estimate of the annual mortality of 7-year and older males, and the denominator expresses survival from age three to age 7, assumed to be the age at which replacement occurs. The comparable rate for adult females is

$$a_2 = \frac{0.11}{1 - 0.11} = 0.123$$

where 0.11 is the annual mortality rate for adult females.

Chapman assumes male recruitment to be an unspecified function of numbers of adult females $f(P)$. Since only the abundance of male recruits is estimated, he uses a ratio, λ, between females and males of age 3, and writes

$$C = f(P)(1 + \lambda) - \left(\frac{a_1}{n} + a_2\right) P$$

and notes that the catch is maximized (a_1, a_2, λ, n constant) if

$$f'(P) = \frac{a_1}{n} + a_2 (1 + \lambda)^{-1}$$

He uses $n = 25$, $\lambda = 1.1$, and thus finds $f'(P) = 0.076$. He calculates the number of females by adjusting estimates of pup abundance by a mean pregnancy rate (0.6). The resulting data is then grouped into three sets of years, 1920-1922, 1950-1956, and 1957-1965. Averages of numbers of adult females and male recruits are used to calculate slopes (rates of change of numbers of recruits with increasing numbers of adult females) and interpolation is used to find a value of P corresponding to 0.076, which Chapman calculates as $P_c = 471,000$.

However, inspection of the plotted data suggests that the maximum might be to the right of this point. Choosing another grouping (Table 3), yields the plot of Figure 9, which suggests the observed slope of about 0.10 continued out beyond $P = 600,000$, which places P_c slightly to the right of that point, if interpolation is used. Hence an element of subjectivity enters the procedure through choice of intervals.

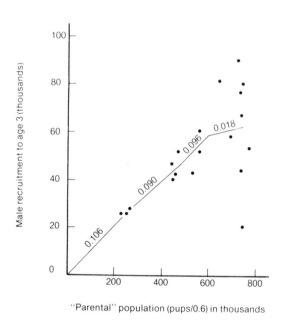

"Parental" population (pups/0.6) in thousands

Figure 9 Stock-recruitment curve for fur seals on St. Paul Island showing the observed slopes.

Table 3 An Alternate Calculation of the Maximum Sustainable Yield Point According to the Data of Chapman (1973a)

Years	Average Number of Pups (thousands)	"Parents" (P=pups/0.6) (thousands)	R_1	Slope	Midpoint
				0.106	125.0
1920-1922	150	250	26.6	0.090	352.9
1962-1965	273.5	455.8	45.1	0.096	527.8
1957-1961	359.8	599.7	58.9	0.018	674.2
1950-1956	449.3	748.8	61.6	—	—

Another view of the same set of data may be obtained by comparing it with a suitably skewed distribution. Richards's (1959) growth function is of some interest here, since it is sufficiently flexible to include curves skewed in both directions. Thus, as remarked above, Pella and Tomlinson (1969) proposed its use to represent a fishery in which the MSY point is to the left of the 50% point. I have not attempted to fit this curve to the fur seal data in any formal way, but have simply manipulated parameters to obtain the representation of Figure 10. The slope of the ascending limb of the curve is just a little greater than 10%, and thus in accord with the observed data. Perhaps the curve is most useful in that it focuses attention on a cluster of points (years) that seem markedly "too high." The same general impression can be obtained from Figure 8, but there the contrast is not quite as dramatic, and is offset by a scatter below the fitted curve.

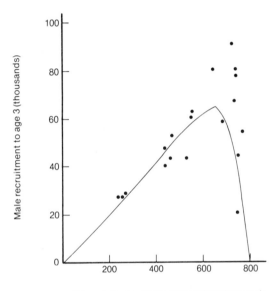

"Parental" population (pups/0.6) in thousands

Figure 10 Comparison of male recruitment to age 3 on St. Paul Island with Richard's curve.

In a sense, searching for an optimal yield point for the fur seal data is currently something of an academic exercise if the population parameters are now nearly those of a stationary population, as suggested by the calculations given previously. Certainly the population did show a healthy rate of growth in the past. The stock-recruitment curve suggests 10%, but may be questioned because it constitutes a considerable oversimplification for a long-lived species like fur seals. However, Kenyon et al., (1954) provide other direct evidence in the form of estimates of pup populations from 1912 to 1924. Rates of growth may be obtained from these data either by fitting an exponential curve directly, using nonlinear methods, or by linear methods on the logarithms of population size. The direct fit (Figure 11) yields an estimated rate of increase as 0.077, with a standard error of 0.0036, while a linear regression of logarithms of the counts has a slope of 0.082 (SE = 0.0033). Both fits show an intriguing tendency for the observed points to oscillate around the trend line.

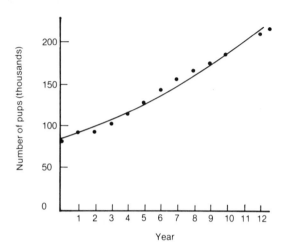

Figure 11   Numbers of fur seal pups born on the Pribilof Islands, 1912 to 1924.

There is thus a question of how to reconcile the early rates of increase with current parameters that suggest a nearly stationary population (see Chapter 5). The available span of reproductive data (1958 to 1974) does not show much evidence of substantial changes, although, as previously noted here, a detailed analysis of the data has not yet been reported. Data (Table 1) on survival in the first years of life do not show much evidence of changes at the lower population levels or in the earlier years. Survival of adult females appears to be the point on which there is the least information, and that based on the uncertain assumption of a stationary population. Hopefully more detailed analyses of the accumulated data will soon be available and will cast some further light on the uncertainties. Meanwhile it may be useful to examine the recent history of the population in terms of the parameter estimates now at hand.

A POPULATION SIMULATION

The only satisfactory index to the abundance of female fur seals on the Pribilofs comes from the long series of pup counts. In early years direct counts of the entire population were made. As numbers increased, various kinds of sampling methods were introduced, with the current method being based on the work of Chapman and Johnson (1968). Any long-term comparison of model and data must thus mainly depend on the pup counts.

The model used here amounts mainly to a simple bookkeeping exercise, using the rates discussed above. Its chief purpose is to attempt an approximate assessment of the compatibility of parameter estimates and the observed recent history of the herd. Various improvements need to be introduced, but these depend on the presently incomplete analysis of reproductive data and the concomitant information on age structure. The main features to be explored are the decline in pup survival discussed above and the impact of harvests of adult females which started in 1956 and continued to 1968. These harvest data (Table 4) were taken from the reports of the North Pacific Fur Seal Commission, which give harvests by age. The Commission reports grouped ages for older animals and these were prorated according to the available age structure from the pelagic sampling.

Survival to age 3 followed a curve of the form shown in Figure 8. The function used was

$$S = S_0 \left[1 - e^{-k(N_{max} - N)} \right]$$

with $S_0 = 0.40$, $k = 0.0333$ and $N_{max} = 300,000$. Maximal survival is thus taken to be slightly higher than that of Table 2. The maximal population size (N_{max}) is also higher than those considered in the table, because the data used there are for St. Paul Island only, since about 20% of the total pups born are found on St. George Island.

Reproductive rates used are those of Figure 1, as are survival rates for age 16 and beyond. Survival rates from age 3 to age 16 are arbitrarily set at 0.93. This choice was made to provide a small positive rate of increase ($r = 0.02$). The simulation begins with an initial age vector constructed according to equations (1) to (3), utilizing the above rates. The initial population (in 1951) is taken as one producing a pup population of about 260,000 individuals. The principal outcome of the simulation is a plot of pup numbers (Figure 12). This is contrasted with a set of estimates of pup numbers, based on several sources. Estimates for 1952 to 1960 are those of Chapman (1973a), as given in Table 1 here (females only), but adjusted by 20% for pups born on St. George Island. Data for 1961 and 1966 are those of Johnson (1975) multiplied by 0.5 to estimate number of females. The estimates for 1967 to 1975 (no estimates are available for 1971) are those from the Fur Seal Investigations Report of the National Marine Fisheries Service (U.S. Department of Commerce) issued in February 1976 (Table 17, p. 40). The 1976 estimate is from a similar report, issued in February 1977, which gives 291,000 pups born on St. Paul Island. These were again multiplied by 0.5 to estimate females, and inflated by 1.2 to account for St. George Island.

Results of the simulation should be regarded as no more than a rather crude test of the hypothesis that the marked drop in pup production beginning about 1966 might be accounted for by a reduction in pup survival and the female harvests of 1956 to 1968. As previously noted, the estimates of pup numbers prior to 1961 are subject to question, since they were based on tagging pups and

Table 4 Harvests of Female Fur Seals on the Pribilof Islands

Age	1956	1957	1958	1959	1960	1961	1962	1963	1964	1965	1966	1967	1968
2	132	0	572	218	19	449	394	699	346	162	0	21	35
3	2183	1150	11393	2016	281	4509	4451	2918	3971	3336	0	935	704
4	6625	5800	8420	7167	562	6798	8501	7915	3942	4056	0	1579	2630
5	4609	11465	3961	3630	964	4436	6139	8518	4159	1490	0	1552	1836
6	2916	6343	3287	2798	600	4723	3354	3970	1822	579	0	1166	1581
7	2120	4031	1200	3323	381	2928	2752	1773	518	348	0	517	1018
8	1745	3550	536	1750	374	2726	2678	2418	230	67	0	537	660
9	1122	3120	462	1146	312	3018	2128	2030	193	56	0	451	757
10	750	1773	574	1053	184	2155	1557	2119	202	59	0	470	584
11	993	1866	128	914	118	2232	2138	2067	197	57	0	459	644
12	941	1768	122	866	111	2114	2026	1959	186	54	0	435	610
13	762	1432	98	702	90	1712	1641	1587	151	44	0	352	494
14	689	1295	89	635	82	1549	1484	1436	137	40	0	318	447
15	658	1236	85	606	78	1479	1417	1370	130	38	0	304	426
16	509	957	65	469	60	1144	1096	1060	101	29	0	235	330
17	339	637	44	312	40	762	730	704	67	20	0	157	220
18	225	423	29	207	27	506	485	468	45	13	0	104	146
19	127	240	17	118	15	287	275	265	25	7	0	59	83
20	78	146	10	71	9	174	167	162	15	4	0	36	50
21	33	63	4	31	4	75	72	69	7	2	0	16	22
22	17	31	2	15	2	37	36	34	3	1	0	8	11
23	17	31	2	15	2	37	36	43	4	1	0	8	11

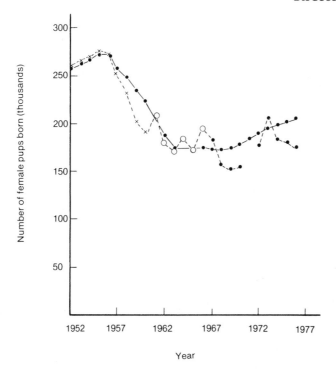

Figure 12. Estimated (dashed line) and simulated (solid line) number of female pups born on the Pribilof Islands. The set of estimates is based on Chapman (1973) (×), Johnson (1975, Table 38) (○), and Fur Seals Investigations Report of the National Marine Fisheries Service issued in 1976 and 1977 (U.S. Department of Commerce) (●).

recovering tags in subsequent male harvests. It later became apparent that biases existed in this data, and corrections were instituted. Beginning in 1961, direct estimates were made by the method of Chapman and Johnson (1968). It needs also to be emphasized that the simulation was started at a pup population level at which a considerable density-dependent effect is established (see Figure 8).

DISCUSSION

There are a number of uncertainties in the above analysis. Important among them is the estimation of adult survival. The simulations suggest that certain kinds of changes in adult survival or in reproductive rates might lead to substantial biases in survival rates estimated from age structure, and that these biases may persist for decades. Other kinds of change may have a much smaller but nonetheless persistent effect. The simulations were based on abrupt changes while more gradual transitions may occur in reality, or an abrupt change may be

Table 5 Age Structure of Female Fur Seals Taken in Pelagic Collections[a]

Age	1958	1959	1960	1961	1962	1963	1964	1965	1966	1967	1968	1969	1970	1971	1972	1973	1974
3	30	30	14	68	71	44	98	138	72	92	55	73	181	148	70	61	41
4	33	65	28	78	107	93	82	198	163	83	148	122	193	213	80	68	105
5	54	79	43	55	94	134	111	62	159	83	58	88	108	91	169	75	89
6	77	82	35	51	55	74	107	100	84	183	73	88	120	99	122	114	83
7	80	100	51	77	71	64	58	65	111	64	107	103	56	87	94	112	133
8	79	114	81	87	75	72	61	89	103	64	59	84	67	57	94	86	127
9	63	75	112	93	56	50	40	46	48	110	62	19	64	42	61	97	105
10	75	67	100	91	76	60	65	27	31	101	62	80	38	68	80	81	98
11	88	68	106	67	70	73	56	49	55	37	61	99	41	38	70	68	54
12	104	53	82	58	74	76	67	41	38	28	62	92	38	61	38	66	32
13	86	39	93	62	44	63	44	22	29	28	37	42	41	30	38	53	29
14	72	49	83	55	50	47	50	27	34	9	40	27	3	11	23	34	35
15	55	61	52	55	41	62	54	38	36	28	47	15	15	15	28	32	16
16	44	48	41	45	38	37	29	33	36	28	40	19	9	11	14	31	25
17	28	25	36	20	34	23	28	27	12	55	33	27	0	8	9	7	13
18	17	19	18	20	19	10	26	22	12	18	17	15	18	11	5	10	3
19	11	11	15	8	11	4	9	5	0	0	16	8	6	0	5	3	6
20	2	3	5	6	8	9	13	5	7	0	11	0	0	4	0	3	0
21	1	5	5	2	2	3	0	3	2	9	5	0	3	0	0	0	0
22	1	3	0	0	2	0	0	0	2	0	5	0	0	4	0	0	0
23	0	1	1	1	0	2	1	3	0	9	2	0	0	0	0	0	0
24	0	1	1	1	1	0	0	0	0	0	0	0	0	0	0	0	6
25	0	1	0	0	0	0	0	0	0	0	0	0	0	0	0	0	0

[a] Adjusted so that each column totals 1000.

reversed after a few years. This is, in fact, what apparently happened in the present case. Pup survival dropped dramatically in the mid-1950s, and then increased again (Table 1). Since there are questions about the estimates of pup production in the 1950s, it may be worthwhile to refer to two other sets of data bearing on this point. One is the male harvest records, which are known to be very accurate. Inspection of the male harvests, arranged by year class, shows that there was a very substantial drop in production from 1952 to 1956. These data also exhibit the same trend as the estimates of pup production, that is, a steady downward progression into the early 1970s, starting in 1959.

Additional evidence as to the impact of the reduced pup survival in the mid-1950s comes from the female age structure data. These data are necessarily influenced by the degree of effort expended in pelagic sampling in a given year, and may also be affected by the locale in which the sampling was done. However, the year classes from 1952 to 1956 seem quite clearly to have been substantially reduced and that reduction is evident throughout the span of the records. Table 5 shows the female age structure data, adjusted to a common base. That is, each year's data has been arranged to total to 1000 animals. An interesting feature of the female age structure data is that it suggests an appreciable reduction in the 1952 age class, in contrast to the male harvests, where the male kill of that age class is one of the largest recorded. However, the female harvests may account for a substantial part of this discrepancy, inasmuch as the year class of 1952 was also the largest contributor to the kill of females (Table 4).

It may be noted from Table 5 that the reduced pup production of the mid-1950s leaves a gap in the productive age classes for more than a decade. It thus seems quite possible that the juxtaposition of two effects may account for the trend of the Pribilof population up into the 1970s. One is the sharp reduction in pup survival in the mid-1950s, suggested here to have been a density dependent phenomenon. This was followed directly by female harvests (1956 to 1968) that further reduced the size of the productive age classes. Such an explanation does not, however, account for the more recent trends in population size.

Pup production appeared to be at its lowest in 1968 to 1970 (Figure 12), and this is corroborated by the age-structure data (Table 5). Since the population as a whole has evidently decreased throughout the 1960s, it seems unrealistic to again invoke the notion of density dependence. However, another factor may well have come into play during that decade. Chapman (1973b) commented that the herd "may be influenced by disturbance which is now greater than at any time since the pelagic operations ceased," and by "competition for food due to large removals of flounders, Alaska pollack, and other species by the large bottom fishery in the Bering Sea and elsewhere along the Pacific Coast of North America" (see also Chapter 14).

LITERATURE CITED

Allen, K. R. 1973. Analysis of the stock-recruitment relation in Antarctic fin whales. Rapp. P.V. Reun. Cons. Int. Explor. Mer 164:132-137.

Allen, R. L. 1975. A life table for harp seals in the northwest Atlantic. Rapp. P. V. Reun. Cons. Int. Explor. Mer 169:303-311.

Chapman, D. G., and D. S. Robson. 1960. The analysis of a catch curve. Biometrics 16:354-368.

Chapman, D. G. 1961. Population dynamics of the Alaska fur seal herd. Trans. 26th North Am. Wildl. Nat. Resour. Conf., pp. 356-369. Wildlife Management Institute, Washington, D.C.

Chapman, D. G., and A. M. Johnson. 1968. Estimation of fur seal pup populations by randomized sampling. Trans. Am. Fish. Soc. 97:264-270.

Chapman, D. G. 1973a. Spawner-recruit models and estimation of the level of maximum sustainable catch. Rapp. P.V. Reun. Cons. Int. Explor. Mer 164:325-332.

Chapman, D. G. 1973b. Management of international whaling and North Pacific fur seals: Implications for fisheries management. J. Fish. Res. Board Can. 30:2419-2426.

Cole, L. C. 1954. The population consequences of life history phenomena. Q. Rev. Biol. 29:103-137.

Eberhardt, L. L. 1977a. Optimal policies for conservation of large mammals, with special reference to marine ecosystems. Environ. Conserv. 4:205-212.

Eberhardt, L. L. 1977b. "Optimal" management policies for marine mammals. Wildl. Soc. Bull. 5:162-169.

Eberhardt, L. L. 1977c. Relationship between two stock-recruitment curves. J. Fish. Res. Board Can. 34:425-428.

Eberhardt, L. L., and D. B. Siniff. 1977. Population dynamics and marine mammal management policies. J. Fish. Res. Board Can. 34:183-190.

Fowler, C. W. 1981. Density dependence as related to life history strategy. Ecology (in press).

Fox, W. W. 1970. An exponential surplus-yield model for optimizing exploited fish populations. Trans. Am. Fish. Soc. 99:80-88.

Gulland, J. A. 1971. The effect of exploitation on the numbers of marine mammals. Pp. 450-467 in: P. J. den Boer and D. R. Gradwell (eds.). Dynamics of Populations. Center for Agricultural Publication and Documentation, Wageningen, The Netherlands.

Johnson, A. M. 1975. The status of northern fur seal populations. Rapp. P.V. Reun. Cons. Int. Explor. Mer 169:263-266.

Kenyon, K. W., V. B. Scheffer, and D. G. Chapman. 1954. A population study of the Alaska fur seal herd. Special Scientific Report. Wildlife No. 12. U. S. Dept. Interior, U. S. Fish Wildl. Serv., Washington, D. C.

Lett, P. F., and T. Benjaminsen. 1977. A stochastic model for the management of the Northwestern Atlantic harp seal *(Pagopilus groenlandicus)* population. J. Fish. Res. Board Can. 34:1155-1187.

Mendelssohn, R. 1976. Optimization problems associated with a Leslie matrix. Am. Nat. 110:339-349.

Pella, J. J., and P. K. Tomlinson. 1969. A generalized stock production model. Int.-Am. Trop. Tuna Comm. Bull. 13:421-490.

Richards, F. J. 1959. A flexible growth function for empirical use. J. Exper. Bot. 10:290-300.

Watson, C. R., M. I. Cochran, J. M. Thomas, and L. L. Eberhardt. 1977. COMP — A BASIC nonlinear least squares curve fitting package. Proc. DECUS Symp., San Diego, California.

ELEVEN

A Decision-Making Framework
for Population Management

LARRY D. HARRIS

IRVIN H. KOCHEL

INTRODUCTION

In many parts of the world, wildlife populations have been pushed to unstable positions. This is especially true of the larger species which not only demand large expanses of contiguous habitat but which create the greatest impact when their numbers exceed upper limits. One of the consequences of this instability is that changes in population size may cause dramatic shifts in the way the species is perceived over very short periods of time or very short distances. Within a decade, or within a few hundred kilometers, a species may change in status from that of a pest to that of an endangered species. This, in turn, leads to a public perception that management is haphazard at best and perhaps misguided (Crutchfield and Pontecorvo, 1969; Talbot, 1977; Caughley, 1977; Holt and Talbot, 1978).

If resource managers wish to improve and demonstrate the rationality of management decisions, there may be merit in examining certain fundamental concepts of formal decision-making. Management schemes must be flexible. Rather than inflexible solutions, there is a need for standard procedures to ensure rational, objective decisions in a highly dynamic environment. Fragmented decision rules that are not tailored to specific local conditions can only serve to compound the problem.

The disciplines of management science and operations research are built around a standardization of rational approaches to making decisions. Although both rely heavily upon modeling, they consist of much more than modeling. Formal decision making is little more than adherence to simple, straightforward procedures which examine the total system of interacting components within which the problem is contained. It consists of answering the following questions:

What are the relevant states of nature (i.e., explicit situations)?

What are the desired states of nature (i.e., an explicit objective)?

What are the alternative courses of action?

What are the consequences of each alternative?

What information is available to relate the magnitude of each action to the likely consequences?

What criterion is to be used to judge the best alternative?

What procedure should be used to direct future courses of action given potentially observed states?

This procedure may seem so simple and straightforward as to appear unnecessary. However, a review of the wildlife science literature shows little evidence that the decisions reached in the management of populations are developed by following such a format. Without explicit statements of objectives, alternative courses of action and precise expressions of the functional relation between actions and consequences, management decisions will continue to appear capricious. Reliable techniques are available to wildlife managers, however, and are developing rapidly, as exemplified by other chapters in this book. Modeling techniques facilitate the testing of numerous management alternatives without relying on direct field observation. A decision-making framework is necessary, however, to ensure that the techniques are applied effectively to the problem at hand. Part of this paper deals with an exploration of the technique of defining the management options which are available. The remainder is an examination of the relative importance of various parameters to the management of populations.

Modeling is no doubt the single most fundamental element of decision theory. Models play a role in forcing a clear statement of the objectives and consideration of alternatives. They provide a scheme for integrating available information. To the degree that a model demands a quantitative expression of the relation between inputs or perturbations and output responses, it forces a distinction between what is known and what is not known. Thus, while clearly identifying what is unkown, it reduces the likelihood of overlooking what is known. To a large extent, the value of a decision-making model is proportional to the number of alternative actions it deliniates and the fidelity of discrimination between the responses that it generates. While models can be too simple and while they often obscure the management options available, they can also be too large, too elaborate or too expensive to serve the practicing manager (e.g., Harris and Francis, 1972).

Models must be designed or chosen on the basis of their applicability to specific situations. This is, in part, the basis for demanding a clearly stated objective at the outset.

One purpose of this chapter is to illustrate certain fundamental principles of the decision-making framework as relevant to large mammals. The first section presents a decision tree as an example of a simple qualitative modeling approach that assists in the articulation and consideration of alternative objectives. The second section illustrates the process of determining the relative importance of various pertinent variables (in this case, population parameters). The third section presents information of use in establishing the nature of the functions interconnecting certain state variables within or related to populations. The three major topics are equally crucial to the development of models for the management of populations and to acceptable decision formulation.

THE DECISION TREE

An elementary form of qualitative modeling that demonstrates only the inter-connections of components and actions is frequently useful for illustrating the options available and for helping to avoid oversights. One elementary technique uses the branches of a decision "tree" to delineate the options one has at various stages and the outcomes likely to result from a given choice. The advantages of such a model are: (1) it gives structure to the process of making decisions and helps the manager approach his particular problem in an orderly, sequential fashion, (2) it requires the evaluation of all possible alternatives and outcomes, desirable or not, and (3) it succinctly demonstrates all of the stages in the process of reaching a particular decision.

For illustration, we develop a hypothetical series of decision nodes relevant to a wildlife manager wishing to increase productivity (Figure 1). The decision nodes in the example are not meant to be all-inclusive nor should it be conclud-ed that any fixed number of options exist at any given node. Nonetheless, the scheme demonstrates the importance of clearly identifying objectives.

The first decision in Figure 1 involves the level of management. Few scientists would consider wildlife populations to be divorced from the larger community or ecosystem context and yet almost all vertebrate management to date has been aimed directly at a single species with little regard for interspecific and total community interactions. Although population management clearly has its place, increased concern is being expressed for community and total ecosystem management strategies (Wagner, 1969, 1977; Harris and Fowler, 1975; Talbot, 1977; Chapters 2, 14, 18, 19). Since neither approach is universally applicable, the first major decision involves the explicit resolution of the community versus the population-level approach. For this example, and in the context of this book, we pursue the population level.

The question of whether to increase, decrease or stabilize a population will often arise (second node, Figure 1). It is even possible for populations in close proximity to be managed with opposite objectives and therefore require com-pletely different approaches. In south Florida, for example, the key deer *(Odocoileus virginiana clavium)* on Big Pine Key is managed as an endangered species, to stabilize its population. On the adjacent mainland, deer are managed for maximum yield. In the surrounding intensively farmed areas, damage by deer can be significant and the population is suppressed accordingly. In spite of the different objectives, neither the general public nor the professionals seem to clearly recognize the distinctly different management approaches required.

Given the decision to increase the population in question, its present density relative to carrying capacity serves as a useful reference point (node 3, Figure 1). Despite the confusion about its definition, populations can occur at, above, or below carrying capacity because of changes in either the populations or the en-vironment. Although most North American management is directed at main-taining populations at what is defined as the carrying capacity, many European populations are managed at densities exceeding what the environment can sup-

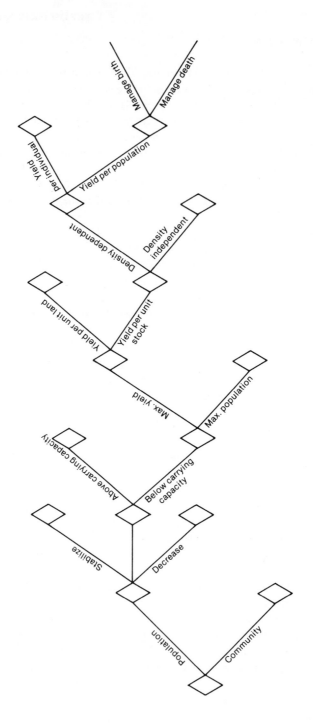

Figure 1 Hypothetical series of decision nodes of a "decision tree" relevant to a wildlife manager wishing to increase productivity.

port naturally (cf. Dagg 1977). Managers run the risk of frustration, if not failure, when the properties of population behavior drawn from populations at one level are indescriminately applied to another.

Much of the wildlife management profession adheres to the belief that yield or production is directly related to population level. In parts of North America this leads to the conclusion that populations should be increased to levels near the carrying capacity of the site whenever possible. Although such a conclusion is supported by theoretical and empirical evidence (Chapter 23) it may not be warranted in all cases. Thus it is important to distinguish between the objective of maximum sustained yield (MSY) and that of maximum population level.

Given that we wish to maximize yield (node 4), an important distinction must be drawn between yield per individual and yield per unit of habitat (node 5). The two production functions differ significantly in relation to large mammal density (and thus grazing intensity). When individuals represent high investment (as in breeding stock) or high returns (as in trophy production) yield per individual may be the performance criterion. Under such conditions, stocking densities should be kept at levels where the production of the individual is maximized. This may be a low density. On the other hand, when investments or returns dictate that production per unit of habitat should be maximized, the optimal stocking density would be considerably higher. Thus production of individuals in the population would be sacrificed to ensure maximum utilization of food and habitat resources by the population.

Since the primary mechanisms responsible for higher population productivities at lower densities revolve around density dependent birth and death responses, it is essential to assess the degree of density dependence (node 6, Figure 1). Density dependence means more than the number of births or deaths produced in response to population density. It means that the number of births or deaths per individual in the population (a specific or relative rate) is a function of density. In this context, the birth and death rates of certain species are clearly more density dependent than others.

The allocation of resources should involve cost-benefit considerations. For this reason critical evaluations of various approaches are necessary. Given that a manager has progressed past all the above decision nodes and the objective remains to substantially increase productivity, at least two approaches remain: to enhance reproduction or to decrease mortality. It seems that to this point, wildlife population, management has been largely directed toward affecting the survivorship function (e.g., bag limits, predator control, and antipoaching). Evidence presented in the next section argues for a greater effort toward managing reproduction.

Additional nodes, additional alternatives at each node, and further explanation of the rationale for the various categories would serve several useful purposes. At one extreme, the scheme could serve to demonstrate the complexity of wildlife management decision-making at all levels. At the other extreme, further elaboration could show how managers need to clearly consider and articulate their objectives and courses of action. We can develop the example no further in the context of our objectives for this paper.

THE RELATIVE IMPORTANCE OF VARIABLES

Birthrate Versus Death Rate

Efforts toward sound management should emphasize variables that have the most significant impact. Whereas a system's status is reflected by the state (or concentration) variables, we affect the system's performance by manipulating the flow (or process) variables. Thus, the ideal decision process involves monitoring the key state variable while directing our limited resources at manipulating the most important rate processes (Forrester, 1968).

These principles seem to apply well to population management. In most cases decisions concerning management are based upon knowledge about population levels, observed sex or age ratios or some correlate such as level of crop damage. Decisions almost invariably affect population survivorship (e.g., changing the hunting season length). It obviously is necessary to evaluate the relative impact of changes in the birthrate versus changes in the death rate. Although both processes are commonly included in population dynamics models, they are frequently combined to generate a single geometric growth parameter such as the instantaneous rate of growth r or the net reproductive rate R_0. Such parameters mask the true importance of the birthrate relative to the death rate. Two forms of evidence presented below suggest that for changes of equal magnitude, most populations are considerably more sensitive to birthrate than to death rate.

Demographers have long known that the birthrate is much more influential than the death rate in determining the age distribution of a population (Chapter 21; Coale, 1957).

Coale's work was based upon demographic data for the Swedish human population in 1860 and 1950. Calculation of the stable age distribution that would result from either set of data proved interesting, but it leaves the effect of increased longevity confounded with the effect of increased fecundity. A more lucid analysis is obtained by calculating the stable age distribution that would result from each of the birth and death rate combinations. The average expectation of life increased from 45.4 years in 1860 to 71.6 years in 1950. Yet, as demonstrated in Figure 2, this 58% increase in life expectancy produces very minor effects compared to the difference resulting from holdng either death rate constant (i.e., either 1860 or 1950) and pairing it with different birthrates. As shown in Chapter 21, population structure is considerably more sensitive to changes in birthrate than changes in death rate.

The second line of evidence is more analytical than empirical and follows the work of Cole (1954) and more recently, Bryant (1971), Charnov and Schaffer (1973), and Bell (1976). Assume that two similar populations representing the extremes of longevity both reproduce during their first year of life. In the first, all the parents die immediately after reproducing for their first and only time as would an annual plant. In the second population, individuals not only live forever, but go on producing offspirng throughout their never-ending lives. It can be easily demonstrated that increasing the litter size of the first population

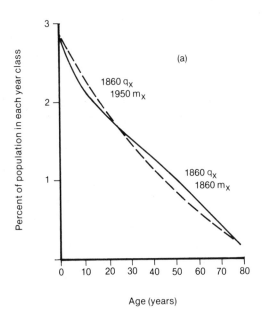

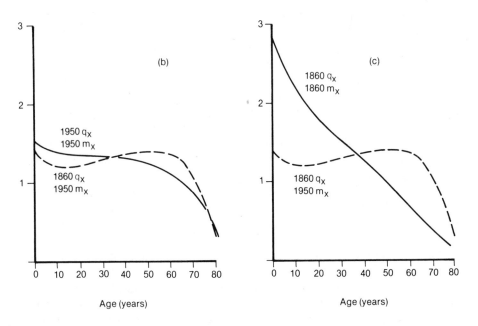

Figure 2 Projected stable age distributions based on Swedish demographic data for 1860 and 1950. Frame *(a)* demonstrates little effect on the stable age distribution from applying the much higher 1950 survival rate (e_x = 71.6 years) to the 1860 birthrate. Frame *(b)* demonstrates little effect from applying the much lower 1860 survival rate (e_x = 45.5) to the 1959 birthrate. Frame (c) demonstrates the dramatic effect of applying the slightly lower 1950 birthrate to the 1860 survival rate.

by one female completely counterbalances the infinite survivorship and the never ending reproduction of the second. This result is not generally recognized. The argument is not as direct for species that do not reproduce during their first year of life, but the importance of early reproduction can not be overemphasized.

An application of the principles outlined above to white-tailed deer will illustrate its significance to decision-making in the management of large mammal populations.

The reproductive rate of deer populations of certain regions in Florida (especially the Everglades) may average less than 1.0 fetus per adult female per year (Harlow and Jones, 1965; Harlow, 1972; C. Loveless, personal communcation). Vigorous white-tailed deer populations in other regions of North America carry as many as 2.47 fetuses per adult female (Nixon 1971). Under such conditions as many as 90% of the females breed and give birth to fawns during their first year of life (Haugen, 1975). This is uncommon in the less productive southern herds (Noble, 1974). These observations indicate that the highly productive northern white-tailed deer herds are represented more by the model given above for an annual species. Females from northern populations seem to have a potential litter size which is almost one female greater than that of the southern herds (50% of 2.47 compared to 50% of 0.9). By inference from the model, the highly productive herd should be able to support very high annual adult mortality and still maintain a growth rate equal to that of a hypothetical herd of low productivity in which all adults survive. Empirical data support the argument since the maximum sustainable harvest rate for productive whitetails appears to be about 40%/year (Andrews and Calhoun, 1968). This may be a low estimate of what can be harvested since in cases where such harvest rates occur they almost invariably include large segments of the young female population (20% of the total harvest of Illinois deer). If mortality were restricted to the adult age classes, increased birthrates should generate greater responses within the population than massive increases in adult survivorship. Arguments such as these become more complex when high infant mortality rates or long birth to reproduction intervals are involved. Even so, simulations suggest that for long-lived ungulates a modest increase in fecundity (e.g., 20%) produces results comparable to those produced by similar increases in infant survivorship (Foose, 1980).

The relevance of exploratory modeling must be considered in relation to its utility in management. As is frequently the case, the model cited above may not have led directly to conclusions for the management of white-tailed deer. Indeed, it was not constructed for any particular species. But when applied to white-tailed deer, the model leads to the conclusion that it is early reproduction and large litter size as characteristic of the northern herds that facilitate such heavy exploitation compared to the less productive southern herds. The implications clearly point toward large potential responses from enhancing fecundity compared to reducing adult mortality. In keeping with the principle of directing our efforts at the most significant rate functions, efforts toward management should be focused on fecundity.

Such management seems reasonable since, as discussed above, very large increases in survivorship are necessary to equal relatively small increases in fecundity. In addition, there seems to be abundant evidence to indicate that high fecundity is a direct function of factors that can be manipulated (e.g., nutrition), Severinghaus and Tanck (1965). Verme (1969) experimented with nutrition and demonstrated that the litter size of white-tailed deer could be increased by 83% while the percent of females pregnant was only increased by 3%.

Birthrate Parameters

Birthrates for populations as a whole are clearly the function of (1) the age of reproductive maturity, (2) litter size and, (3) litter frequency. Many simple population models are unable to discriminate between these three biological functions and yet considerable evidence suggests that they are not equally important (Chapters 2, 20-23). For example, Smith (1954) illustrated the relative importance of the mean generation time compared to total offspring (R_0). Such analyses reveal that the growth rate of populations is much more sensitive to the generation time than to total offspring.

This is verified by data from ungulates. Ransom (1967) compared the ovulation and conception rates of a population of white-tailed deer in Manitoba to several other productive herds. He concluded that the fertility of the adults compared favorably with the highest levels of fertility reported for white-tailed deer in general. Indeed, the pregnancy rate for the older age classes of the herds in Manitoba was greater than that for a herd in Iowa as reported by Haugen (1975) and the ovulation rate was nearly as high. Yet, the population in Manitoba is described as having a "relatively low annual increment" (Ransom, 1967) while the Iowa herd is highly productive. The primary difference seems to involve the fact that 74% of the Iowa fawns conceived and averaged 1.2 fawns each. A very low percentage of Manitoba fawns became pregnant. The low productivity of many southern deer herds may be explained similarly. For example, Noble (1974) concludes that few, if any, does in the population in Mississippi experience estrus before they are at least 1 year old.

Like age of first reproduction, the interval between reproductive cycles also directly alters the mean generation time. For this reason, we believe it is the next most sensitive parameter of population growth while litter size appears least important of the three (see chapter 23 for perspective involving age structure).

Differential Mortality Rates

Caughley (1966) generalized the mammalian mortality rate schedule as a U-shaped function of age. This has led to the general belief that even though mortality is highly variable with age, the shape of the curve is not particularly variable between sites or circumstances. Models built solely upon this generality may err in their representation of reality. Not only is the pattern of mortality dynamic, it appears that the different patterns have greatly different significance to the underlying populations.

Given the forces of natural selection, mortality rates could be expected to vary as an inverse function of genetic value. The reproductive value (Fisher 1958) describes the future value of a present individual. Since animals do not achieve their highest reproductive value (V_x) until the onset of reproduction, both younger and older individuals have lower reproductive values than prime aged individuals (Figure 3). The curve of reproductive value as a function of age is essentially the inverse of Caughley's generalized mortality schedule.

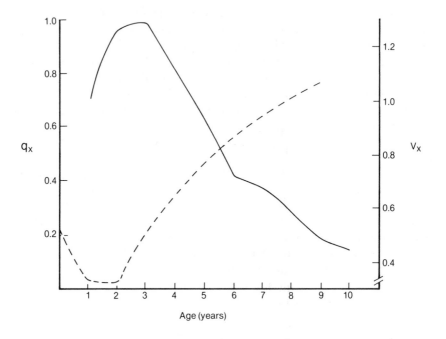

Figure 3 The mammalian mortality rate, q_x (dashed line; Caughley, 1966), and reproductive value, v_x (solid line), as a function of age.

Data on ungulate mortality resulting from predation by coevolved predators appears to confirm this relationship (Pimlott et al., 1969; Hornocker, 1970; Mech and Frenzel, 1971; Schaller, 1972; Kruuk, 1972; Mech, 1977). Additional support comes from the observation that male mortality rates are generally greater than those of females (Cowan, 1950; Green, 1950; Robinette et al., 1957; Taber and Dasmann, 1954), since reproductive value applies only to the females in a population. It should be noted that this differential mortality rate is not simply a reflection of behavioral or ecological patterns or of human activities such as hunting; it may also appear *in utero*. The sex ratio at birth tends to favor females in stressed populations while favoring males in vigorously growing populations (Robinette et al., 1957; Trivers and Willard, 1973). Overall, it appears that mortality rates are inversely related to reproductive values in natural communities. Thus, at least under certain conditions, the population segments with the highest reproductive value appear to reflect the lowest mortality rates.

In contrast to the above are mortality patterns involving circumstances more independent of the influence of natural selection. Under highly stressful or catastrophic conditions the pattern described above may not apply. For example, in at least four cases of documented catastrophic mortality the ratio of female to male carcasses was nearly 2-1. Massive die-offs of sika deer (*Cervus nippon*, Christian et al., 1960), African elephant (*Loxodonta africana*, Corfield, 1973), reindeer (*Rangifer tarandus*, Klein, 1968) and red deer (*Cervus elephas*, Mitchell and Staines, 1976) yielded carcass sex ratios of (female to male) 65:35, 65:35, 60:40, and 83:17, respectively. Since the sex-ratio prior to the mortality was not reported in any of these cases, the evidence for significant differential mortality is inconclusive. Moreover, the reindeer data may be inadmissible since virtually the entire population was decimated. Also, Child (1972) reports the details of a massive wildebeest die-off, which seem to run counter to our hypothesis. Nonetheless, even if mortality occurred in the same frequency as the sex ratio of the live population it would be inconsistent with the much more general trend of greater male mortality. Only a few sets of the more appropriate data on age-specific mortality under different circumstances exist (e.g. Table 1). They are equally inconclusive but suggest strong differences from the pattern that derives from predator-induced mortality.

Finally, mortality induced through hunting should bear the least resemblance to natural patterns. Data for white-tailed deer from Michigan (Eberhardt, 1960) and New Hampshire (Laramie and White, 1965) are readily available. Neither data set shows any significant difference between the frequencies of year classes in the live population and the hunter kill (Figure 4).

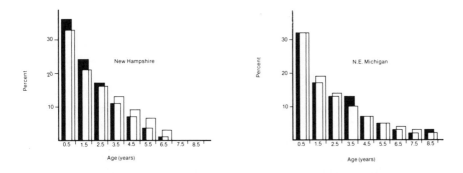

Figure 4 Percentage composition by age of the populations (black) and of the harvests (light) for two white-tailed deer populations.

Although inconclusive in themselves, we believe these data counter the proposition that ungulate mortality patterns can be generalized for individual conditions (but see Chapter 22). More importantly, there is doubt that resulting changes in sex and age composition resulting from differential mortality always boosts the reproductive value of the population as maintained by Caughley (1976). There is obviously variety in the ways mortality operates. Under certain circumstances animals of low reproductive value are removed first and populations are minimally impacted. Under more catastrophic conditions, animals of high reproductive value also succumb with a significant impact on the populations. As MacArthur (1960) has concluded: "the same reproductive value may be removed (from any population) by taking few individuals of high reproductive value or many individuals of low." Modeling efforts that do not consider the nature and relative intensity of different forms of mortality may not be at all representative of the real world (see Chapters 20, 23). We now turn our attention to the nature of the relations between state variables such as density and sex ratio and the response variables birthrate, death rate, and yield.

Table 1   Cumulative Proportion of Female Carcasses Occurring in Percentile Classes After Massive Die-offs of Three Large Mammals.[a]

	Percentile Class									
	10	20	30	40	50	60	70	80	90	100
Sika deer	0.29	0.41	0.67	—	—	—	—	—	—	1.00
Red deer	0.25	0.50	0.54	0.54	0.58	0.67	0.67	0.80	0.96	1.00
African elephant	0.38	0.44	0.47	0.53	0.60	0.67	0.79	0.84	0.93	1.00

[a]To achieve comparability, the life spans of 10 years for sika deer (Christian et al., 1960), 17 years for red deer (Mitchell and Staines, 1976), and 60 years for the African elephant (Corfield, 1973) were divided into 10 percentile classes.

YIELD CURVES AND COMPENSATORY CHANGES

Despite a vast literature on the subject, few quantitative or experimental data exist to show the relation between man-induced harvesting and other mortality rates, such as predation or disease. Second, with one exception (Silliman and Gutsell, 1958), no experimental study has investigated a vertebrate's response in productivity to different levels of harvesting. Understanding the nature of both of these responses is critical to accurate modeling, decision-making, or sustained yield harvesting.

There is both mathematical and biological evidence that birthrates, and thus productivity, decrease as population levels increase. The mathematical rationale has historically relied heavily on the logistic growth model. Such models produce the conclusion that maximum rates of population change occur at the inflection point of a density dependent growth curve. To the extent that productivity is indicated by population changes, maximum productivity is predicted by the logistic model to occur at 50% of saturation density. Work with invertebrates such as unicellular algae (Ketchum et al., 1949), Daphnia (Slobodkin and Richman, 1956), blowfly larva (Nicholson, 1954), fruit flies (Fugita and Utida, 1953), and other insects (MacLagen, 1932) tend to corroborate this hypothesis. Experimental harvesting of guppies *(Lebistes reticulatus)* also supports the hypothesis (Silliman and Gutsell, 1958). However, Fowler (1981, Chapter 23) has shown that most long-lived species tend to be most productive at levels close to saturation density while short-lived species are most productive at levels below 50% of saturation density.

Also, as demonstrated by Fowler (Chapter 23), the shape of a yield curve may be determined by the sex of harvested animals. In particular, it appears that a harvest of males only will often give rise to productivity being greatest at populations which are more numerous than when a harvest of both sexes occurs.

Such responses are easily demonstrated expirically as in the following experiment. Fourteen replicate populations of a closed genetic strain of wild-caught mice *(Mus musculus)* were used. Each population was established with individuals taken from one original population. The indoor pens in which they were reared were approximately 3 m² with *ad libitum* food, water, and nest boxes provided. Temperature and light were under moderate control. The populations were followed closely for 18 months at which time it appeared that all were oscillating around a common level (35 ± 2.7 individuals, Figure 5). We arbitrarily defined saturation density K as the initial mean density of 35 and reduced 12 of the populations to fractional levels of K. Four populations were assigned to each of three densities: $0.35K$, $0.50K$, $0.65K$. Two populations were maintained as controls. Two of the four populations assigned to each density were randomly committed to a male-dominated harvesting schedule and two to an equal-sex harvesting regime. Monthly harvests over a full year removed the number of individuals necessary to return each population to its prescribed density. Different measures of productivity were calculated to be expressed as yield per pen, yield per individual, and yield per female. The incidence of death, mange, and agonistic behavior were closely monitored.

Productivity of the populations of different density showed highly significant differences $(P < 0.01)$ with peak productivity occurring at 50% K. While the productivity of the populations of $0.50K$ and $0.65K$ were not significantly different from each other, the productivity of the two extreme populations $(0.35K$ and $1.0K)$ was greatly different and averaged an order of magnitude less than that of the two medium-density populations as shown in Figure 6.

A highly significant $(P < 0.01)$ interaction between density and the sex of the harvest was noted. The populations exposed to male-dominated harvests invariably attained greater productivity at $0.65K$. The populations exposed to equal-sex harvests were invariably most productive at 50% of saturation density (Figure 6). In all cases only one of the three densities gave rise to peak productivity; productivity at the other three densities was only about 10% as great. Even though the point of maximum productivity occurred at different densities,

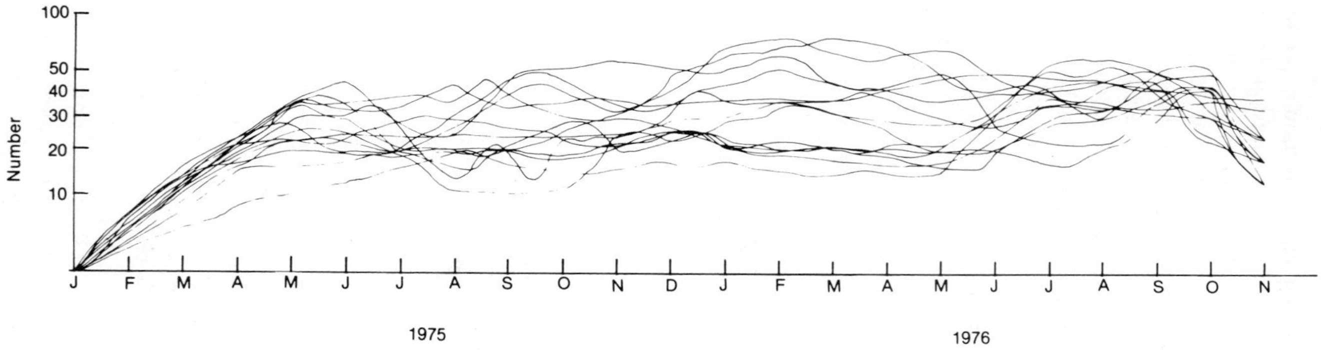

Figure 5   Population size over time for 14 captive populations of wild-caught mice *(Mus musculus)*.

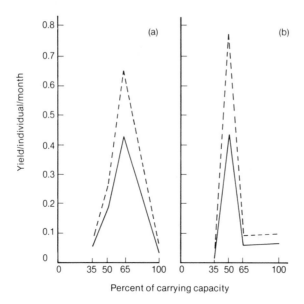

Figure 6 Yield per month from harvested mouse populations as a function of relative density. Yields per individual in the population (solid line) and per female in the population (dashed line) are shown when the harvest is primarily on males *(a)*, and when both sexes are harvested *(b)*.

the two schemes based on the sex of the harvest produced no significant difference in overall specific productivity.

Both birthrate and death rate were highly responsive to population density (Figure 7); losses due to natural mortality were directly proportional to density ($Y = -26.3 + 1.2x$, $r^2 = 0.98$). The response of the birthrate was nearly identical to the yield response described above (Figure 7).

The compensatory aspect of natural mortality and reproduction is clearly illustrated by these results. In all cases, the level of natural mortality was inversely related to the number of individuals cropped. Cropping may have been substituting for natural mortality in addition to a natural decline due to density dependence. Since the birth rate was also enhanced in the intermediate levels of these populations, the total productivity of harvested populations was substantially greater than unharvested ones.

The frequency of agonistic behavior was also inversely related to population density. Except for the population kept at $0.35K$, the incidence and severity of mange was also inversely related to density. Agonistic encounters and mange were probably prime contributors to the lowered birthrate and or the high natural mortality at high density.

In summary, our work with mice supports the work of Fowler (Chapter 23, 1981) in several ways. First, as a species with an intermediate life span, mice showed maximum productivity at levels above 50% of the saturation level (when expressed in numbers and not as specific rates). This is in contrast to in-vertebrates, which show such dynamics at 50% or less at saturation densities. Se-cond, the productivity of populations subjected to a male harvest peaks at higher densities than those from which both sexes are harvested but which are otherwise similar. These empirical observations, in combination with the explanations provided by more theoretical approaches (Chapter 23) emphasize the need to consider such underlying principles in managing large mammal populations.

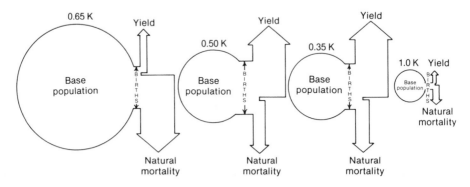

Figure 7 Interrelation of population density, birthrate, natural mortality and yield at four different harvest rates. Area of circle is proportional to the population level at which the populations were maintained ($k = 35 \pm 2.7$ animals).

DISCUSSION

Although the significance of the results above is apparent, their application to free-ranging populations must be made with caution. Were we modeling the ex-perimental population of mice we would incorporate a highly nonlinear, density-dependent birthrate which would vary over an order of magnitude and peak at some median density level. This birth rate would not only be sensitive to population density but sex ratio and social stability. The birthrate would not be mediated as much through litter size, as it would be by age at which the first lit-ter is produced and the frequency of production of litters. A nonlinear, density-dependent death rate would also be necessary to describe the experimental populations. This response would be somewhat more complex than the birthrate response because of the two distinct forms of mortality (harvest and natural).

Individuals removed in our experimental work were representative of the population's age and social structure whereas individuals succumbing to natural deaths were not. These two different types of mortality produced different im-pacts on the parent populations with high rates of natural death being associated with lower productivity and high cropping removals perpetuating high produc-tivity. To a large extent, then, harvesting (even in a random manner) appeared to substitute for natural death.

At the level of the total population, yield integrates the differences between birth and natural mortality. Each, in turn, is a function of density and the type of removal. The sex of the harvest is obviously important as are the age and social structure of the harvest (see Chapters 21, 23). Clearly, the tradeoffs are significant and need to be considered in research and management which involves populations.

It is through the application of techniques used in management science and operations research that we are forced to look at relationships such as those discussed above. Through such approaches the relevant variables and relationships are identified. As above, having identified the important aspects of the system which we propose to manage, details concerning the nature of internal structure and relationships are then established. The process is one of obtaining bounds for a system which is then represented in sufficient detail to provide relevant insights of use in the process of manipulation. The management of populations of large mammals must be based on such approaches.

LITERATURE CITED

Andrews, R. D., and J. C. Calhoun. 1968. Characteristics of a white-tailed deer population in Illinois. Ill. Dep. Conserv. Game Manage. Rep. No. 1. 8 pp.

Bell, G. 1976. On breeding more than once. Am. Nat. 110:57-77.

Bryant, E. H. 1971. Life history consequences of natural selection: Cole's result. Am. Nat. 105:75-76.

Caughley, G. 1966. Mortality patterns in mammals. Ecology 47:906-918.

Caughley, G. 1970. Population statistics of chamois. Mammalia 34:194-199.

Caughley, G. 1976. Wildlife management and the dynamics of ungulate populations. Adv. Appl. Biol. 1:183-246.

Caughley, G. 1977. Analysis of Vertebrate Populations. John Wiley and Sons, Inc., New York.

Charnov, E. L., and W. M. Schaffer. 1973. Life history consequences of natural selection: Cole's result revisited. Am. Nat. 107:791-793.

Child, G. 1972. Observations on a wildebeest die-off in Botswana. Arnoldia (Rhod.) 5:1-13.

Christian, J. J., V. Flyger, and D. E. Davis. 1960. Factors in the mass mortality of a herd of Sika deer, *Cervus nippon*. Chesapeake Sci. 1:79-95.

Coale, A. J. 1957. How the age distribution of a human population is determined. Cold Spring Harbor Symp. 22:83-89.

Cole, L. 1954. The population consequences of life history phenomena. Q. Rev. Biol. 29:103-107.

Corfield, T. F. 1973. Elephant mortality in Tsavo National Park, Kenya. E. Afr. Wildl. J. 11:339-368.

Cowan, I. M. 1950. Some vital statistics of big game on overstocked mountain ranges. Trans. N. Am. Wildl. Conf. 15:581-588.

Crutchfield, J. A., and G. Pontecorvo. 1969. The Pacific Salmon Fisheries — A Study of Irrational Conservation. Johns Hopkins Press, Baltimore. 220 pp.

Dagg, A. I. 1977. Wildlife Management in Europe. Otter Press, Waterloo, Ontario.

Eberhardt, L. 1960. Estimation of vital characteristics of Michigan deer herds. Report 2282, Game Div., Mich. Dep. Conserv., Lansing. 192 pp.

Fisher, R. A. 1958. The Genetical Theory of Natural Selection. Dover, New York.

Foose, T. J. 1980. Demographic management of endangered species in captivity. Intl. Zoo Yrbk. 20:154-166.

Forrester, J. W. 1961. Industrial Dynamics. M. I. T. Press, Cambridge. 464 pp.

Fowler, C. W. (1981). Density dependence as related to life history strategy. Ecology (in press).

Fujita, H., and S. Utida. 1953. The effect of population density on the growth of an animal population. Ecology 34:488-498.

Green, A. U. 1950. The productivity and sex survival of elk, Banff National Park, Alberta. Can. Field Nat. 64:40-42.

Harlow, R. 1972. Reproductive rates in white-tailed deer of Florida. Q. J. Fla. Acad. Sci. 35:165-170.

Harlow, R., and F. K. Jones, Jr. 1965. The white-tailed deer in Florida. Tech. Bull. No. 9. Florida Game and Fresh Water Fish Commission, Tallahassee.

Harris, L. D., and N. K. Fowler. 1975. Ecosystem analysis and simulation of the Mkomazi Reserve, Tanzania. E. Afr. Wildl. J. 13:325-346.

Harris, L. D., and R. Francis, 1972. AFCONS: A dynamic simulation model of an interactive herbivore community. Tech. Rep. No. 158. Grassland Biome U.S. International Biological Program, Ft. Collins, Colorado 88 pp.

Haugen, A. O. 1975. Reproductive performance of white-tailed deer in Iowa. J. Mammal. 56:151-159.

Holt, S. J., and L. M. Talbot. 1978. New principles for the conservation of wild living resources. Wildl. Monogr. 59. 33 pp.

Hornocker, M. C. 1970. An analysis of mountain lion predation upon mule deer and elk in the Idaho primitive area. Wildl. Monogr. 21. 39 pp.

Ketchum, B. H., L. Lillick, and A. C. Redfield. 1949. The growth and optimum yields of unicellulan algae in mass culture. J. Cell. Comp. Physiol. 33:267-279.

Klein, D. R. 1968. The introduction, increase, and crash of reindeer on St. Matthew Island. J. Wildl. Manage. 32:350-367.

Kruuk, H. 1972. The spotted hyena. University of Chicago Press, Chicago.

Laramie, H. A., and D. L. White. 1965. Some observations concerning hunting pressure and harvest on white-tailed deer. N. Hamp. Fish Game Dep. Tech. Circ. No. 20. 55 pp.

MacArthur, R. H. 1960. On the relation between reproductive value and optimal predation. Proc. Natl. Acad. Sci. 46:143-145.

MacLagan, D. S. 1932. The effect of population density upon rate of reproduction with special reference to insects. Proc. R. Soc. London 111:437-454.

Mech, L. D. 1977. Wolf-pack buffer zones as prey reservoirs. Science 198:320-321.

Mech, L. D., and L. D. Frenzel. 1971. Ecological studies of the timber wolf in Northeastern Minnesota. USDA For. Serv. Res. Paper No. 52. 62 pp.

Mitchell, B., and B. W. Staines. 1976. An example of natural winter mortality in Scottish red deer. Deer 3:549-552.

Nicholson, A. J. 1954. Compensatory reactions of populations to stresses and their evolutionary significance. Aust. J. Zool. 2:1-8.

Nixon, C. M. 1971. Productivity of white-tailed deer in Ohio. Ohio J. Sci. 71:217-225.

Noble, R. E. 1974. Reproductive characteristics of the Mississippi white-tailed deer. Mississippi Game and Fish Commission, Game Div., Jackson. 58 pp.

Pimlott, D. H., J. A. Shannon, and G. B. Kolenosky. 1969. The ecology of the timber wolf in Algonquin Provincial Park. Dept. Lands and Forestry, Ontario. 92 pp.

Ransom, A. B. 1967. Reproductive biology of white-tailed deer in Manitoba. J. Wildl. manage. 31:114-123.

Robinette, W. L., J. S. Gashwiler, J. B. Low, and D. A. Jones. 1957. Differential mortality by sex and age among mule deer. J. Wildl. Manage. 21:1-16.

Schaller, G. B. 1972. The Serengeti Lion, a Study of Predator Prey Relations. University of Chicago Press, Chicago.

Severinghaus, C. W., and J. E. Tanck. 1965. Productivity and growth of white-tailed deer from the Adirondack region of New York. N. Y. Fish Game J. 11:13-27.

Silliman, R. P., and J. S. Gutsell. 1958. Exploitation of fish populations. Trans. N. Am. Wildl. Conf. 22:467-471.

Slobodkin, L. B. and S. Richman. 1956. The effect of removal of fixed percentages of the newborn on size and viability in populations of Daphnia pulcaria. Limn. Ocean. 1:209-236.

Smith, F. E. 1954. Quantitative aspects of population growth. Pp. 277-294 in: E. Boell (ed.). Dynamics of Growth Processes. Princeton University Press, Princeton, New Jersey.

Taber, R. D., and R. F. Dasmann. 1954. A sex difference in mortality in Columbian black-tailed deer. J. Wildl. Manage. 18:309-315.

Talbot, L. M. 1977. Wildlife quotas sometimes ignored in the real world. Smithsonian 8:116-122.

Trivers, R. L., and D. E. Willard. 1973. Natural selection of parental ability to vary the sex ratio of offspring. Science 179:90-92.

Wagner, F. H. 1969. Ecosystem concepts in fish and game management. Pp. 259-206. in: G. M. Van Dyne (ed.). The Ecosystem Concept in Natural Resource Management. Academic Press, Inc., New York.

Wagner, F. H. 1977. Species vs. ecosystem management: Concepts and practices. Trans. N. Am. Wildl. Nat. Resour. Conf. 42:14-24.

Verme, L. J. 1969. Reproductive patterns of white-tailed deer related to nutritional plane. J. Wild. Manage. 33:881-887.

TWELVE

A Management Perspective
of Population Modeling

THOMAS M. POJAR

INTRODUCTION

Mathematical representations of a population's dynamics, by necessity, have in common some estimate of mortality, natality, and density. Given these three variables as they represent a population at any point in time would permit projections of the population's rate of increase (or decrease) and, therefore, population size at some future date. As simple as this concept is, its application can be quite complicated when dealing with wild populations and data usually characterized by poor precision and questionable accuracy. Although disturbing, it is not surprising that the basis for some decisions involving wildlife management are reduced to gross oversimplifications.

The development of a population model demands examination of all data, relations among elements, and theory associated with the system (Guynn et al., 1976). The model-builder's principal challenge is to incorporate the information and theory into a model that will mimic the real-world system. In short, viewing a population through the structure of a model forces examination of pertinent aspects of the population and reduces the reliance on oversimplified rationale in management. Through modeling, the potential and limitations of populations can be broadly defined for purposes of management and planning.

There seems to be an optimal level of complexity for models that are management oriented. Digernes et al. (1973) developed a model that used only major population parameters to evaluate various cropping routines. These authors suggest that an ideal model would be more holistic and include, besides the basic population parameters, climatic, biotic (forage, etc.), and geographic data. Medin and Anderson (1979) developed a mule deer model that included a degree all of these except geographic parameters. Walters et al. (1975) describe a model that simulates population dynamics of caribou and biomass dynamics of their major food species. These aspects of the overall system are related through a foraging submodel that considers snow depth, seasonal migration, and total area of useable winter habitat. This model was useful in exploring hypotheses concerning food supply and hunting as limiting factors of this population, but the authors concede that, "In retrospect, it appears that much simpler population models could be used for caribou herd management in the forseeable future. Prediction based only on gross annual increment and overall natural mortality give nearly the same (management) results..." (Walters et al. 1975).

241

The foregoing are examples of what Levins (1966) calls the "brute force" approach to modeling which is an attempt to produce a faithful, one-to-one reflection of certain aspects of the system's complexity. As Levins points out, carrying this approach to the extreme results in attempting to deal with too many parameters, many of which are only vaguely defined and nearly impossible to measure. He favors the approach in model-building which sacrifices this degree of precision to realism and generality, which is the philosophy that Walters et al. (1975) alluded to in the analysis of their caribou model. The strength of the more general models is that they deal with fewer parameters which can be measured and compared with simulation results. Simplifying assumptions and omission of vague or difficult-to-measure parameters result in output that is potentially less precise but more realistic.

An example of a relation that is difficult to define, but is included in many models, is the attempt to relate the organism to its environment through a density driven feedback mechanism affecting natality and/or mortality (e.g. Allen and Kirkwood, 1978; Connolly et al., 1971; Gross et al., 1973). The mechanisms that limit population growth provide a constant reserve for study and speculation by ecologists (Wynne-Edwards, 1965), although it is easily accepted that limitation does occur (Krebs, 1978; Stubbs, 1977). Because population attributes (such as size, composition, natality, and mortality rates) can be estimated, one approach is to assume that the animals can be used as "the integrators of their environment" (Larkin, 1977, p. 2). The resulting mathematical function relating animal and environment ignores the mechanisms but accepts the principle. It is difficult to define that relation in many populations due to imprecise population parameter estimates, lack of long-term data (Dolbeer et al., 1976), or possibly because threshold population levels have not been reached (Craighead et al., 1974). This is a major dilemma in pragmatic model-building and the approach of assessing "population condition" suggested by Hanks (Chapter 3) may offer the best solution. Small changes in a density dependent function can make large differences in long-term simulation results and consequently major differences in management recommendations (Connolly, 1978; Craighead et al., 1974). Fortunately, most pragmatic models are used to project only short-term events, such as year-to-year harvest strategies; therefore, the impact of an inaccurate density dependent function is lessened. In the model by Gross et al. (1973) this problem is partially evaded by providing the option of using a density dependent reproductive function, observed or assumed reproductive function, observed or assumed reproductive rates, or a combination of both, which allows some flexibility in dealing with this phenomena, depending on the data available.

Since the early 1970s there has been a great deal of effort put into the development of computerized population models for a variety of wildlife species including Black-footed ferret (Clark, 1976), Coyote (Connolly and Longhurst, 1975; Knowlton, 1972; Sheriff et al. (unpublished), Blackbird (Dolbeer et al., 1976), Ring-necked pheasant (Taylor, 1978), Red fox (Zarnoch et al., 1974), Bear (Bunnell and Tait, 1978; Craighead et al., 1974), Grasshopper (Gyllenberg,

1974); Mallard (Walters et al., 1974) and Whale (Allen, 1973). In addition, a general cervidae model by Gross et al. (1973) and species oriented models for caribou (Walters et al., 1975), moose (Karns and Snow, 1979; Reuterwall and Ryman, 1978), deer (Anderson et al., 1974; Cooperrider, 1974; Medin and Anderson, 1979; Riffe, 1970; and Walls, 1974), and elk (Fowler and Barmore, 1979) have been published. Normally, a model is built for a particular population of interest. The model is built around the data from that population and in some sense, fit to the observations, Then, depending on the objective of the modeling effort, a variety of simulations are run using various sets of inputs that have some relevance to actual manipulation of the real-world system and various types of sensitivity analyses may be conducted. To advance the utility and reliability of a model, it should be tested using data sets that are independent of those used in the model's construction. Drummond (1976) contends that frequently this approach will point to the need for further model modification and that "In general, the greater number of data sets accommodated by a model, the more reliable it becomes."

MANAGEMENT—WITH OR WITHOUT MODELING

Managers of wildlife populations most often attempt to affect the size or composition of a population through regulation of the harvest. Persons who influence the decision-making process behind this management include a full spectrum of educational levels and backgrounds, including politicians, landowners, biologists, and administrators. Each group requires a different level of understanding of the functions of a population. Biologists want as much detail as possible while politicians may want only a superficial overview.

With the diversity of potential users (and critics) it is obvious that instilling the use of models in the decision-making process will be difficult. Gross (unpublished) described some aspects of developing a population model that is both relevant and understandable to persons involved in management. He points out that the computational processes must be as simplistic and explainable as possible. This point seems to be especially important when managers are grappling with the question of the validity of the model. Managers must understand the mechanics of the model in order to accept it. Inexplicable processes breed suspicion. For this reason, and as mentioned in Chapter 21, complex mathematics are undesirable even if some programming and computer time are saved.

As Garfinkel et al. (1972) indicate, there must be continued interaction between those producing the models and the ultimate client for the model. This interaction is important during the conceptual, developmental, and output stages (Innis, 1972) of model development to enhance its acceptability and utility. Walters et al. (1975) used this approach successfully. The ultimate client or primary user for management-oriented models, at least in Colorado, is the field-level manager, the person who collects the data, initiates the management recommendations, and, most importantly, bears the brunt of the consequences. He is continually forced to make decisions based on insufficient evidence.

(Coleman (1972) states that it is common and unavoidable for biomedical researchers and physicians to make major decisions based on insufficient evidence. He contends that decision-making in biomedical research is improved through the model-building process, even though it "must frequently use descriptions of component parts that are no better than qualitative or intuitive in nature." Although decisions involving wildlife management are not as crucial (at least in terms of human life) as those of a physician, they are important professionally and to society. If, in fact, population modeling is to be of assistance in management, the modeling effort should be geared to the primary user in terms of understandability and utility.

As mentioned in Chapter 21, a critical step in introducing population modeling into the decision-making process is to demonstrate to the primary user its relevance to the solution of real-world problems. Gross (unpublished) concludes, modeling examples dealing with abstract or "clean" problems do little to dispel the "so what?" attitude of managers. Interest in model application is usually marginal until the modeling effort is directed to a particular manager's specific problems. He has an intrinsic and vested interest in population's that are well known to him and is much more open to demonstrations of relevance of the model to real-world problem-solving.

Most management of ungulate populations in the intermountain region of the United States in the past has been based on the manager's mental model of the system. But, mental models are characteristically inconsistent, incomplete and never completely explicit. Consider an actual example of a manager's mental model of a mule deer population which consisted of 10,000 animals, a buck-doe-fawn ratio of 40-100-70 (thus only 1905 bucks) and a consistent annual reported harvest of 4000 bucks. A logical model, whether computerized or not, is frequently the best way to demonstrate that this population could not withstand the reported harvest for one season, not to mention on a sustained basis. It should be stressed that the foregoing example, although possibly an extreme case, represents the sort of situation in which a management model must function if the model is to be of utility in the real world of game management.

The exposure of inadequacies and fallacies of such mental models is the basic target of logical models. Such faulty concepts are changed by analyzing data through the structure of the model which provides managers with a more realistic assessment of the complexity of the system. Although simulation models for populations are incomplete (in the sense that all components of the real world system cannot be included) they are capable of handling, in an orderly and consistent fashion, many more components of the population than mental models and are therefore much more realiable.

IMPLEMENTATION OF POPULATION MODELS IN WILDLIFE MANAGEMENT

The following section describes the introduction of a model for simulating populations to management personnel in Colorado. The model (ONEPOP) was developed by Gross et al. (1973) specifically with the wildlife manager/decision-maker in mind. The purpose of developing this model was "to increase the capability of managers to effectively and efficiently husband big game resources" (Gross, unpulbished). Development of the model was directed mainly toward simulating the population dynamics of large ungulates. Thus far, in Colorado, it has been used for the management of elk *(Cervus elaphus)*, mule deer *(Odocoileus hemionus)*, pronghorn antelope *(Antilocarpa americana)*, and bighorn *(Ovis canadensis)* populations.

It is not the purpose of this chapter to describe or critique the model, but rather to explicate its application and utility in managing large ungulate populations. Additional documentation and validation of the model (ONEPOP) can be found in Bartholow (1977), Roelle (1977), and Williams (1977).

The Herd Concept in Modeling

Perhaps one of the most important contributions of modeling to Colorado's game management is the implementation of the concept of herd management. Herd management involves the realistic definition of geographic boundaries of a population, based on the criterion that the area encompass the herd's year-round range. The object is to delineate populations so as to minimize emigration and immigration so that these unpredictable and difficult-to-quantify phenomenon do not have to be considered in the simulated population. Colorado is already divided into game management units (GMUs) used for hunter distribution and sampling units for hunter-harvest surveys. Since harvest data are an essential part of the simulation, we attempted to locate herd boundaries to coincide with GMU boundaries, if at all possible. Divided GMUs would require an estimate of the proportion of harvest in each segment, which, by necessity, would be based on conjecture with no quantitative means to accommodate potential year-to-year fluctuation. Thus, division of GMUs is undesirable since it would be a critical factor in the simulation results if a significant proportion of the total herd harvest was involved.

The initial herd unit boundaries were delineated by biologists at the regional level. Their bases for boundary locations was empirical information from several years' experience with aerial counts of their region and fragmentary banded animal information. Concrete evidence for placement of herd boundaries, of course, would be most difficult to obtain; therefore, it is understandable that some herd boundaries are still under debate.

Developing a Simulation

Simulations were developed for each of 33 elk, 35 deer, 16 pronghorn antelope, and one bighorn sheep population, based on (1) information collected on each population, (2) information collected elsewhere in Colorado for the same species, and (3) information from published literature. These initial simulations provided a feasible demographic hypothesis of the dynamics of the population.

Depending on the quality and quantity of real data available for each population, the simulation mimicked the real-world population to varying degrees. In each case, the initial simulation was reviewed with regional and field-level managers who had some responsibility in that herd's management. These "modeling workshops" served several purposes. First, it was usually the first exposure of field personnel to modeling, so the input data and the model were explained in detail. Second, the output of the initial simulation was reviewed and compared with available empirical data and the mental models of those associated with the herd. During this phase there was extensive discussion regarding the quality and quantity of data available, many hypotheses were explored through the model, and much conjecture related to the population's dynamics was aired. The objective of these workshops was to develop a simulated population that mimicked the real-world population to the extent that it could be used for planning and management.

The end result of the modeling workshops was usually a reasonable representation of the population with which various harvest strategies and their consequence could be evaluated. The available data for the population were organized in a standardized fashion by the model and all information regardless of source, was available for critical review in terms of its contribution toward making the simulation more realistic. In addition, considerable reorientation was accomplished because the population had to be viewed as an entity rather than population fragments associated only with field personnel's arbitrary boundaries of responsibility.

Results of these workshops varied with the extent of the data base available. The extremes ranged from a mule deer population that only had an estimate of harvest available, to another mule deer population that had estimates of density, natality, and mortality that were obtained in a statistically sound fashion. The workshop dealing with the first, resulted in a very tentative simulation based on a population that could sustain the reported harvest and remain at a hypothesized density. Much of the input data were adapted from real data collected elsewhere in Colorado. An estimate of natality and density obtained directly from this population was necessary to reduce the amount of speculation in the original simulation. These estimates were obtained by field personnel using the method of Bowden and Anderson (1975). A second workshop was then held incorporating the new information into the simulation which was subsequently used to affect management. The modeling effort was directly responsible for providing the incentive to collect the necessary data.

In the second case above, where significant amounts of population data had been collected as the result of an ongoing research project (Gill, 1970; Roper 1973; Freddy, 1976), the modeling effort concentrated on providing a correspondence between simulated values and observed values (henceforth referred to as "alignment" between compared values). This still required subjective evaluation of some data points because of suspected bias and poor counting conditions. For example, the observed drop in the deer population from 1973 to 1974 (Figure 1) was not accompanied by any other evidence that the herd had

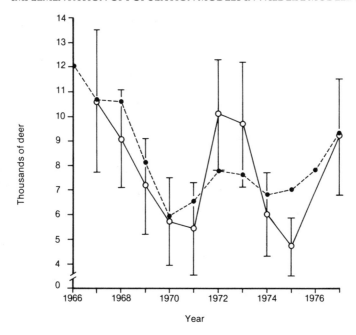

Figure 1 Comparison of observed (solid line) and simulated (dashed line) population size of a mule deer population with a good data base using the 1977 observed value as the strong alignment point (see text). Ninety percent confidence limits around observed values are shown.

experienced a 38% decline. The observed total mortality for the winter of 1973-1974 was 6.4%, and there was no empirical evidence of excessive winter mortality in terms of reports of a die-off by either local residents or field personnel. Research personnel who conducted the census suggested that the 1974 count was conservative because snow conditions affected the pattern of distribution of deer upon which the sampling scheme was based (Freddy, 1975). Similar conditions existed in 1975, and counting conditions were so poor in 1976 that a census was not conducted. Nearly ideal counting conditions existed in 1977, and a good census was obtained. Therefore, the point estimate of the 1977 postseason population size received special emphasis in fitting the simulated to the observed value (hereafter called an alignment point, which means a value in the simulation that has a corresponding observed value). The observed fawn-doe ratios (Table 1) were used as alignment points for recent years, while the mortality rates in the simulation were adjusted to allow the simulated population size to track the observed population as nearly as possible without adjusting mortality annually. Observed mortality rates were largely ignored as alignment points because of the high variability and potential inaccurcy (due to decomposition and scavenging of carcasses).

By viewing the data set through the model it was concluded that estimates of overwinter mortality were less reliable than estimates of population size and

Table 1   Comparison of Observed and Simulated Population Parameters of a Mule Deer Population in Colorado[a]

Year	Population Size		Fawns-100 Does		Overwinter Herd Mortality	
	Obs.	Sim.	Obs.	Sim.	Obs.	Sim.
1966	NC[b]	12,043	NC	82	NC	15.1
1967	10,600	10,712	90	84	NC	14.9
1968	9,112	10,654	85	81	16.0	14.4
1969	7,206	8,124	77	98	7.0	15.4
1970	5,730	5,916	41	41	15.8	12.4
1971	5,429	6,567	61	61	8.5	13.7
1972	10,087	7,807	78	78	0.9	14.7
1973	9,670	7,649	88	88	6.4	15.9
1974	6,005	6,805	75	74	4.3	15.2
1975	4,677	6,998	70	70	2.5	14.5
1976	NC	7,809	80	80	NC	14.8
1977	9,182	9,363	93	93	NC	—

[a]For game management unit D-9.
[b]NC = No count made.

natality. As a consequence, the systematic search for dead deer on the winter range has since been discontinued, thus streamlining the program for collecting data.

As demonstrated by this example of fitting or "simulation alignment," even a conscientious, statistically sound effort to collect pertinent data does not eliminate the necessity for subjective value judgments. Until the 1977 postseason count was made, the strongest alignment point for population size was the count from 1973, which was the last year good counting conditions existed. The simulated population, in numbers, aligns well with that observed through 1973, but departs drastically from 1973 through 1977 (Figure 2). The decision to use the count from 1977 as the strongest alignment point involved several considerations. Counting conditions were considered to be as good as for any previous year; intuitively it seemed very unlikely there were more deer in 1977 than in the mid 1960s (simulated population, Figure 2); and all but one of the simulated points in Figure 1 are within the 90% confidence limits of the estimate, while all of the simulated points after 1973 are outside the confidence limits in Figure 2.

The process of developing a population simulation demands close scrutiny of available data and consideration of the principles of population dynamics. The acceptance of this concept may earn population modeling an influential position in the management of large ungulates. The actual impact of the population simulation model on management decisions may depend more on the uniqueness of the population and its habitat characteristics than on the quality of the base available as will be demonstrated in the following examples.

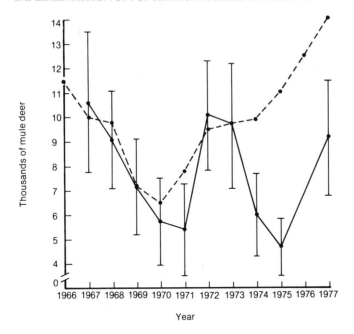

Figure 2 Comparison of observed (solid line) and simulated (dashed line) population size of the mule deer population shown in Fig. 1 using 1973 observed value as the strong alignment point. Ninety percent confidence limits around observed values.

Utility of Simulation for Management

Examples of Impact on Mule Deer Management

In the mule deer population represented by the nearly complete data base, the effect of the population model on management was minimal. Attempts to regulate the harvest have largely been unsuccessful because "weather, deer distribution, hunter participation, and hunter success may all be too variable to allow predetermined harvest objectives to be achieved within desired limits" (Freddy 1975). With certain weather conditions (heavy, early snow) causing deer concentration, hunters have historically been able to harvest on the order of 30 to 40% of the population. These conditions seem to occur 2 or 3 years in every 10 years. Assuming that cover and inaccessability of portions of the range allow a sufficient number of breeding bucks to survive, then the only critical management would be to set a maximum number on antlerless harvest to protect the herd during those years of high vulnerability.

In another deer population with a much less extensive data base, the impact of simulation was more profound. This population is located near metropolitan areas where hunting pressure could be excessive. Harvest was regulated in the past by issuing a limited number of either-sex permits and an unlimited number of bucks-only permits. The observed postseason fawn-doe ratio has been very low in recent years, ranging from 35 to 45 fawns per 100 does. The simulated over-winter natural mortality rate of 13 to 15% resulted in a postseason population size in the neighborhood of 7000 animals in 1975, which fit the hypothesized density, according to the managers. Based on this simulated population, with natural mortality held constant and the postseason fawn-doe ratio set at 45-100, the population was projected through 1980 (Table 2). The number of yearling and adult females remained stable without any harvest, which means the number of females dying of natural causes nearly equals the number of yearling females added to the population each year. If so, any harvest of antlerless animals would cause a decline in the producing segment of the population (Figure 3). However, if a buck-doe ratio near 20-100 is satisfactory, then, as described in Chapter 23, the number of antlered animals harvested can be a large portion of the number of yearling males added to the population each year. The projection in Table 2 carries the reported 1976 harvest through 1980, showing the effects on the simulated population.

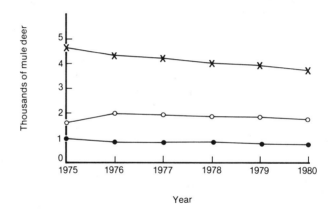

Figure 3   Effect of a limited either-sex harvest of 653 antlered and 200 antlerless animals on a simulated mule deer population. Numbers of does ($\times$), fawns ($\bigcirc$), and bucks ($\bullet$) are shown.

Owing to the apparent lack of response of this herd to conservative seasons since 1970, there was considerable pressure by some managers to close the hunting season in 1976 and "let the herd come back." Through the simulation, it was demonstrated that, as discussed in Chapter 23, this strategy would only build the antlered side of the population, since the female segment was stabilized by natural losses, not hunting (Figure 4). Unless the objective was to increase the buck-doe ratio, there seemed to be no need to close the season. The resulting decision was to issue an unlimited number of bucks-only permits in 1976 which resulted in a reported yield of 653 bucks and 7071 recreation days (Colorado Division of Wildlife, 1976).

Table 2   Projected Population Values with a Bucks-Only Harvest of a Mule Deer Population in Colorado[a]

| Year | Simulated Postseason Values | | | | Reported Values[b] | | |
	Yearling and Adult Bucks	Yearling and Adult Does	Fawns	Bucks-100 Does	Harvest Bucks	Harvest Does	Fawns-100 Does
1975	965	4609	1603	21	612	0	35
1976	792	4506	2106	18	653	0	45
1977	821	4542	2043	18	653	0	45
1978	859	4556	2058	19	653	0	45
1979	901	4557	2067	20	653	0	45
1980	942	4555	2072	21	653	0	45

[a]For game management unit D-10.
[b]1976 values are projected through 1980.

The simulation model was an influencing factor in the preceding decision. For decisions involving relatively coarse resolution, such as deciding whether to close the season or to have a bucks-only hunt, simulation techniques are adequate to suggest a reasonable alternative.

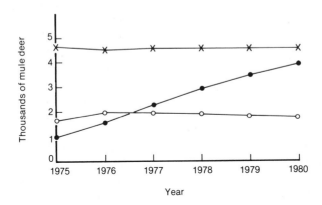

Figure 4   Effect of a closed hunting season on the same simulated mule deer population present in Fig. 3. Numbers of does ($\times$), fawns ($\bigcirc$), and bucks ($\bullet$) are shown.

Since this herd is located near metropolitan areas, there is considerable demand for management to provide maximum recreation. More specific management and adequate monitoring of this population is essential if prudent herd utilization is to be accomplished in the future. An experimental deer inventory research project was initiated in 1976 with the objective "to design a sampling scheme based on the means, variances, and deer density strata . . . capable of estimating mean deer numbers $\pm 15\%$ at $\alpha = 0.05$" (Anderson 1977). In addition, buck-doe-fawn ratios will be sampled using the statistical and sampling procedures described by Bowden and Anderson (1976). If that level of precision in estimating density is accomplished and regular estimates are made of population numbers and buck-doe-fawn ratios, then more specific management of this herd will be feasible.

Examples of Impact on Pronghorn Antelope Management

Pronghorn antelope are quite vulnerable to hunting because they occupy open, unforested areas. Harvest is controlled in Colorado by issuing a limited number of permits specific to age and sex depending on the area. Under any of the specific regulations the manager must determine how many permits should be issued to accomplish the herd objectives. Since the harvest can be more strictly controlled on this species, the opportunity presents itself for more specific management. Data for these populations include basic species life history information taken from the literature, trend counts (counts used as indices for determining changes over time), and estimates of herd structure. Attempts are made to cover 100% of the area in conducting trend counts in early spring, using either fixed wing or helicopter aircraft. Herd structure is estimated through flights made in late summer following recommendations by Bear (1969). The major shortcomings of the trend counts are that, many times, portions of the total area do not get counted, and of course, the ratio of observed animals to the total population is unknown. Counts are adjusted upward, using Table 1 in Caughley (1974) as a rough guide, combined with pilot and observer judgment concerning the counting conditions. Normally the count is assumed to represent 75 to 90% of the total.

The major problems associated with attempting to collect aerial trend and herd-structure data, both on an annual basis, are shortages of time and funding. Since the optimal time for collecting herd structure data and trend data occur at different seasons, total aerial coverage twice a year is necessary. Logistically, this presents the serious problems of unpredictable flying conditions and limited flying time; this has been the cause on numerous occasions of incomplete coverage of herd units. In an effort to overcome this, personnel in the Southeast Region have recently begun to conduct aerial trend counts on half of the herd units and herd-structure counts on the other half one year, alternating the procedures the next year. Unless there is strong empirical evidence to the contrary, the simulated values will be used for years in which field data are not collected. If this is successful, great savings in time and money will be realized.

In 12 of the simulated pronghorn populations, the simulations are used extensively in establishing and evaluating proposed management strategies. In one example, simulation was used to communicate through all levels of management the need to change from a limited either-sex harvest to a harvest where the number of bucks to be harvested can be specified separately from the number of "does" (includes all fawns). The management objectives were to (1) maintain the posthunt population between 2000 and 2200 animals; (2) reduce the buck-doe ratio from mid-40s to the mid-20s; and (3) if possible, maintain a harvest of 230 animals annually. Simulation was used to evaluate various management schemes in an attempt to meet all three objectives (Table 3). Continuing the either-sex harvest at the 1975 level resulted in a steady decline in population size and numbers of mature females (the producing segment of the population), while the buck-doe ratio remained in the mid-to-upper forties (Table 4). The bucks-only harvest (Table 5) allowed the population to grow too large and the buck-doe ratio to get too low over the five-year projection period. The number of permits issued for 1976 was intended to harvest 200 bucks and 25 does. The actual reported harvest was 206 bucks and 28 does. The effect of continuing this harvest rate through the projection period meets all of the original objectives (Table 6). This harvest strategy will be followed until field data indicate the population is not responding in the predicted fashion or until the objectives are changed.

Table 3   Historic Attributes of a Simulated Pronghorn Population

| | Harvests[a] | | | Results[b] | | |
Year	Antlered	Antlerless	Total	Population Size	Bucks-100 Does	Adult Females
1970	252	115	367	3141	45	1154
1971	170	75	245	3201	46	1181
1972	156	73	229	2752	49	1183
1973	198	78	276	2399	43	1103
1974	171	72	243	2326	43	937
1975	152	79	231	2204	46	880

[a]Actual reported harvest.
[b]Simulated values.

Table 4   Pronghorn Population Projection with Either-Sex Harvest

| | Harvest | | | Results | | |
Year	Antlered	Antlerless	Total	Population Size	Bucks-100 Does	Adults Females
1975	152	79	231	2204	46	881
1976	156	73	229	2184	46	881
1977	156	73	229	2171	46	863
1978	156	73	229	2118	47	836
1979	156	73	229	2068	47	814
1980	156	73	229	2038	47	806

Table 5   Pronghorn Population Projection with Bucks-Only Harvest

	Harvest			Results		
Year	Antlered	Antlerless	Total	Population Size	Bucks-100 Does	Adult Females
1975	152	79	231	2204	46	880
1976	230	0	230	2182	37	935
1977	230	0	230	2200	29	974
1978	230	0	230	2194	24	995
1979	230	0	230	2203	20	1024
1980	230	0	230	2245	17	1073

Table 6   Pronghorn Population Projection with Specified Harvest

	Harvest			Results		
Year	Antlered	Antlerless	Total	Population Size	Bucks-100 Does	Adult Females
1975	152	79	231	2204	46	880
1976	206	28	234	2178	40	914
1977	206	28	234	2180	34	931
1978	206	28	234	2154	31	934
1979	206	28	234	2139	28	944
1980	206	28	234	2154	25	971

The most common use of simulated pronghorn populations is to assist in setting the number of permits to be issued. However, they have also been used to evaluate effects of a trapping-and-transplant operation. The original management plan was to trap animals from a particular population, without curtailing the hunter harvest, for four to five consecutive years to maximize returns on labor invested in building the trap. After seeing the predicted effect of five years of trapping on the simulated population, the plan was altered to include only two years of trapping and transplanting (Figure 5).

Examples of Impact on Elk Management

One final example involves an elk population (Table 7). The original simulated set of parameters was designed to fit the counts presented in Figure 6 and matches that trend reasonably well through 1972. The observed trend reflected the conservative harvests of 1971 and 1972, but increased by 20% in 1973, concurrent with a 70% increase in harvest. The simulated population could not sustain the reported 1976 harvest. Conditions for counting after 1973 were poor, so any opportunity for determining the response of the real-world population to the relatively heavy harvests of 1973 to 1976 was lost. Field personnel were unable to

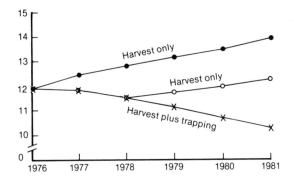

Figure 5 Effect of trapping and harvest on the Haswell pronghorn population (see text).

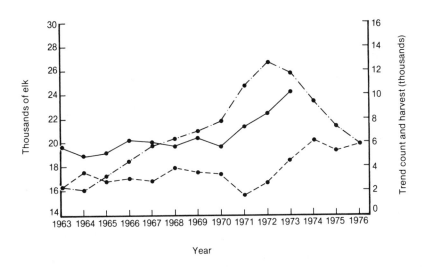

Figure 6 Simulated number of animals (dash-dot line), trend counts (solid line), and harvest (dashed line) of an elk population.

detect any decline in the population. Other indirect indices such as hunter success or effort per kill gave no evidence of decreasing availability of animals. In view of this somewhat meager evidence, the hypothesis that the simulated population was too small and that, in fact, the real population could sustain the reported harvest was explored. A simulation with decreased natural mortality allowed the population to expand. This simulated population still declined when harvested at the 1976 level (Figure 7). This left the managers in a dilemma

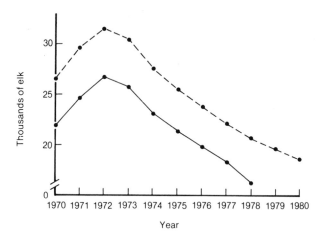

Figure 7 Comparison of test (dashed line) and current (solid line) simulated response to the continuation of a harvest of 3299 antlerless animals, as in 1976.

because they felt the original simulation matched the real population well, at least through 1972. Yet there was no real evidence that future harvest should be reduced to prevent the projected population decline. Consequently the harvest required to stabilize the population at the postseason 1976 level was calculated using a special subroutine in the model (the subroutine calculates the harvest, by sex and age, necessary to maintain a specified population size and structure. All other parameters in the simulation are left unchanged). The harvest so calculated for 1977 indicated that the antlerless harvest (which is totally controlled by permits) should be reduced by 30%. The antlered harvest, which can only be controlled on part of the area, would need to be decreased by only 7.6% to maintain the prescribed bull-cow ratio. The manager's decision was that the number of antlerless permits would be reduced by 19.4% and the number of antlered permits by 6.2% on the portion of the area where control is possible. The reported antlerless harvest for 1977, compared to 1976, was down by 17.4% and the antlered harvest up 6.5% (Table 8). The overall hunter success was 30%, which is higher than any value reported between 1971 and 1976.

In this case the model does not mimic the real-world population well enough to base management decisions on it totally. The managers chose to lean heavily toward the calculated harvest indicated for antlerless animals but did not accept it completely. It appears the population can sustain a somewhat higher harvest than the simulation indicates. The hypothesis that the simulated population should be larger has been tentatively rejected because of the relatively good agreement between observed and predicted values through 1972 and little empirical evidence of a declining population. A reduction in natural mortality in recent years due to mild winters may be the next best hypothesis. The utility of the simulation would be greatly enhanced with a good trend count, or better still, with an objective, quantitative estimate of population size.

Table 7   Historic Attributes of an Elk Population in Colorado (E-6)

Year	Postseason Population Size		Harvest		
	Simulated	Trend	Bulls	Cows and Calves	Total
1963	16,219	5,807	1783	422	2205
1964	16,056	4,818	2501	1001	3502
1965	17,181	5,253	1993	765	2758
1966	18,405	6,297	1873	1142	3015
1967	19,835	6,055	1737	1090	2827
1968	20,299	5,700	2184	1710	3894
1969	20,957	6,371	2276	1269	3545
1970	21,800	5,560	2543	903	3446
1971	24,718	7,430	1039	585	1624
1972	26,713	8,445	1800	850	2650
1973	25,834	10,149	3146	1365	4511
1974	23,194	NC[a]	3492	2643	6135
1975	21,409	NC	3481	1881	5362
1976	19,925	NC	3481	2299	5780

[a]NC = no count made.

Table 8   Comparison of 1976 reported harvest with prescribed, anticipated, and reported harvest of 1977 for elk population[a]

Harvest				
Antlered	% Change[b]	Antlerless	% Change[b]	
3481	0	2299	—	Reported 1976 harvest based on hunter survey
3216	−7.6	1566	−31.9	Prescribed by DESIRED DENSITY routine of ONEPOP
3481[c]	0.0	1853	−19.4	Anticipated, based on number of permits issued
3709	+6.5	1900	−17.4	Reported 1977 harvest based on hunter survey

[a]For game management unit E-6.
[b]Compared to reported 1976 harvest.
[c]No change from 1976 because of anticipated overall increase in hunting pressure.

DISCUSSION

Simulation is a means of developing a better understanding of complex systems. In the management of wild animal populations, decisions are made that are presumed to affect the system, that is, the population and its habitat. Wildlife managers are frequently forced to make these decisions based upon limited information from the real system plus their personal experience with the system. Learning from experience should be a process of reinforcement. In wildlife management detecting changes resulting from particular decisions is difficult. Changes in management are usually of a very conservative nature, so that the population response, if any, is well beyond detection with our current methodology. Therefore, the manager is left without reinforcement—either positive or negative—for his actions, and he is inhibited from learning from experience.

Simulation as a tool for learning provides instantaneous reinforcement. An inaccurate combination of population size, natality, and mortality in a simulation may result in extinction of the simulated population. A bucks-only harvest versus a closed season may allow the total simulated population to increase, but does it accomplish other of the manager's objectives? An infinite number of trials can be made with immediate results which enhances the understanding of the system and consequently the ability to wisely manage. Strauch (1976) states: "The very nature of simulation ensures increased understanding, for one cannot replicate the structure of a system, manipulate the parts of that structure, and judge its validity on the basis of real-world data without becoming increasingly aware of the complex interactions that occur within it."

New understanding of a system is a significant benefit of the simulation process. The end product of simulating a population is a hypothesis concerning its dynamics, and as Coleman (1972) states: "If a hypothesis is to be of value it must be tested." Testing of these hypotheses involves measuring the performance of the real population against the simulated population. Development of a simulated population has utility for management but the full potential of simulation will not be realized until practical performance measures (Gross, 1972) are developed and used.

Conservation agencies face a broad spectrum of impediments associated with management of wildlife populations. Identifying and collecting data that are relevant to decision-making and adequate to detect a meaningful change in the magnitude of a parameter is undoubtedly the greatest challenge. But, nonuse or the misuse of existing population information is the least excusable shortcoming of current management. The modeling of populations can synthesize relevant information into usable form for decision-making and management.

LITERATURE CITED

Allen, K. 1973. The computerized sperm whale model. Rep. Int. Whaling Comm. 23:70-74.

Allen, K., and G. P. Kirkwood. 1978. Simulation of southern hemisphere Sei whale stocks. Rep. Int. Whaling Comm. 28:151-157.

Anderson, A. E. 1977. Experimental deer inventory—Northeast Region. Colorado Division of Wildlife Federal Aid Project W-38-R. Game Research Report Part 2. Pp. 227-250.

Anderson, F. M., G. E. Connolly, A. M. Halter, and W. M. Longhurst. 1974. A computer simulation study of deer in Mendocino County, California. Tech. Bull. 130. Agric. Exp. Sta. Oregon State University, Corvallis. 72 pp.

Bartholow, J. M. 1977. Fort Niobrara Refuge: Big game management modeling. M.S. Thesis, Colorado State University, Fort Collins. 224 pp.

Bear, G. D. 1969. Evaluation of aerial antelope census technique. Colorado Division of Wildlife Game Information Leaflet No. 69. 3 pp.

Bowden, D. C., and A. E. Anderson. 1975. Evaluation of herd structure methodology. Colorado Division of Wildlife Federal Aid Project W-38-R. Game Research Report Part 2. Pp. 475-499.

Bowden, D. C., and A. E. Anderson. 1976. Evaluation of herd structure methodology. Colorado Division of Wildlife Federal Aid Project W-38-R. Game Research Report Part 2. Pp. 505-554.

Bunnell, F. L., and D. E. N. Tait. 1978. Bears in models and reality—Implications to management. In: C. J. Martinka (ed.). Bears—Their biology and management. Fourth International Conference of Bear Research and Management, Kallispell, Montana.

Caughley, G. 1974. Bias in aerial survey. J. Wildl. Manage. 38:921-933.

Clark, T. W. 1976. The black-footed ferret. Oryx 13:275-280.

Coleman, T. G. 1972. Simulation is helping biomedical research. Simulation Today No. 8. Pp. 29-32 in: Simulation, Vol. 19, No. 4.

Colorado Division of Wildlife. 1976. 1976 Colorado big game harvest. Department of Natural Resources, Division of Wildlife, Denver. 209 pp.

Connolly, G. E. 1978. Predator control and coyote populations: A review of simulation models. Pp. 327-345 in: M. Bekof (ed.). Coyotes Biology, Behavior, and Management. Academic Press, Inc., New York. 384 pp.

Connolly, G. E., and W. M. Longhurst. 1975. The effects of control on coyote populations: A simulation model. University of California, Davis, Division Agricultural Science, Bulletin No. 1872. 37 pp.

Connolly, G. E., W. M. Longhurst, F. M. Anderson, and A. N. Halter. 1971. A computer simulation model for evaluating deer hunting strategies. Calif.-Nev. Wildl. pp. 1-9.

Cooperrider, A. 1974. Computer simulation of the interaction of a deer population with northern forest vegetation. Ph.D. Thesis, State University of New York, College of Environmental Science and Forestry, Syracuse. 220 pp.

Craighead, J. J., J. R. Varney, and F. C. Craighead. 1974. A population analysis of the Yellowstone grizzly bears. Montana Forest and Conservation Experiment Station, School of Forestry, University of Montana, Missoula. Bulletin No. 40. 21 pp.

Digernes, T., H. Haagenrud, R. Langvatn, E. Reimers, T. Skogland, K. Bo, E. Gaare, A. Sorenssen, S. Skjeneberg, and J. Jenssen. 1973. A simulation model for cervide herds. A preliminary report (unpublished). 42 pp.

Dolbeer, R. A., C. R. Ingram, and J. L. Seubert. 1976. Modeling as a management tool for assessing the impact of blackbird control measures. Pp. 35-45 in: Proceedings of the 7th Vertebrate Pest Control Conference, University of California, Davis.

Drummond, D. C. 1976. Systems modelling: A tool for ecologists. Proc. N. Z. Ecol. Soc. 23:51-59.

Fowler, C. W., and W. J. Barmore. 1979. A population model of the Northern Yellowstone elk herd. Pp. 427-434 in: R. M. Linn (ed.). Proceedings of the 1st Conference on Scientific Research in the National Parks, Vol. 1. National Park Service Transactions and Proceedings Series, No. 5.

Freddy, D. 1975. Middle Park deer study—Experimental harvest regulations. Colorado Division of Wildlife Federal Aid Project W-38-R. Game Research Report Part 2. Pp. 209-240.

Freddy, D. 1976. Middle Park deer study—Experimental harvest regulations. Colorado Division of Wildlife Federal Aid Project W-38-R. Game Research Report Part 2. Pp. 259-282.

Garfinkel, D., J. McLeod, M. Pring, and D. DiToro. 1972. Application of computer simula-tion to research in the life sciences. Simulation Today No. 5. Pp. 17-20 *in*: Simulation, Vol. 19, No. 1.

Gill, R. B. 1970. Middle Park deer study—Population density and structure. Colorado Divi-sion of Wildlife Federal Aid project W-38-R. Game Research Report Part 3. Pp. 311-336.

Gross, Jack E. 1972. Criteria for big game planning: Performance measures vs. intuition. Trans. N. Am. Wildl. Nat. Resour. Conf. 37:246-259.

Gross, J. E. (no date). Management agencies, ivory towers, and models: Applied operations research in wildlife management (unpublished manuscript). 18 pp.

Gross, J. E., J. E. Roelle, and G. L. Williams. 1973. Program ONEPOP an information pro-cessor: A system modeling and communication project. Progress Report, Colorado Cooperative Wildlife Research Unit, Colorado State University, Fort Collins. 327 pp.

Guynn, D. C., Jr., W. A. Flick, and M. R. Reynolds. 1976. Mathematical modeling and wildlife management: A critical review. Proceedings of 30th Annual Conference, Southeastern Association of Fish and Wildlife Agencies. Pp. 569-574.

Gyllenberg, G. 1974. A simulation model for testing the dynamics of a grasshopper popula-tion. Ecology 55:645-650.

Innis, G. 1972. Simulation of ill-defined systems: Some problems and progress Simulation Today No. 9. Pp. 33-36 *in*: Simulation, Vol. 19, No. 6.

Karns, P. D. and W. J. Snow. 1979. A computer simulation model for studying the popula-tion dynamics of moose in northern Minnesota. Minnesota Department of Natural Resources, Grand Rapids (unpublished report). 24 pp.

Knowlton, F. F. 1972. Preliminary interpretation of coyote population mechanics with some management implications. J. Wildl. Manage. 36:369-382.

Krebs, C. J. 1978. A review of the Chitty hypothesis of population regulation. Can. J. Zool. 56:2463-2480.

Larkin, P. A. 1977. An epitaph for the concept of maximum sustained yield. Trans. Am. Fish. Soc. 106:1-10.

Levins, R. 1966. The strategy of model building in population biology. Am. Sci. 54:421-431.

Medin, D. E., and A. E. Anderson. 1979. Modeling the dynamics of a Colorado mule deer population. Wildl. Monogr. No. 68. 77 pp.

Reuterwall, C., and N. Ryman. 1978. A computerized model for prediction of moose popula-tion dynamics at different hunting policies. Department of Genetics, University of Stockholm, Sweden. Publication SNV PM 1022, 19 pp.

Riffe, J. E. 1970. Computer simulation of deer populations. M.S. Thesis, Pennsylvania State University, University Park. 98 pp.

Roelle, J. E. 1977. Refuge management modeling: The National Bison Range. Ph.D. Disser-tation, Colorado State University, Fort Collins. 311 pp.

Roper, L. A. 1973. Middle Park deer study—Experimental harvest regulations. Colorado Division of Wildlife Federal Aid Project W-38-R. Game Research Report Park 2. Pp. 147-162.

Sheriff, S. L., A. T. Cringan, and M. I. Dyer. (no date). A coyote population model for testing management strategies. Colorado State University, Ft. Collins (unpublished manuscript). 12 pp.

Strauch, P. G. 1976. Modeling in the social sciences: An approach to good theory and good policy. Simulation Today No. 39. Pp. 153-156 *in*: Simulation, Vol. 26, No. 1.

Stubbs, M. 1977. Density dependence in the life-cycles of animals and its importance in K-and r- strategies. J. Anim. Ecol. 46:677-688.

Taylor, M. W. 1978. A simulation model for ring-necked pheasants. Nebraska Game and Parks Commission, Lincoln. Technical Series No. 3, 55 pp.

Walls, M. L. 1974. A dynamic white-tailed deer population simulator and lessons from its use. M.S. Thesis, Virginia Polytechnic Institute and State University, Blacksburg. 167 pp.

Walters, C. J., R. Hilborn, and R. Peterman. 1975. Computer simulation of barren-ground caribou dynamics. Ecol. model. 1:303-315.

Walters, C. J., R. Hilborn, E. Oguss, R. M. Peterman, and J. M. Stander. 1974. Development of a simulation model of mallard duck populations. Can. Wildl. Serv. Occas. Pap. No. 20. 35 pp.

Williams, G. L. 1977. Simulation modeling of a big game at Wichita Mountains Wildlife Refuge. Ph.D. Dissertation, Colorado State University, Fort Collins. 242 pp.

Wynne-Edwards, V. C. 1965. Self-regulating systems in populations of animals. Science 147:1543-1548.

Zarnoch, S. J., R. G. Anthony, and G. L. Storm. 1974. Computer simulation of population dynamics of Red Foxes. Trans. Northeast Sect. Wildl. Soc. 31st NE fish and Wildl. Conf. Pp. 183-204.

THIRTEEN

Application of Population Models to Large Whales

K. RADWAY ALLEN

INTRODUCTION

Apart from concern over the future of whale stocks and fear that the risks of further destructive overexploitation have not been totally obviated, there are a number of reasons why the large whales form a particularly interesting subject for the study of population dynamics.

They are the largest group of mammalian species that have been subject to extensive and continuous exploitation. They consist of a variety of species, many of which are closely related and often similar in much of their essential biology. Many of them are worldwide in their distribution, and their populations are often very large. They have provided opportunities to compare the biological differences in their responses to exploitation, and to study the interactions of similar species occupying niches that overlap, particularly with respect to their food supply. The great differences in the histories of the exploitation of the various species also provide opportunities for comparative study.

Fortunately, the basic data regarding the exploitation of the majority of stocks are unusually complete, as compared with those relating to most exploited animals. Such series of data often cover long periods of time. For most of the principal fisheries we have good data back to about 1868 on the numbers of animals taken annually. Since about 1930 these data also cover the sex and size and often the reproductive condition of most of the animals taken.

Since whaling came under continuing international regulation with the establishment of the International Whaling Commission (IWC) in 1947, a number of methods of regulating whaling operations have been used, but the most important has been the fixing of catch limits for each season. Under the so-called "new management procedure," which the Commission put into effect in 1975 (IWC, 1977), these limits are determined by applying rigid formulas based on the assessed condition of the stocks. The need to provide the Commission with the information required under these rules has greatly stimulated attempts to carry out detailed assessments of each of the stocks, and this has led, in turn, to a rapid increase in attempts to build quantitative models of whale populations.

Models are used in two distinct ways in carrying out whale stock assessments. The first is in estimating current and original stock size. The second is in determining both the relative stock size at which maximum sustainable yield, in either numbers or weight, is obtainable, and the equilibrium yield at this and other stock levels.

The purpose of this chapter is not to describe the models now in use in any mathematical detail, but to deal with their essential structure, and to draw attention to some of the problems involved in their construction which may have interesting parallels in the study of other large mammals.

DEVELOPMENT OF MODELS

The development of whale population models has drawn on the techniques and concepts used in modeling both fish and mammal populations. Rather naturally, most of the early work was based on fairly simple techniques, and these had largely been developed in the study of fish populations. There were several reasons for this. First, these techniques could often be applied even where little biological data regarding the animals were available. Second, having been developed for fish, they were applicable to other marine animals which are much more difficult to observe, or even to count directly, than most terrestrial species. Third, fisheries models have generally been developed with the specific aim of assessing the yields that can be obtained from a population and determining the conditions under which these yields will be available. This, of course, has been the main reason for modeling whale populations, and will continue to be so as long as they are regarded as a resource. This objective does not apply to the same extent in the modeling of terrestrial mammals since relatively few of these are being managed as resources to be harvested on a sustained yield basis.

From these early stages, development of whale population models has been in the direction of increasing their complexity in an attempt to simulate more closely the internal dynamics and external relationships of the real populations. Thus, increasing incorporation of the characteristic features of the whales themselves has caused the more sophisticated models now being used to become progressively more like those used in the study of terrestrial mammals, and less like the classical fish populations models.

Further general discussion of the development and use of models in connection with whale stock assessments can be found in Chapman (1974a), Allen and Chapman (1977), and Allen (1980a).

The simplest models used for whale populations had only two components: the recruited population and the net recruitment added to it each year. Probably the first application of such a model to the management of whales was in the studies carried out by the Committee of Three Scientists, established by the IWC in 1961 to make an independent assessment of the Antarctic baleen whale stocks (Chapman, 1964). This committee used the technique developed by Schaefer (1954) to estimate the amount of recruitment to the Antarctic fin whale population in successive years, and hence, using assumptions as to the net recruitment rate, calculated the size of the recruited population. A further extension of this model, also due to Schaefer, assumes that the net recruitment rate is linearly related to population size. From this it follows that the curve of sustainable yield against population size is a symmetrical parabola, and the Committee of Three Scientists used this relationship to estimate the maximum sustainable yield of fin whales and the population level at which this occurred.

Perhaps the most obvious development from this simple model is to separate net recruitment into its components of gross recruitment, comprising the actual additions to the population, and the natural mortality that constitutes the removals. Separation of the two components draws attention to the fact that while the natural mortality is a direct function of the current population, the recruitment is derived from a parent population existing at some earlier time. If the amount of recruitment is considered as a function of this parent population, it is necessary to allow for this lag effect in modeling the population. These lag effects may be very significant in populations undergoing rapid depletion, because the recruitment may remain at a high level, since it is derived from an earlier larger population, while the amount of natural mortality decreases with the current population size. To make provision for lag effects in even simple models it becomes necessary to know the ages at which animals are recruited to both the breeding and the exploited components of the population. Where these ages are different, the breeding and exploited components of the population are of different numerical size and have to be separated in the model.

Models that separated the processes of recruitment and natural mortality, and also allowed for the lag effect in recruitment, were used by the Committee of Three Scientists in its early study of southern hemisphere baleen whales in addition to the simpler models referred to above. This separation has been maintained in most subsequent models.

Distinction between the processes of recruitment and natural mortality, together with the use of the models in the estimation of the size of the exploited or recruited component of the population, should lead logically to a distinction within the models between the exploited and the breeding components. Only if the ages at recruitment and at first parturition for the females and at sexual maturity for the males are the same is this distinction unnecessary. In the earlier work, attention was concentrated largely on fin and blue whales in which the difference between ages at maturity and at recruitment is small, and it was felt unnecessary to distinguish between the population components in this respect. Later, however, when sei whales were heavily exploited and assessment of their stocks was necessary, it became essential to develop models that could provide for the relatively large difference in the ages for this species (Doi and Ohsumi, 1970); these have been used in most recent studies.

The process of recruitment itself has also been separated in many models into two components, the number of young born and the number subsequently surviving to enter the exploited or mature population.

The management procedure instituted by the IWC in 1975 bases the regulations for each stock on the current size of the stock relative to the maximum sustainable yield (MSY) level and fixes catch limits as a proportion of the MSY. This has made it necessary to develop models that can be used to provide estimates both of MSY population levels and of the yield at various population sizes. The concept of sustainable yield implies that a population must be capable of remaining in equilibrium when the appropriate catch is regularly taken at a given population level. Models of such populations must contain functions that cause the net recruitment rate to increase as the population decreases. Thus, factors affecting recruitment directly, for example, pregnancy rate, must be negatively density dependent; or those operating in the opposite direction, for example, natural mortality rate or female age at maturity, must be positively density dependent. Any number of functions may be made density dependent in an inherently stable model.

In the Schaefer model referred to above, net recruitment is density dependent; but more recently a variety of different density-dependent relationships operating on different processes have been proposed for whale models. These have included arbitrary curves (Doi et al., 1970; Ohsumi, 1972) or mathematical functions (Allen, 1976, see also Chapter 23).

The earliest population studies concentrated on baleen whales largely because the main emphasis in the industry at that time was on these animals. Baleen whales have a relatively simple social structure, the herds consisting of approximately equal numbers of both sexes, and there is little difference in the rate of harvesting of the two sexes. The rise in the importance of the sperm whale fisheries in the late 1960s made more detailed assessments of populations necessary for management, and about 1968 more sophisticated models began to be developed (IWC, 1969; Ohsumi and Fukuda, 1972). In this species the sexes must be treated separately, both because there is a harem structure and because the great difference in the size of the sexes leads to quite different harvesting rates. In many sperm whale fisheries exploitation has been concentrated almost entirely on males. Simple yield-curve models therefore could not be applied, and models had to be developed that simulated the social structure and provided for separate treatment of the sexes (Allen and Kirkwood, 1977a).

In the last few years more complex models have, in turn, been developed for baleen whales. These treat the sexes separately although those now in use assume that the production of young is a function only of the female population. The most advanced baleen whale model yet in use (1978) is that evolved at the IWC Scientific Committee meeting in 1977 (IWC, 1978).

Still more recently a beginning has been made in the development of multispecies models, which provide for competition for food between species of baleen whales. The present models are, however, still essentially single-species models, in the sense that only one species is treated as dependent and the others are regarded as external competitors with independent population levels (IWC, 1978).

CURRENT BALEEN WHALE MODELS

The whale population models that have, up to the present, been applied in practice are of the block or compartment type. In these the population is divided into a series of groups (e.g., juveniles, mature females, recruited males), and functions are incorporated in the model determining the number of individuals transferred annually between pairs of compartments, or between compartments and the external world. In the existing models the numbers of compartments incorporated are quite small, but further development on the same lines leads to matrix or cohort models in which each age group represents a separate compartment. Only a few preliminary experiments with matrix models have been carried out (Allen and Kirkwood, 1977b; Smith, 1977). This is partly because we still have very little good data for whales on the age specificity of such parameters as mortality and pregnancy rates. The existing data, however, suggest that the variations with age are not great in relation to the uncertainties of the absolute values or to the likely range of density-dependent effects.

The essential features of the existing baleen whale models in their present form (Figure 1) are that the sexes are treated separately and that within each sex three components are generally recognized. These are (1) immature and unexploited; (2) *either* immature and exploited, *or* mature and unexploited; and (3) mature and exploited.

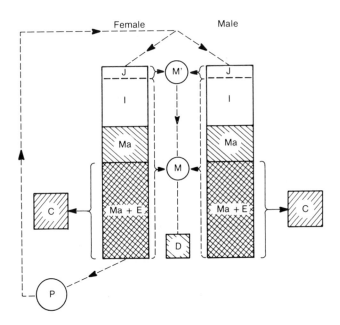

Figure 1 Schematic structure of current baleen whale models. *J*, juveniles subject to juvenile mortality rate; *I*, immature; *Ma*, mature; *E*, exploitable; *C* catch; *D*, natural deaths; *M'* and *M*, natural mortality in juveniles and older animals, respectively; *P*, pregnancy.

In some models the immature and unexploited group is further subdivided into a juvenile group, with a separate natural mortality rate, and an immature group with the same mortality rate as the older animals.

In models currently used, as distinct from cohort models, recruitment and maturity are usually taken as occurring at a fixed age, so that for all animals of a given sex in one population, maturity occurs before, after, or at the same time as recruitment. Thus, for each sex in a particular model only one of the immature exploited or mature unexploited groups will occur. These groups are therefore alternatives and which of them occurs depends on the characteristics of the species concerned. Thus, fin whales are recruited to the fishery before they become mature, whereas the opposite commonly applies to sei whales.

In these models the number of animals in each compartment is changed annually using the basic equations:

$$N_{2,j} = (N_{2,j-1} - C_{2,j-1})(1 - M) + R_{1,j} - R_{2,j}$$

or

$$N_{2,j} = (N_{2,j-1} - C_{2,j-1})e^{-M} + R_{1,j} - R_{2,j}$$

In this equation, $N_{2,j-1}$ and $N_{2,j}$ are the numbers of animals in compartment 2 at the beginning of years $j - 1$ and j; $C_{2,j-1}$ is the catch from compartment 2 in year $j - 1$; $R_{1,j}$ and $R_{2,j}$ are the numbers of animals transferred at the beginning of year j from compartments 1 and 2 to compartments 2 and 3, respectively. The catches are usually known and if necessary can be allocated between the compartments according to an appropriate formula. Recruitment into the youngest compartment, that is, the beginning of the first year class, is based on the pregnancy rate and the number of animals in the parent female mature stock in the appropriate year. In some models the juvenile or immature unexploited compartment is omitted, and recruitment is then directly into the second compartment. In this case recruitment has to be calculated from the pregnancy rate and mature female population in the appropriate year, adjusted according to the age at recruitment and the natural mortality rate for the young age groups.

Recruitment into older compartments is usually calculated as the number of animals entering the next younger compartment in the appropriate earlier year, modified by the mortality that takes place during passage through the compartment. The number leaving the compartment is then the number entering multiplied by $(1 - M)^t$, or e^{-tM}, where t is the number of years in the lower compartment with an adjustment if necessary for any catch.

In such models the key parameters are adult natural mortality rate M, juvenile natural mortality rate M', pregnancy rate p, age at recruitment to exploited stock t_r, and age at maturity t_m. All of these may be varied with time, either using observed time series or by making them density dependent.

Age at recruitment is determined to a large extent by the factors that control the minimum size of animals normally taken. These factors include not only minimum size limits imposed by regulations, but also size selection by the industry for economic or other reasons as well as the location and season of operations. Such factors may lead to irregular changes in the age at recruitment (Allen, 1980b). It is possible also that age at recruitment may be affected by density-dependent changes in growth rate, although such effects have not yet been identified.

The natural mortality rate of adults at the unexploited stock level may be estimated from the age structure, although problems may arise if the unexploited stock has been changing in size. At other stock levels, natural mortality rate cannot be measured, since it is not possible to separate it from fishing mortality. It does not seem practicable therefore to determine whether the natural mortality rate is subject to density-dependent changes. The juvenile mortality rate is almost impossible to measure directly, since the unexploited animals to which it applies are not available for study. At the unexploited level however it can be calculated by means of the balance equation:

$$M' = 1 - \left(\frac{2M}{p\,(1 - M)^{(t_m - t_j)}} \right)^{\frac{1}{t_j}}$$

if M and M' are annual mortality rates, or

$$M' = \frac{1}{t_j} \left[\ln\!\left(\frac{P}{2(1 - e^{-M})} \right) - M(t_m - t_j) \right]$$

if M and M' are instantaneous rates, on the assumption that the population is in equilibrium, and recruitment is therefore equal to matural mortality. In these equations t_j is the age up to which the juvenile mortality rate operates. It is commonly set arbitrarily at 2 years.

Pregnancy rate and age at maturity have both been demonstrated to vary as a result of changes in the size of the population, not only of the primary species but also of competitors. Both parameters have changed for fin and sei whales in the Antarctic from about 1930 onward. This was while fin whales were being harvested, but before the exploitation of sei whales began (Gambell, 1976).

In most of the current models the density-dependent function is the modified logistic equation:

$$X(D) = \left[X(0) - X(D^*) \right] \left[1 - \left(\frac{D}{D^*} \right)^{n+1} \right] + X(D^*)$$

where $X(D)$ is the value of the density-dependent parameter X when the independent variable of the density-dependent effect has the value D; $X(0)$ and $X(D^*)$ are the values of X when D has the limiting values 0 and D^*; and n is an arbitrary exponent. If X is internally density dependent, D will be a measure of some aspect of the size (e.g., numbers or biomass) of the whole population, or some component of it. The function may also be used to represent dependence on the magnitude of other aspects of the ecosystem, for example, the total baleen whale biomass (Allen and Kirkwood, 1978). This form has been used because, as well as being convenient to manipulate, it has the characteristic that at low population, or levels of the variable D, the parameter tends to become independent of population change, which appears to be biologically plausible, if not general for large mammals (Chapters 12, 14, 20, 23).

In an internally density-dependent population in which the net rate of recruitment is determined by functions of this form applying to the annual rates of gross recruitment or natural mortality or both, it can be shown that the relative MSY population level, or the ratio of the population at maximum sustained yield to that at equilibrium (D'/D^*), is determined entirely by the value of n and is given by

$$\frac{D'}{D^*} = \frac{1}{(n+2)} \frac{1}{(n+1)}$$

It occurs at 50% of unexploited level if $n = 0$, at 60% for $n = 1.4$, and 70% for $n = 4.0$.

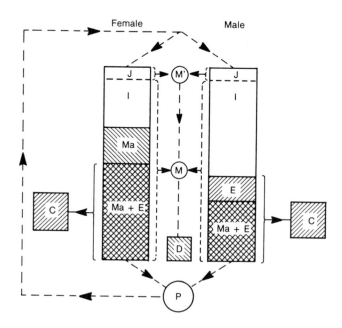

Figure 2 Schematic structure of current sperm whale models. J, juveniles subject to juvenile mortality rate; I, immature; Ma, mature; E, exploitable; C, catch; D, natural deaths; M' and M, natural mortality in juveniles and older animals, respectively; P, pregnancy.

CURRENT SPERM WHALE MODELS

The essential difference between the sperm whale model (Figure 2) and the baleen whale models is that pregnancy rate and therefore recruitment is influenced by both sexes in the former (Allen and Kirkwood, 1977; Allen and Chapman, 1977). Because of the harem structure, only socially mature males are involved in reproduction, and since their age at maturity—about 25 years—is much greater than the female age at maturity—10 years—the number of these animals in the total unexploited population is only about half that of the mature females. Since harems normally consist of about 10 to 15 breeding females to a single bull there is still a large "excess" of socially mature males in the population. Most of these apparently surplus animals are found at any one time in much higher latitudes, remote from the breeding herds, which live mainly in tropical and subtropical waters. The role of these surplus males is little known. In particular we have little information on the degree of interchange between them and the males with the harems, either during or between breeding seasons, and therefore on the extent to which they are essential to the reproduction process. The geographical separation suggests that little interchange is likely to occur during a single breeding season. The current model assumes that at least some of these males serve a useful purpose and incorporates a component of reserve males that must be present, in addition to the harem males, for reproduction to be fully efficient. If r is the ratio of reserve males to males with harems, and H is the average size of a harem, the ratio of mature males to mature females in the population should therefore be not less than $(1 + r)/H$. The model assumes that if the ratio falls below this value then pregnancy rate will decline linearly. In present IWC studies the reserve ratio is set at 0.3 and the mean harem size at 10. Since it is likely that the actual average ratio of females to mature males in the harem is closer to 15:1, the model provides approximately one reserve male for each male in a harem.

APPLICATION OF MODELS

Models are used for two distinct purposes in studies aimed at providing a basis for the management of whale stocks. The first is to obtain estimates of the past and present sizes of the populations. The second is to forecast the population size, relative to the unexploited level, at which the MSY can be taken and to calculate the MSY itself as a proportion of the unexploited or MSY population size.

Although both direct counts and mark-and-recapture methods have been used in obtaining estimates of whale populations, the most widely used methods have involved setting up a model of the population, applying the known catches, and determining the combination of original population size and catchability coefficient q, which will give the best fit to observed changes in the indices of abundance, particularly the catch-per-unit effort (Chapman, 1974a; Allen and

Chapman, 1977; Allen, 1980). The quality of the estimates obtained is, however, very dependent not only on the reliability of the indices of abundance, but also on the structure of the model and the values and density-dependent ranges of the vital parameters used in it. The problems of obtaining measures of catch and effort that will provide a good index of the changes in the real abundance of a whale population have been causing increasing concern to the Scientific Committee of the IWC in the last few years (IWC, 1980).

When models are used in this way to obtain estimates by simulation of past events, it is not necessary that they should include any density-dependent effects. Appropriate constant values for the various vital parameters may be used, or if the period of exploitation is relatively short it may even be assumed that the absolute amount of recruitment remains unchanged (Chapman, 1974b). If parameters are believed to have changed during the period of the simulation, either observed time series of values or density-dependent relationships can be used. In a computerized model, it is possible to include both of these procedures as alternatives (Allen and Kirkwood, 1978).

However, when models are used to estimate the MSY population level relative to the unexploited level, they must incorporate density-dependent functions. The nature of these functions and their associated parameter values determine the relative MSY level and the MSY as a proportion of population size. Thus, models that are satisfactory as a basis for estimating population may be of no use in determining relative MSY level, and therefore in classifying stocks under the IWC procedure. It is important, however, that where models are being used in estimating the size of a population and then in classifying it, they should be, if not identical, at least compatible with respect to the rates of recruitment and mortality that they produce. There is an almost endless variety of mathematical functions that can be used to define the density-dependent relationships in these models, but historically we have had no good data on their real shape. To obtain this we would need to determine parameter values quite accurately at several different population levels. There seems to be little hope that we will be able to obtain data of this kind directly from the populations for many years. Thus, the relative MSY levels currently used in the models are to a great extent arbitrary, as is the current use of the modified logistic curve. In the earlier modeling 50% of the unexploited stock size was commonly used, but since the introduction of the "new management procedure" most models used by the IWC have been based on a 60% MSY level. Recent work has, however, suggested that even higher MSY levels may be biologically realistic (Chapters 10, 14, 23). Considered as a basis for management the higher MSY appear more conservative in the sense that they impose protection on the stocks to a higher population level. However, they also imply, for a given range over which net recruitment can vary with population size, a higher MSY (see Chapter 23). The higher MSY levels are therefore also less conservative in the sense that they permit a higher catch to be taken if this is being calculated as a proportion of MSY, as under the present procedure.

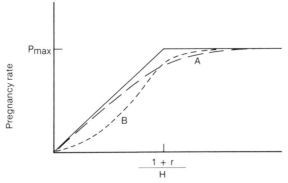

Figure 3   Relationship between pregnancy rate and ratio of male to female in mature populations in the existing model using a discontinuous function (solid line) and in models using continuous functions (broken lines), curve A being a monotonic curve and B a sigmoid curve.

In sperm whales the dependence of the pregnancy rate on both the mature male and mature female populations creates a special problem in the determination of sustainable yields. The relationship between pregnancy rate and ratio of male to female mature populations in the existing model is shown by the solid line in Figure 3, and the maximum sustainable yield of males is obtained when the ratio is equal to $(1 + r)/H$. In such models the MSY level and yield itself are much more sensitive to the values of r and H than to the other vital parameters.

Unfortunately we do not have as yet any good evidence either as to whether the actual relationship is appropriately described by a function of this kind or as to the necessary proportion of reserve males. A discontinuous function of the kind now used seems to be an unlikely description of a biological relationship. It could, however, be easily replaced by any one of a number of continuous functions such as those shown by broken lines in Figure 3. The MSY level can be calculated for any specific form of the function. A curvilinear function would however contain additional parameters which would be still more difficult to estimate. We may also note that a sigmoid curve, such as B, would result in a population model much more sensitive to catches exceeding the sustainable yield level than a monotonic curve like curve A.

Some functions have been proposed based on attempts to model the behavior of the breeding animals (Beddington and May, 1980; Botkin et al., 1980) but these have not yet been found to simulate observed events more successfully than the model described above (IWC, 1979).

COMPARISON WITH OTHER LARGE MAMMALS

From the preceding discussion, it appears that one of the most important problems involved in placing management of sperm whales as an exploitable resource on a sound basis is that of understanding the influence of the relative size of the mature male population on reproduction (see Chapters 4, 6, 9, and 23 for discussions of this in other species). Basically there are probably at least two ways by which this can be approached.

One is by comparative studies of the pregnancy rates in various exploited populations in which the sex ratio is at different levels. For example, there is already evidence that in the eastern Indian Ocean and in the western North Pacific (IWC, 1979) there have been reductions in the pregnancy rate, which may have resulted from a reduced ratio of socially mature males to mature females.

The other approach could be based on studies of the actual behavior of the animals, particular of the socially mature males. If techniques can be developed by which the movements of individual animals can be followed (e.g., by radio tags), we may be able to gain understanding of the nature of their migrations and of the part that the apparently surplus males in the higher latitudes play in the actual breeding process.

Comparison with large terrestrial mammals in which breeding behavior is more easily observed may help to throw light on these problems. Many of these have harem structures, so that there are large numbers of sexually mature males that do not take an active part in the breeding process at any one time. Frequent struggles for mastery between the males holding the harems and the unemployed animals are a common feature of such social structures (see Chapters 4 and 6). In sperm whales, however, it seems that a large proportion of the nonbreeding males are geographically so remote from the breeding herds that their attempts to displace the harem masters must be much less frequent. It would be interesting to know whether there are large terrestrial polygynous mammals in which the geographical separation between breeding herds and unemployed males is as great as it is is sperm whales.

Another problem of great importance in the understanding of whale population dynamics, and on which light might be thrown by comparison with more easily observable terrestrial species, concerns the early mortality. Recent evidence seems to suggest that in whales the rate of this mortality is sometimes not much greater than the adult mortality rate. This result may appear surprising at first sight, since it is commonly assumed that young animals of most kinds are more susceptible to injury, predation, and accidental causes of death and should therefore be expected to have a higher mortality rate. It has, however, been suggested that among whales, and perhaps particularly sperm whales, a highly developed social structure leads to a level of maternal care that effectively protects the young animals to the extent that they may indeed have even greater chances of survival than older animals. It would be particularly interesting to know whether such a situation exists in large terrestrial animals. If this is the case, it then becomes important to know whether or not the juvenile mortality rate is influenced by changes in population size, and if so in what direction.

LITERATURE CITED

Allen, K. R. 1976. A more flexible model for baleen whale populations. Rep. Int. Whaling Comm. 26:247-263. App. Report and papers of the Scientific Committee.

Allen, K. R. 1980a. Conservation and Management of Whales. University of Washington Press, Seattle.

Allen, K. R. 1980b. Size distribution of male sperm whales in pelagic catches. Rep. Int. Whaling Comm. Sperm Whales: Special issue: 51-56.

Allen, K. R., and D. G. Chapman. 1977. Whales. Pp. 335-358 *in*: J. Gulland (ed.). Fish Population Dynamics. John Wiley and Sons, New York.

Allen, K. R., and G. P. Kirkwood. 1977a. Further development of sperm whale population models. Rep. Int. Whaling Comm. 27:104-105.

Allen, K. R., and G. P. Kirkwood. 1977b. A sperm whale population model based on cohorts (SPCOH). Rep. Int. Whaling Comm. 27:268-271.

Allen, K. R., and G. P. Kirkwood. 1978. Simulation of southern hemisphere sei whale stocks. Rep. Int. Whaling Comm. 28:151-157.

Beddington,, J. R., and R. M. May. 1980. A possible model for the effect of adult sex ratio and density on the fecundity of sperm whales. Rep. Int. Whaling Comm. Sperm Whales: Special Issue: 75-76.

Botkin, D. B., D. S. Schimel, L. S. Wu, and W. S. Little. 1980. Some comments on density dependent factors in sperm whale populations. Rep. Int. Whaling Comm. Sperm Whales: Special Issue: 83-88.

Chapman, D. G. 1964. Special Committee of Three Scientists, final report. Rep. Int. Whaling Comm. 14:39-84.

Chapman, D. G. 1974a. Estimation of population parameters of Antarctic baleen whales. Pp. 336-351 *in*: W. E. Schevill (ed.). The Whale Problem; A Status Report. Harvard University Press, Cambridge, Massachusetts.

Chapman, D. G. 1974b. Estimation of population size and sustainable yield of sei whales in the Antarctic. Rep. Int. Whaling Comm. 24:82-90.

Doi, T., and S. Ohsumi. 1970. On the maximum sustainable yield of sei whales in the Antarctic. Rep. Int. Whaling Comm. 20:88-96.

Doi, T., S. Ohsumi, K. Nasu, and Y. Shimadzu. 1970. Advanced assessment of the fin whale stock in the Antarctic. Rep. Int. Whaling Comm. 20:60-87.

Gambell, R. 1976. A note on the changes observed in the pregnancy rate and age at sexual maturity of some baleen whales in the Antarctic. FAO, ACMRR/MM/SC37:1-20.

International Whaling Commission. 1969. Report of the IWC-FAO working group on sperm whale stock assessment. Appendix 5. Rep. Int. Whaling Comm. 19:60-62.

International Whaling Commission. 1977. Chairman's report of 27th meeting. Rep. Int. Whaling Comm. 27:6-12.

International Whaling Commission. 1978. Report of the special meeting on southern hemisphere sei whales. Rep. Int. Whaling Comm. 28:335-343.

International Whaling Commission. 1979. Report of the sub-committee on sperm whales. Rep. Int. Whaling Comm. 29:65-74.

International Whaling Commission. 1980. Report of the Scientific Committee to the 31st meeting of the Commission. Rep. Int. Whaling Comm. 30:42-59.

Ohsumi, S. 1972. Examination of the recruitment rate of the Antarctic fin whale stock by use of mathematical models. Rep. Int. Whaling Comm. 22:69-90.

Ohsumi S. and Y. Fukuda. 1972. A population model and its application to the sperm whale in the north Pacific. Rep. Int. Whaling Comm. 22:96-110.

Schaefer, M. B. 1954. Some aspects of the dynamics of populations important to the management of the commercial marine fisheries. Bull. Inter-Am. Trop. Tuna Comm. 1:26-56.

Smith, T. D. 1977. A matrix model of sperm whale populations. Rep. Int. Whaling Comm. 27:337-342.

FOURTEEN

Evaluation of Marine
Mammal Population Models

DOUGLAS G. CHAPMAN

INTRODUCTION

Models have been widely used to assist in the management of exploited wildlife populations since World War II, although some of the earlier attempts date back to the years before World War II. For example, the earliest fisheries models are due to Baranov (1918) and Thomson and Bell (1934), and the first consideration of marine mammal yield was by Hjort et al. (1933). Theoretical ideas were being developed at the same time by Lotka (1925), Voltera (1931), and Gause (1934).

Most of the early models were designed to provide estimates of maximum sustainable yield (MSY), that is, the maximum number or weight of animals that might be exploited each year while leaving the population unchanged. Such models assume that there is a population level that has this property regardless of the age and/or sex structure of the exploitation. These MSY models usually fall into one of two categories. The first of these will be referred to as structural and the second as empirical.

The empirical model requires estimates of the "yield" of the population at various levels of the population (as from historical data). To these data a model is fitted and a maximum yield level estimated. In the structural model various processes that go into population growth are considered. The yield is derived as a function of these processes. However, it is usually necessary to determine the responses of the elemental processes to exploitation before a maximum sustainable yield can be estimated. These responses may be determined empirically.

Before reviewing some of the marine mammal population dynamics models that have been published and used, it is useful to suggest a classification. First we note that there are single-species models, multispecies models, and ecosystem models. Most models deal with a single species. In some applications of these models, it is assumed that changes in population do not affect the environment, which is taken to be fixed. More sophisticated models may attempt to deal with two or more interacting species, particularly if both are or have been exploited. Finally, and ideally, one should consider models that treat the whole ecosystem as is required in the United States under the Marine Mammal Protection Act. The complexity of the ecosystem and the lack of information on the interrelationships and on the parameters of the multiplicity of elements in the system have limited efforts in this direction.

It is useful to distinguish among three other categories of models as developed within the framework of single species. These are growth models, management yield models, and perturbation models. To define these we need some notation. Let $N(t)$ be a measure of population size, either of the whole population or perhaps some component, expressed as a function of time t. Then a population growth model is simply an estimation of the function $N(t)$. In practice we may be dealing with a population that has been substantially reduced by exploitation. In this case we may be interested in the growth from the present exploited population level N_p to the maximum level N_m.

Management yield models are often used for determining the maximum rate of change of the population as a function of N. The first derivative $N'(t)$ is, of course, the increment of the population. It is assumed that if this increment is harvested by man, then $N(t)$ will be unchanged and that in fact this can be done on a sustainable or continuing basis. However, it is obvious that the population does not consist of a number of interchangeable elements but has age and sex structure. Both affect the growth function and are affected by any removal. In particular the level of N that maximizes $N'(t)$ will depend on the structure of the removed portion of the population as described in Chapters 20 and 23. For example, harvesting the zero age class (as in harp seals) is different from harvesting adults of both sexes (as in large baleen whales).

These issues are of importance in the management of marine mammals. As published in the U.S. Federal Register, a marine mammal population is at optimum level if its population equals or exceeds the level of maximum net productivity. If the growth model is of logistic form, then maximum net productivity occurs at the level of population which is 50% of the unexploited level. If the population were to be exploited, then it is customary to assume that the MSY level would be also at this 50% level, but this is not necessarily the case.

The level that provides maximum sustainable kill will depend both on the age and sex classes in the kill, but also on the factors that bring about the population's response to exploitation (as discussed in Chapter 23), and the number of males required for breeding. It is common to assume that there are two factors that respond to exploitation—immature mortality rate and pregnancy rate. Presumably pregnancy rate responds to the size of the adult stock, but it can also depend on the male-to-female ratio. The immature mortality rate may depend on the number of adults or the number of immatures or both. The importance of the sex ratio to the kill is obvious and has been widely discussed in the polygynous species such as the northern fur seal and the sperm whale.

Consideration may be given to the effect of perturbations on severely depleted populations, and specifically the probability of such populations becoming extinct. This issue has become a matter of interest with respect to the Alaskan bowhead whale population. One aspect of the problem has been studied by Beddington (1978), namely, the expected probability of extinction under a management policy that seeks to continue harvest at the MSY level in the presence of exogenous perturbations. This analysis shows that given such pertur-bations and no correction in the harvest policy, extinction is virtually assured. What is, perhaps, more difficult, is attempting to determine the probability of extinction of a depleted stock, even in the absence of any harvest. We do not know the effect of the perturbations in the environment on the vital parameters. Many who express concern for whales would like to know whether blue whales (for example) will become extinct after their severe overexploitation in most oceans. At this time it is probably not possible to answer this question in any scientific manner. Thus, while perturbation models have been studied rather extensively in theoretical ecology in recent years, their application to real populations, particularly ones as difficult to study as marine mammals, does not yet seem fruitful. Models of the effect of perturbations will not be considered further in this chapter.

Finally, population dynamics may refer to the fluctuations about some equilibrium. This may be the equilibrium known as the maximal or unexploited population level, or it may refer to the equilibrium sought by a manager who is attempting to maintain the population at the MSY level through harvest. The particular concern about fluctuations of the equilibrium may be the determina-tion of the probability of the population going to extinction or the nature of the equilibrium. It is known that if birth and death rates are constant and equal, as would be expected in the equilibrium situation, variations increase in time (Ken-dall, 1948). However, as oscillations increase, feedback may alter birth or death rates and hence dampen the fluctuations.

Before leaving the topic of population models in general, it is worth pointing out that there are many studies of marine mammals which are identified as including or relating to population dynamics, but which do not in fact include or make possible a population model in the sense used in this chapter. Many of these deal with current information on specific population parameters (e.g., pregnancy rate, mortality rate) which, of course, are vital information for struc-tural models, provided it is available over several population levels. Other population dynamics models may refer to variations of the population over time, for example, between times of the day or of the tide or between seasons. Such information not only is of importance for behavioral studies but also may be a key piece of information in standardizing or validating population counts. However such population dynamics studies are not directly applicable to long-term growth or yield models.

GROWTH MODELS

Because many of the species of marine mammals were severely overexploited during the nineteenth century and have subsequently had a respite from harvesting, we have a number of cases where there have been observations of the rebuilding of the population. Unfortunately, in many cases, observations of population levels have been sporadic and of doubtful accuracy so that no model can be fitted, or, if fitted, the resulting models are of questionable value. Another problem in developing such models is the patchy distribution of many of the marine mammals. This is well exmplified by the Alaskan sea otter and the northern elephant seal. These remnant populations were initially localized, perhaps on a single island. As this local population builds, one of two things may happen. Density-dependent mechanisms may come into play to reduce population growth, or part of the population may emigrate to another location. Still a third alternative is that both of these happen in part, and it is difficult to distinguish the relative roles.

Northern Fur Seal

The depletion of the northern fur seal population *(Callorhinus ursinus)* in the late nineteenth century is well known. After the Fur Seal Treaty of 1911 northern fur seals first received total protection followed by a managed harvest for about 50 years. Despite many extensive studies, it has never been possible to count all elements of this population, though complete counts were made of pups from 1912 to 1916 and again in 1922, with partial counts in 1917-1921 and 1924. These partial counts were extrapolated through the application of an exponential growth model to give an estimate of the total pup population. These counts and some estimates are shown in Table 1. No counts of pups were attempted after 1924 until after World War II. Thereafter other methods of estimating the size of this segment of the population were utilized. These are reviewed in Kenyon et al. (1954), and the more recent shearing and sampling methods are discussed in Chapman and Johnson (1968). Other elements that have been counted include adult males, both those holding harems and so-called "idle" males. The size of both of these components of the herd are, in large part, the reflection of management policy.

In the early part of the recovery of fur seal population a number of subadult males, called a "breeding reserve," were deliberately spared. The number so spared was gradually reduced in the succeeding years and the practice discontinued by 1934. The subadult male kill reflects this management procedure as well as the variations in year-to-year survival. After 1951 the harem bull count remained at its apparent maximum value for about a decade and then fell rapidly. This decrease may have been caused in part by a deliberate attempt to harvest a greater proportion of the subadult males through longer killing periods and/or a broader allowable size range within which animals could be killed. This practice was adopted at the same time that the reduction of females was carried out so that the effects of the two programs are confounded. A new period of growth of the number of harem males began about 1973.

While there have been a series of estimates of pups as well as of the age 3 male recruitment since the late 1940s, these data apply to the population as it has fluctuated around equilibrium values, partly due to natural causes and partly due to man-imposed factors. Since critical data on the growth of the herd from 1924 to 1947 is missing, it is difficult to determine whether the growth model is symmetrical, that is, of logistic form, or whether it is skewed.

Other Species

Many other pinniped populations were depleted during the eighteenth and nineteenth centuries. Since the time of commercial extinction some have received some protection, and in any case most that were not totally exterminated have begun to rebuild. Not only have populations rebuilt at residual locations, but many populations have spread to colonize areas that had become totally depopulated. Unfortunately, for many populations that were seriously depleted and have now rebuilt to maximal levels (or close to such levels, for example, the Alaskan sea otter or the walrus in the north Pacific), population data are too scanty or unreliable to be used in a growth model. In other cases, while unregulated harvesting has been terminated, there has been continuing harvest sufficient to prevent normal growth of the population. Often data (even on the harvest) are lacking or uncertain. These considerations again apply to the Pacific walrus.

Table 1   Estimates of Population Size for Various Species Over Time as Taken from the Literature Indicated[a]

Northern elephant seal (Ref. 1)

Species	1950	1957	1960	1965	Early 1970s 1972
Baja California	6600	12,650	14,400	13,295	23,250
California	100	600	600	775	7,910
Total	6700	13,250	15,000	14,070	31,430

San Miguel Island pups (Ref. 2)

1958	1964	1968	1972	1973	1975	1976
80	796	1624	2482	3088	3547	4014

Northern fur seals (Ref. 3) — Pribilof Islands (All figures in thousands)

1912	1913	1914	1915	1916	1917	1918	1919	1920	1921	1922	1924
82.0	92.9	93.2	103.5	117.0	128.0	142.9	157.2	167.5	176.7	188.9	208.4

Guadalupe fur seal (Ref. 4)

	1954	1955	1957	1964	1965	1966	1967	1968	1975	1976	1977
Summer	32	92					198	314	80	355	1073
Winter	14	72		252	285				254		
Spring			107	240	211	372		148			470

Southern fur seal (Ref. 5) — Bird Island, pups

1957	1958	1959	1960	1961	1962
5350	6800	8300	9400	9900	10,200

Sea otter — California (Ref. 6)

1938	1947	1950	1955	1957	1959	1963	1966	1969	1972	1973	1974	1975
310	530	660	800	880	1050	1190	1260	1390	1530	1720	1730	1760

Amchitka (Ref. 7, 8)

Ref.	1931	1936	1937	1939	1943
7	1000		1761	1870	3420
8		804	1261	1335	3417

California sea lion[a] (Ref. 9)

1927	1928	1930	1936	1938	1946	1947	1958	1961	1965	1969	1970
1450	1867	1411	3177	3882	7338	3660	12,619	18,363	17,169	17,451	18,047

Right whale (Ref. 10)

	1969	1970	1971	1972	1973	1974	1975	1976
South Africa, adults	43	55	65	96	69	98	108	148
South Africa, calves	17	13	19	26	21	31	31	23

Gray whale — Baja California counts (Ref. 11)

1952	1953	1954	1959	1960	1962	1964
827	912	1315	1509	1455	1193	1052

California, shore estimates (Ref. 12)

1952/1953	1954/1955	1957/1958	1959/1960
2919	3603	4417	6069

[a]Counts before 1946 were boat counts and pups were excluded; counts in 1946 and 1947 were by mixed methods; some records were lost in 1947; counts since then have been by aerial survey and have included pups.

[b](1) Leopold and Gogan (1976); (2) Delong and Johnson (in prep.) ; (3) Kenyon et al. (1954); (4) Fleischer (1978); (5) Bonner (1968); (6) Woodhouse et al. (1977); (7) Lensink (1960); (8) Kenyon (1969); (9) Ripley et al. (1962), Carlisle and Aplin (1971); (10) Best (1976, 1977), Anon. (1978); (11) Hubbs and Hubbs (1967); (12) Gilmore (1960), Rice (1961).

The northern elephant seal illustrates well the problems of determining growth models for a pinniped population that originally occupied a quite extensive range. Aside from the scanty data with various corrections necessary for censuses taken at different times of the year and for missing components of the population, we have a stock with one primary population source (Guadalupe Island). This parent population has dispersed first to neighboring areas in Baja California and then to the Channel Islands. In the California Channel Islands, San Miguel Island has been the primary population center, but it is not clear whether the newer colonies, for example, Ano Nuevo or the Farallon Islands, have been colonized primarily from the San Miguel stock or whether there is still input from the Guadalupe and/or other Baja California stocks.

Treated as a total stock (Table 1), the exponential growth is about 6% per year, though the California component has a much larger rate of increase. Part of the larger rate of increase in California undoubtedly is due to immigration. The interesting biological question involves the extent to which migration determines the difference. If the increased growth in the northern subpopulation were primarily self-generated, it would be reflected either in more pups per adult (i.e., higher birthrate) or a higher survival rate, most likely of juveniles and subadults. The little available data on pup-to-adult ratios (summarized in Leopold and Gogan 1976) do not support the former possibility. Any higher survival rate among young cannot be distinguished from immigration of the same component from the main population in Baja California.

The Guadalupe fur seal has recently been studied by Fleischer (1978) who presented a table of all available population estimates by season (see Table 2). The exponential growth rates using estimates made in the same season are as follows: summer 0.091 ($r^2 = 0.521$); winter 0.067 ($r^2 = 0.728$); spring 0.066 ($r^2 = 0.597$). The average of these values, 0.075, is that shown in Table 1.

Estimates of exponential growth for the California sea lion (Table 2) are similar to those for the Guadalupe fur seal and the total elephant seal population as discussed above.

The estimated sea otter growth rates are rather confusing. In California where the most observations have been made and where data are perhaps more reliable, the estimated growth rate is relatively low. Here the population is expanding northward and southward from the original center. There undoubtedly have been some reductions in population growth through human disturbance and illegal killing. In the Alaskan area, particularly on the Aleutian chain, population growth is more complicated; as one island becomes overpopulated, migrants must move to adjoining islands which, in some cases, involves long distances. The only island for which reasonable data are available is Amchitka. Two sets of estimates from the island (Lensink, 1960; Kenyon, 1969) suggest quite different growth rates, both of which are substantially larger than estimates from California. The Alaskan area supports higher densities of otters per square mile of habitat than California, and this may be reflected in higher growth. Data on otter density for California area given in Woodhouse et

Table 2 Estimated exponential growth rates of several rebuilding marine mammal populations

Species and Population	Time Period	Exponential Growth Rate	Number of Points	r^2
Mirounga angustirosis				
Northern elephant seal, Baja California	1950-1972	0.063	5	0.894
Northern elephant seal, California	1950-1972	0.179	5	0.911
Northern elephant seal (total)	1950-1972	0.071	4	0.998
Northern elephant seal pups, St. Miguel Island	1958-1976	0.203	7	0.921
Callorhinus ursinus, Pribilofs, pups	1912-1924	0.082	12	0.992
Arctocephalus townsendi, Guadalupe fur seal, Guadalupe Island	1954-1977	0.075	7	—
Arctocephalus tropicalis gazella, Bird Island, South Georgia	1957-1962	0.128	6	0.899
Enhydra l. tris L.				
Sea otter, California	1938-1975	0.045	13	0.980
Sea otter, Amchitka	1931-1943	0.122	4	0.878
Sea otter, Amchitka	1936-1943	0.192	4	0.951
Zalophus californianus, Sea lion, southern California	1927-1946	0.087	6	0.948
	1927-1938	0.088	5	0.872
Eubalaena australis				
Right whale, South Africa, adults	1967-1976	0.154	8	0.878
Right whale, South Africa, calves	1067-1976	0.092	8	0.561
Eschrichtus gibbosus				
Gray whale shore counts	1952/1953-1959/1960	0.104	7	0.999
Gray whale aerial censuses	1952-1959	0.080	4	0.753

al. (1977) as about 12 per square mile of habitat. Similar density data for Alaska are given by Kenyon (1969). Densities vary widely over subarea and over time, but some densities as high as 30 to 40 per square mile of habitat are shown there. It is to be noted that Woodhouse defines habitat as areas to 20 fathoms depth, while Kenyon considers all areas to 30 fathoms. If both areas were measured to the same depth, Alaska's densities would be much higher.

Growth rates comparable to those for the sea otter, based on Lensink's data, are also calculated from Bonner's data for the southern fur seal. One explanation for the rapid growth rate of this krill-eating seal is the decline of competition for food with the depletion of the large baleen whales (Laws, 1977). If this is indeed correct, it demonstrates both the importance of exogenous factors and the plasticity of the population parameters.

The estimated growth rates as shown in Table 2 range from 0.045 to 0.216 with a mean of 0.11. By omitting those populations thought to be experiencing immigration and determining a single figure for each population, the mean of such population estimates if 0.10. Visual inspection of Table 1 would indicate that California sea lions and southern fur seals have experienced exponential growth up to population levels greater than one-half that of equilibrium level. The data for California sea otters and San Miguel elephant seals show similar tendencies, although their interpretation is not as clear. Data from the remaining populations shown in Table 2 seem to show no clear indication of a decrease in growth rate which would be indicative of density-dependent changes. To more critically examine for such possibilities, yield models may be used as discussed in the next section.

The two whale stocks for which there are data appear to have similar or even somewhat higher exponential growth rates. The California gray whale has clearly leveled off since the period used in the computation, though it is difficult to do much with the data in the way of developing a model for the complete growth function. The shore counts at San Diego were distorted by boat traffic, certainly by the 1960s (Rice, 1965). There were gaps in the aerial counts between 1954 and 1959; certainly the aerial counts appear to be stationary between 1959 and 1964 which suggests that the exponential growth comes to a halt very rapidly and quite close to the carrying capacity of the environment. The growth rate of the right whale along the coast of south Africa is surprisingly high, although, as Best (1976) has stated, the increase in calves may be the more reasonable measure of growth. It is also possible that this population is growing by immigration as well as through natural increase.

YIELD MODELS

Seal Populations

The establishment of the North Pacific Fur Seal Commission in 1957 led to extensive studies on all populations of *Callorhinus ursinus* in the North Pacific

been managed more rationally, though much remains to be learned about the biological parameters of this population. Only recently have estimates of the number of pups born been made that are reasonably reliable (Shaughnessy, 1975). The total population is estimated on a rather crude basis involving extrapolation from the number of pups.

Many of the dynamic properties proposed for a model of the south African fur seal (Shaughnessy and Best, 1975) are based on data from the northern fur seal for lack of better values. One of the parameters assumed to change with population size is the pregnancy rate. This is taken to be a linear relationship using a *conjectural* upper limit for the Pribilof herd. The other parameter assumed to be density dependent is the pup mortality which again has been estimated from Pribilof data. Natural mortality after about eight months (the midpoint of the age of exploitation of pups) is assumed not to be density dependent.

An analysis of this model utilizing these assumptions indicates that the maximum sustainable yield is achieved when 35% of the pups are taken annually. The yield and population decline rapidly at higher rates.

In view of the uncertainty as to whether these parameter values (and particularly their changes in response to exploitation) apply to the South African fur seal, it is important to test the sensitivity of the model to changes in the values used. So far there has been no opportunity to test the model predictions in the actual operations of the harvest.

The use of models discussed above is subject to many problems, even though there are no basic disagreements regarding the general approach. The situation is different with respect to the Atlantic harp seal *(Pagophilus groenlandicus)*. There are three generally distinct stocks—an eastern herd that summers near Spitsbergen, a central group between Greenland and Spitsbergen, and a western stock that is exploited in the Gulf of St. Lawrence and off Labrador and Newfoundland. A great deal of effort has been given to the study of this western stock, a recent source of controversy. As a result several models have been developed to provide estimates of the maximum sustainable yield (see chapter 7).

It is generally agreed that the present population is below the level which would give a maximum sustainable yield, but the present population level and the sustainable yield are subject to disagreement. There are also disagreements on some of the basic biological parameters. One of the models was developed by Allen (1975) who used a Leslie matrix approach. With the parameter values he used, the dominant eigenvalue of the projection matrix is 1.09 (R. Allen, personal communication—the value given in the paper, 2.03, is incorrect). Several other models have been developed by Capstick and Ronald (1976), Lett and Benjaminsen (1977), and Winters (1978). A recent study has been carried out for the U.S. Marine Mammal Commission (Beddington and Williams, 1979).

One other seal model should be mentioned. Smith (1973) produced a model for the ringed seal *(Phoca hispida)* of the Canadian eastern Arctic. This is an important biological study which gives much information on the biological parameters of this population. Smith also uses a Leslie matrix model from which he concludes that "the estimated total take is 7.2% of the total estimated population size of 70,684. This is the best estimate for the maximum sustainable yield, and it obviously errs on the safe side as it is based on the assumption that the population growth is 0." No evidence is given that population growth is 0, but even if it were he would have only shown that 7.2% is the present sustainable yield. A sustainable yield can occur at any level of the population, and without further information it cannot be assumed to be the *maximum* sustainable yield.

In summary it may be said that while several seal populations have been extensively studied and models of varying degrees of sophistication have contributed insight to the dynamic processes involved, they have certainly not solved the problem of managing for MSY. Many questions have been left unresolved.

Cetacean Populations

The two cetacean stocks for which the most information is available (and which have been modeled) are the blue and fin whales of the southern oceans. For both stocks there are estimates of the original (preexploitation) population size and of the population through the period of exploitation. From these it is possible to compute estimates of yield using the method of Schaefer, (Schaefer, 1954) modified by taking into account the lag between birth and recruitment to the exploited population. The resulting relationships are exhibited in Figures 1 and 2 for blue and fin whales, respectively. The former is taken from Chapman et al. (1966), the latter from Allen (1972).

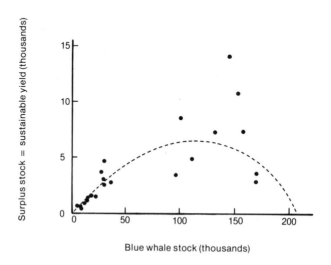

Figure 1   Surplus production model for Antarctic blue whales as from Chapman et al. (1966).

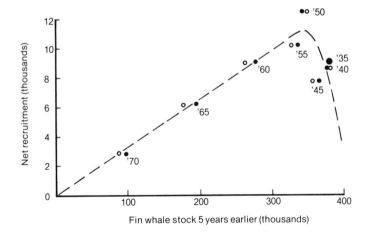

Figure 2   Surplus production model for Antarctic fin whales as from Allen (1972). (○) 4 year recruitment; (●) 5 year recruitment.

Neither of these figures suggests that the logistic is the appropriate model. The equation

$$Y(P) = aP^m(K - P)^n \tag{1}$$

may be fitted to the data sets. Here K is the unexploited population size, $Y(P) =$ yield, and P is the size of the exploited population. This is a generalization of a model considered by Allen (1972). The level of the population giving MSY is

$$\left(\frac{m}{m+n}\right)K$$

Using $K = 400$ for fin whales and $K = 210$ for blue whales, the level of each population yielding MSY (MSYL) may be calculated from equation (1). These along with values calculated by the nonparametric method of Chapman (1973) are shown below along with the ratio of MSYL to the unexploited population (in parentheses).

	Fin Whale	Blue Whale
Equation (1)	330 (0.82)	114 (0.54)
Nonparametric model	322 (0.80)	134 (0.64)

A different method of analysis of the fin whale stock has been given by Ohsumi (1972). He develops two empirical relationships between pregnancy rate and population level, and two relating age of maturity to population level using curves fitted by eye. He also considers two possible responses of mortality rate to population level. These responses are then used to calculate yield as a function of size of the population for the 2^3 or eight possible cases. These lead to estimates of the MSYL at 47 to 57% of the unexploited population level. This range is much lower than the values derived from the Schaefer-type analysis. The two different approaches need to be reexamined to determine the reasons for the substantially different estimates. However, this needs to be part of a broader reanalysis of the fin whale data, taking into account recent improvements in the method of determining population estimates. Furthermore, both approaches suffer from the defect of treating the whole population in the southern oceans as a single stock, whereas it is now generally accepted that there are several different stocks for each species with perhaps some low level of genetic mixing.

Porpoise Populations

Several populations of the eastern tropical Pacific have been subjected to a fairly intense exploitation through incidental catches by the Yellowfin tuna purse seine fishery since the late 1950s. The populations primarily impacted are the spotted dolphin *(Stenella attenuata)*, the spinner dolphin *(Stenella longirostris)* of which there are two races known as eastern and whitebelly, and the common dolphin *(Delphinus delphis)*, although several other species have been incidentally caught.

Population analyses of these stocks have been conducted (Anon., 1976, 1979). Because the kill of porpoises was "incidental" (i.e., the fishery was directed at the tuna), data from the early years of the exploitation are almost completely lacking. Since 1971 much biological and kill data have been collected, but these fail to provide the comparative data needed for a population yield model.

Even so, a model has been constructed with the objective of estimating the level of the current population relative to the unexploited. This involved using data from another unexploited *Stenella* population, that off the Pacific coast of Japan. Some other conclusions were reached by analogy with other marine mammal population dynamics models. Because of the lack of data it was not feasible to construct a model such as those cited above for the northern fur seal and the two southern whale stocks. Since the level of kill has been sharply reduced since 1973, the present level of data collecting should yield information relative to changes in the biological parameters. In view of the lack of data, models must be built by analogy with better studied species. Such models are useful first approximations, but it will be a long time before critical questions they have raised can be answered.

MULTISPECIES AND ECOSYSTEM MODELS

Changes in parameters of reproduction in whales were referred to previously. The review by Gambell (1973) called attention to the changes in such parameters with respect to the sei whale prior to its substantial exploitation. This led to consideration of the impact of reduction of the major whales (blue and fin) on other species of cetaceans and on the pinnipeds feeding in the southern oceans. Laws (1977) made crude estimates of the changes in krill consumption due to the reduction of whales. Unfortunately there are deficiencies in our present estimates of food consumption by many of the species for which krill is a major dietary source.

Estimates of the whale stock and of the changes over the past 50 years are comparatively good, though most estimates are of the exploited stocks. Any complete analysis involving changed levels of krill predation must include the younger age groups. These represent a much smaller proportion of the cetacean biomass but presumably have higher food consumption per unit of body weight. Information concerning present population levels of other marine mammals in the Antarctic is much less adequate, and past changes are totally conjectural. Little is known about the past abundance or levels of food intake for fish and invertebrate predators on krill.

A further complication at this level is illustrated by Laws' (1977) paper. It is not clear what area should be considered, partly because the krill distribution is not fully known. Laws refers to the area south of Antarctic convergence (approximately 50⁰ south), and he partitions part of the sei and sperm whale stocks on the basis of their distribution north and south of the convergence. However, all of the species are captured as well in the latitudinal zone from 40⁰ to 50⁰ south, though the proportion of food taken in this zone as contrasted to that taken further south is uncertain.

Thus, while the role of krill in the ecosystem is of major importance, the partitioning of the krill between its users now and in the past is very uncertain. Nor can it be predicted with any certainty what will happen if there is a direct harvest of krill by man. It seems certain that in the long run this would remove a share that would otherwise be taken by some of the present krill consumers; but which species would be primarily affected is unknown. Further discussion of these issues is contained in Chapter 2.

An interaction on a limited scale was studied by the Scientific Committee of the International Whaling Commission at its 29th meeting (Anon., 1978) in which a multiple regression of apparent sei whale pregnancy rate on crude estimates of the exploited stock sizes of blue, fin, and sei whales was produced. Two models were fitted to this multiple regression, a linear model and a nonlinear model using a power of the estimated sei whale population. In both cases the regression coefficient for the fin whales was nonsignificant, and this component was dropped from the model. Biologically this is puzzling since, in general, it is the blue whales that have been found much further south than sei whales. Thus, from location and feeding habits, interaction between fin and sei whales seems more reasonable than that between blue and fin whales.

It should be pointed out that the correlation between blue and fin whale population sizes is very high so that the model cannot distinguish well between these two causative variables, and we may have a statistical artifact. Also the data on sei whale pregnancy rates are suspect. These data were calculated from information reported to the Bureau of International Whaling Statistics from catches. A more careful examination of pregnancy data given by Masaki (1977) shows that (1) observed pregnancy rate in catch increases from north to south in the southern oceans (i.e., by zone); (2) the pregnancy rate decreases during the whaling season; (3) the observed pregnancy rate varies from area to area; and (4) the observed pregnancy rate may vary with the nationality of the catching operation due to diligence in searching for and reporting the fetus. Most of the early sei catches were taken at the land station at South Georgia located in Area II between 54⁰ and 55⁰ south latitude. On the other hand, all recent catches have come from pelagic operations scattered over all areas and zones but with emphasis varying over time. An analysis of Masaki's data shows that the trend in pregnancy rates during the 1960s is downward in two areas, slightly upward in three others. None are significantly different from zero. The exact changes in sei whale pregnancy rates over the last 30 years can only be determined by an analysis that takes into account the factors cited above.

Over the past decade many attempts have been made to construct ecosystem models, particularly terrestrial models, and indeed the International Biological Program focused much of its attention on this effort. The number of models developed for marine ecosystems is much smaller, and most of these do not involve marine mammals. Two models that do, have been produced, one by Green (1977) and the other by Laevastu and Favorite (1977). Both are system-level models, Green having developed ecosystem models for both the Ross Sea and the Southern Ocean.

Green's mathematical model is a set of linear differential equations with time varying coefficients. These are mathematical representations of the flow (of carbon or biomass) between compartments, in which it is assumed that the donor compartment is controlling the flow, although some consideration is also given to flow controlled by recipient. The latest revised model contains 24 compartments together with the environment.

The Laevastu and Favorite computer model of the eastern Bering Sea ecosystem is mathematically less sophisticated but involves more complex information. For example, the model is a four-dimensional system. The Bering Sea is divided first into a two-dimensional grid with squares 95.25 km in each direction. Less attention is paid to the third dimension (depth), although the model can focus on it. The fourth dimension is time, with time units being a week or month. This is in effect a bookkeeping or input-output model that keeps track of the transfer of biomass from unit to unit and requires an extensive data input in the way of consumption and growth functions, and so forth.

It is clear that such model construction has value to those who construct them. As such, they are helpful in developing better understanding of the ecosystem. To the nonparticipant they can shed light on problems of data deficiencies and research needs. Such recommendations are made in the papers referred to above and several other chapters in this book (6, 12, 21). Before they can be used as management tools or as a contribution to basic science, it is clear that the input data must be reviewed carefully for its accuracy in the light of present knowledge; and where data is lacking or of doubtful reliability, extensive sensitivity analyses must be performed. Green has done some such testing, but in view of the number of compartments and coefficients, much more testing is needed. For example, in the Southern Ocean model there are 24 compartments and 105 transfer coefficients. If one were to consider high, low, and median values for each of these, this could imply $3^{129} = 3.5 \times 10^{61}$ possible combinations! Such an overwhelming number must be reduced and an efficient sampling scheme devised to explore the remaining combination. This in turn will lead to more intensive exploration of particular variables or combinations thereof. Such studies can serve to sharpen and refine the research recommendations that have come from the ecosystem models, but it is probably true that at this stage only in rare situations can the model be used as a mangement tool.

Finally, it is pertinent to ask what should be the recommendation of those of us who have been involved in modeling to assist in management decisions. Do we continue to develop and base recommendations on single-species models; do we insist on ecosystem or multispecies models; or do we inform the decision-makers that the data base is inadequate to construct adequate models? We have seen the results of management when little or no science was brought to bear on the decision-making processes. The International Whaling Commission in the 1950s is an example.

One of the problems of the model builder is to try to convey to the decision-maker the degree of uncertainty in the scientific analysis. This uncertainty occurs as statistical variability in parameter estimates and as uncertainty about the basic structure of the model. There are undoubtedly some cases where the scientist has hidden such assumptions and perhaps also the statistical uncertainty from the decision-maker. Clearly this is wrong. The scientist has to make clear the uncertainty of the models. Also he or she should strive to achieve the best model possible, seeking a balance between simplicity and completeness.

We may be able to embed simple one-species models within more complex ecosystem models and through theoretical studies learn how much the simple model may give results that are not consistent with those of the ecosystem model. Sensitivity studies will be essential both for the models and such comparisons, but, as indicated earlier, sensitivity studies are a subdiscipline in which we are only beginning to learn our way. Naturally, as model builders and as scientists, we are hardly likely to conclude that we should abandon using input from such activities in the decision-making process.

LITERATURE CITED

Allen, K. R. 1972. Further notes on the assessment of Antarctic fin whale stocks. Rep. Int. Whaling Comm. 22:43-53.

Allen, R. L. 1975. A life table for harp seals in the Northwest Atlantic. Rapp. P. V. Reun. Cons. Int. Explor. Mer 169:303-311.

Anon. 1976. Report of the workshop on stock assessment of porpoises involved in the Eastern Pacific yellowfin tuna fishery. Southwest Fisheries Center Admin. Rep. No. LJ-76-29. La Jolla, California 54 pp. + 7 appendices.

Anon. 1978. Report of the Scientific Committee. Rep. Int. Whaling Comm. 28:49-50.

Anon. 1978. South Africa. Progress report on cetacean research. Rep. Int. Whaling Comm. 28:117-118.

Anon. 1979. Report of the Status of Porpoise Stocks Workshop (August 27-31, 1979, La Jolla, California). Southwest Fisheries Center Admin. Rep. No. LJ-79-41.

Bararov, F. 1918. On the question of the biological basis of fisheries (in Russian). Nauchn Issled. Ikhtiologicheskii Inst. Izv. 1:81-218.

Bartholomew, G. A., and C. L. Hubbs. 1960. Population growth and seasonal movements of the northern elephant seal. Mammalia 24:313-324.

Beddington, J. R. 1978. On the risks associated with different harvest strategies. Rep. Int. Whaling Comm. 28:165-167.

Beddington, J. R., and H. A. Williams. 1979. The status and management of the harp seal in the Northwest Atlantic. Report to the U.S. Marine Mammal Commission. Contract #MM1301062-1.

Best, P. B. 1976. Status of whale stocks off South Africa, 1974. Rep. Int. Whaling Comm. Report and papers of the Scientific Committee of the Commission, 1975, pp. 264-285.

Best, P. B. 1977. Status of the whale stocks off South Africa, 1975. Rep. Int. Whaling Comm. 27:116-121.

Bonner, W. N. 1968. The fur seal of South Georgia. Br. Antarctic Surv. Sci. Rep. 56:1-81.

Capstick, C. K., and K. Ronald. 1976. Modelling seal populations for herd management. FAO scientific consultation on marine mammals. Bergen, Norway, ACMRR/MM/SC/77. 19 pp.

Carlisle, J. G., Jr., and J. A. Aplin. 1971. Sea lion census for 1970 including counts of other California pinnipeds. Calif. Fish Game 57:124-126.

Chapman, D. G. 1961. Population dynamics of the Alaska fur seal herd. Trans. N. Am. Wildl. Conf. 26:356-369.

Chapman, D. G. 1966. A critical study of the Pribilof fur seal population estimates. Fish. Bull. U.S. Fish Wildl. Serv. 63:657-669.

Chapman, D. G., and A. M. Johnson. 1968. Estimation of fur seal pup population by randomized sampling. Trans. Am. Fish. Soc. 97:264-270.

Chapman, D. G. 1973. Spawner recruit models and estimation of the level of maximum sustainable catch in fish stocks and recruitment. Rapp. P. V. Reun. Cons. Int. Expor. Mer 164:325-332.

Chapman, D. G., K. R. Allen, and S. J. Holt. 1966. Special Committee of Three Scientists Final Report in International Commission on Whaling. Rep. Int. Whaling Comm. 14:39-92.

DeLong, R. L., and A. M. Johnson. (In prep.). Increase in the northern elephant seal population of San Miguel Island, California. National Marine Mammal Laboratory, Northwest and Alaska Fisheries Center, Seattle, Washington.

Fleischer, L. A. 1978. The distribution, abundance and population characteristics of the Guadalupe fur seal, *Artocepahlus townsendi*, (Merriam, 1897). M.S. Thesis, University of Washington, College of Fisheries, Seattle.

Gambell, R. 1973. Some effect of exploitation on reproduction in whales. J. Reprod. Fert. Suppl. 19:533-553.

Gause, G. F. 1934. The Struggle for Existence. Hafner, New York.

Gilmore, R. M. 1960. A census of the California gray whale. U.S. Fish Wildl. Serv. Spec. Sci. Rep. Fish No. 342, 30 pp.

Green, K. A. 1977. Antarctic Marine Ecosystem Modelling: Revised Ross sea model, general southern ocean budget, and seal model. Report to the U.S. Marine Mammal Commission, NTIS §PB-270 375, Springfield, Virginia. 111 pp.

Gulland, J. A. 1974. Distribution and abundance of whales in relation to basic productivity. Pp. 27-52 in: W. E. Schevill (ed.). The Whale Problem. A Status Report. Harvard University Press, Cambridge, Massachusetts.

Hjort, J., G. John, and P. Otterstad. 1933. The optimum catch HVAL RADETS SKRIFTER (Scientific Results of Marine Biological Research), No. 7, Essays on population, pp. 92-127.

Hubbs, C. L., and L. C. Hubbs. 1967. Gray whale censuses by airplane in Mexico. Calif. Fish Game 53:23-27.

Ichihara, T. 1972. Maximum sustainable yield from the Robben fur seal herd. Bull. Far Seas. Fish. Res. Lab. 6:77-94.

Kendall, D. G. 1948. On the generalized "birth-and-death" process. Am. Math. Statis. 19:1-19.

Kenyon, K. W. 1969. The sea otter in the eastern Pacific Ocean. U. S. Bur. Sport Fish and Wildl. N. Am. Fauna 68:1-352.

Kenyon, K. W., V. B. Scheffer, and D. G. Chapman. 1954. A population study of the Alaska fur seal herd. U.S. Fish Wildl. Serv. Spec. Sci. Rep. Wildl. No. 12, 77 pp.

Laevastu, T., and F. Favorite. 1977. Preliminary report on dynamical numerical marine ecosystems model (Dynamics II) for eastern Bering Sea. Information Report U.S. Dept. of Commerce. Northwest and Alaska Fisheries Center, Seattle, Washington. 81 pp.

Laws, R. M. 1977. Seals and whales of the southern seas. Philos. Trans. R. Soc. London Ser. B. :81-96.

Lensink, C. J. 1960. Status and distribution of sea otters in Alaska. J. Mammal. 41:172-182.

Leopold, A. S., and P. J. P. Gogan. 1976. A review of the population ecology of the northern elephant seal (Mirounga angustirostris). Unpublished report.

Lett, P. F., and T. Benjaminsen. 1977. A stochastic model for the management of the northwestern Atlantic harp seal (Pagophilus groenlandicus) population. J. Fish. Res. Board Can. 34:1155-1187.

Lotka, A. J. 1925. Elements of Physical Biology. Williams and Wilkins, Baltimore.

Masaki, Y. 1978. Yearly change in the biological parameters of the Antarctic sei whale. Rep. Int. Whaling Comm. 28:421-430.

Nagasaki, F. 1961. Population study on the fur seal herd. Tokai Reg. Fish. Lab. Spec. Publ. No. 7, 60 pp.

Ohsumi, S. 1972. Examination of the recruitment rate of the Antarctic fin whale stock by use of mathematical models. Rep. Int. Whaling Comm. 22:69-90.

Ohsumi, S. 1976. Population assessment of the California gray whale. Rep. Int. Whaling Comm. Report and papers of the Scientific Committee of the Commission, 1975, pp. 350-357.

Rice, D. W. 1961. Census of the California gray whale. Nor. Hvalfangsttid. 50:219-225.

Rice, D. W. 1965. Offshore migration of gray whales off southern California. J. Mammal. 46:504-505.

Rice, D. W., and A. A. Wolman. 1971. The life history and ecology of the gray whale (Eschrichtus robustus). Am. Soc. Mammal Spec. Pub. 3. 142 pp.

Ricker, W. E. 1954. Stock and recruitment. J. Fish. Res. Board Can. 11:559-623.

Ripley, W. E., K. W. Cox, and J. L. Baxter. 1962. California sea lion census for 1952, 1960 and 1961. Calif. Fish Game 48:228-231.

Schaefer, M. B. 1954. Some aspects of the dynamics of population important to the management of commercial marine fisheries. Bull. Inter-Am. Trop. Tuna Comm. 1:27-56.

Shaughnessy, P. D. 1975. The status of seals in South Africa and Southwest Africa. Document 33-407, UN/FAO/ACMRR, Ad Hoc III/34.

Shaughnessy, P. D., and P. B. Best. 1975. A simple population model for the South African fur seal *(Arctocephalus pusillus pusillus)*. Unpublished manuscript.

Smith, T. G. 1973. Population dynamics of the ringed seal in the Canadian eastern Arctic. Fish. Res. Board Can. Bull. 181. Ottawa. 55 pp.

Thompson, W. F., and F. H. Bell. 1934. Biological statistics of the Pacific halibut fishery. (2) Effect of changes in intensity upon total yield and per unit of gear. Rep. Int. Fish. (Pac. Halibut) Comm. 8. 49 pp.

Voltera, V. 1931. Lecons sur la Theorie Mathematique de la Lutte pour la Vie. Hermann, Paris.

Winters, C. H. 1978. Production, mortality and sustainable yield of Northwest Atlantic harp seal *(Pagophilus groenlandicus)*. J. Fish. Res. Board Can. 35:1249-1261.

Woodhouse, C. D. J., R. K. Covan, and L. R. Wilcoxin. 1977. A summary of knowledge of the sea otter *(Enhydra lutris)* in California and an appraisal of the completeness of biological understanding of the species. U.S. Dept. of Commerce, NTIS PB-270 374, Springfield, Virginia. 71 pp.

FIFTEEN

Elephants and Their Habitats: How Do They Tolerate Each Other

HARVEY CROZE
ALISON K. K. HILLMAN
ERNST M. LANG

INTRODUCTION

It has become clear in the past few years that the ability to plan for the management of large tropical herbivores is not merely a matter of attaching real values to the transitional probabilities in the life of the animal. Only in three populations in the whole of the African continent—Manyara in Tanzania, Amboseli in Kenya, and Addo in South Africa—are we close to knowing the living elephant's life history in any detail. Moreover, in several spheres, such as erosion processes and primary production events, we are beginning to suspect that high-amplitude changes in local climate over short periods are contributing heavily to the variation in semiarid ecosystems (Dunne, 1980; McNaughton, 1976).

The same may be true of large mammal population dynamics. The effect of the climatic perturbations (rather than the "cycles" of Norton-Griffiths, 1975; Cobb, 1976) is mitigated through the changes in resource availability. Thus, increases in buffalo populations appear to be controlled by the rate of juvenile mortality which may in turn be directly related to the quality of the resource base in any particular year (Sinclair, 1973). The ability of primary production to cause an almost immediate effect on a season's secondary production can result in a situation in which, over a five-year period, wildebeest suffer a 10% mortality due to drought and, in good seasons, a net population increase of 16% (Croze, 1980).

Such a quick recovery may not be surprising for small, medium, or even some large herbivores. But such changes are not so clearly characteristic for the upper end of the scale, such as would be the case for elephants. Large body size and slow growth suggests that the response of a population of such species to environmental change will be necessarily slow. It is necessary for the managers of East African ecosystems to investigate this hypothesis. It has led, for example, to the suggestion that negative feedback processes in elephant reproduction will be too slow to produce a compensation in population size before habitat destruction occurs; and, moreover, that after any resulting dramatic decrease in population size, recovery through reproduction will be too slow or even unable to reestablish the population (e.g., Laws and Parker, 1968). If this is true, then the only recourse is for managers to step in and artificially control numbers before the habitat is destroyed and the population reduced to the point of no return.

If elephants are considered not as singular oversized species, but as a species with body size at one end of a continuum (Western, 1979), it would be expected that their rates of growth and reproduction will be comparable to those of other species on a weight-for-weight basis. Indeed, elephants would be surprisingly ill-adapted for their environment if this were not so. They have lived since the Pleistocene in an unstable environment, in which the droughts followed by floods of the past two decades are probably not exceptional. It would thus seem reasonable to suggest that elephants are able to respond to their environment in ways to ensure the long-term survival of both the elephants and their habitat.

Evidence that lends considerable support to this suggestion falls into three categories. First, field observations indicate that both elephant populations and their habitats are able to recover quite rapidly from periods of low turnover. Second, the wide ranges of elephant population parameters which have been reported in the literature suggest to us a considerable potential flexibility to adjust to local long- and short-term habitat conditions. Two of the key parameters that vary significantly are age at first conception and juvenile survival. Slobodkin (1961) and Laws (Chapter 2) point out that age at first reproduction may be extremely important in contributing to population growth in large mammals. This may be especially true of slow-growing animals that produce just one young, such as man. Croze (1972) suggested that the ability to produce reproductive stock quickly might also be an adaptive quality for elephant populations, particularly in ecosystems subjected to environmental extremes. And, from simulation models, Hanks and McIntosh (1973) and Fowler and Smith (1973) deduced that mortality, particularly of the youngest age classes, may be important in affecting elephant population growth. Third, new evidence from known-aged animals shifts the time perspective for elephant growth curves and suggests that body growth in elephants may be more rapid than previously thought.

The crucial link between early maturity and primary production has not yet been made with field data. However, there are some tempting indications which will be presented below. Using a variable projection matrix model (Smith, 1973) we will look at theoretical short-term change in elephant population growth rate as a function of change in age at first maturity and calf mortality.

Finally, we will discuss the management implications of the hypothesis that elephant populations may respond to changes in local habitat conditions before irreversible trends set in, either in the elephants or in their habitats.

THE APPROACH AND THE NATURE OF THE EVIDENCE

Habitat Recovery

Increasing human populations and increasing agricultural land use have considerably reduced the area available to wildlife since the turn of the century. The more recent effect of harassment by illegal ivory hunting has also tended to concentrate elephants into smaller ranges of relatively greater safety.

The obvious corollary to a decrease in a species' range is a decrease in its resource base. For a wide-ranging species like the elephant, this means that the animal's flexibility to buffer the effects of local resource depletion by moving elsewhere is lost (see Chapter 2). During the mid-1960s, the question arose concerning whether or not increased demand on reduced habitats by elephants would result in a self-induced progressive trend toward deterioration from which neither the elephant or the habitat could recover (Glover, 1963, Glover and Sheldrick, 1964; Laws, 1970; Laws and Parker, 1968; Pienaar and van Nierkerk, 1963). In examing the theoretical effects of a permanently reduced carrying capacity and cumulative resource deficit, Laws (1969b) concluded that one possible outcome of this process would be the conversion of much of the habitat to desertlike conditions. Such concepts were picked up and magnified by the popular press (e.g., Beard, 1965), resulting in reactions that were often emotional but not without some ecological basis.

These predictions, which were most relevant to Tsavo National Park in southern Kenya, rested largely on the premise that the changes being observed in ecosystems like that of Tsavo were changes involved in moving away from a state of equilibrium. However, the gradual reduction of entropy through succession to a climax habitat in equilibrium is largely dependent on a reliable supply of water, which removes the uncertainty of the next season's production of plants and animals. As such, this type of equilibrium habitat is almost exclusively confined to high rainfall areas, temperate zones, or tropical highlands. Since uncertainty is the rule in semiarid ecosystems, it is expected that species which evolved there—both animal and plant—would have genomes prepared to deal with alternating and often irregular events, especially those involving primary production. The ability to deal with such variability should manifest itself independently of the cause of the change in the resource base, be it drought or excessive population density.

"Overcrowding" is a term that gives rise to semantic problems since it is not always adequately defined. It implies an overloading of the resource base by a high density of animals so that there are insufficient resources to support individual animals. "Insufficient," in turn, must be defined in terms of loss of condition, increase in mortality, or decrease in fecundity (see Chapter 3). A protein deficit during the long dry season and the consequent loss of herbivore condition is a frequent, if not annual, event in semiarid ecosystems (e.g., Croze et al., 1980, Chapter 3). The test of overcrowding must include the observation of chronic deleterious effects, even during periods marked by precipitation.

The crucial question (and one which sounds disturbingly dangerous to the temperate-zone ecologist) is not how to stop changes in animal numbers or habitat structures but what are the limits to, and rates of recovery from, observed changes.

During the peak of the drought in the early 1970s, more than 6000 elephants died in the Tsavo ecosystem (Corfield, 1973). It is now with the benefit of hindsight that we can point to the current conditions at Tsavo and question the basis on which the hypothesis of inevitable and nearly total destructive change was made. Here we will only generally and rather briefly discuss the habitat in Tsavo, and leave it to others with firsthand data to provide the hard facts (e.g., Agnew, 1968; T. Corfield, in prep.; W. van Wijngaarden, in prep.).

Tsavo today (1978) certainly is not a desert. The death of large numbers of elephants due to starvation and poaching in the last few years, however, does not seem to provide the total explanation for this observation. Precipitation has increased and has appeared to reverse any trend toward becoming a desert.

Even before the exceptional precipitation of 1977 (40% higher in November through April alone than the 10-year annual average), it was apparent that *Commiphora* and *Melia* trees were regenerating. Increased precipitation has brought about a marked recovery of grass cover, with up to 70% (visual ground estimate in April 1978) around Aruba dam, the center of what was predicted to be the area most detrimentally affected. The cover is made up largely of annuals, but perennial grasses are conspicuously becoming reestablished. In an April 1978 aerial survey of the 43,000 km Tsavo ecosystem, one-third of the area was observed to have a canopy cover of less than 10%. Most of the low-canopy area consisted of grasslands with a grass cover ranging from 50 to 90%. Less than 3% of the entire ecosystem had a grass cover of under 20% (Douglas-Hamilton, 1978). The soil loss during the drought period in the early 1970s, when the ground in some places was almost bare, was not exceptional (T. Dunne and W. van Wijngaarden, personal communication). It had been projected that a loss of grazers, hartebeeste especially, would be observed, but this has not been substantiated (Cobb, 1976 and personal communication).

Finally, the hypothesis that overcrowding through restricted range has occurred in Tsavo must be examined. Large-scale surveys of Tsavo and the surrounding areas indicate there was a virtually continuous distribution of elephants, from Tsavo East to the Tana River, as late as 1976 (J. Allaway and the Kenya Wildlife Conservation and Mangagment Department, unpublished data).

The physiognomy of the habitat in Tsavo, however, has changed. In many places (not ubiquitously) the habitat of the Tsavo ecosystem has changed from bushland to wooded and bush grassland. Biomass has shifted from large-bodied slow-growing things like elephants and bushes to smaller, faster-producing things like grazers and grass. The net loss to the system has been negligible; production and nutrient turnover has increased. If plants can respond to such changes in the physical environment in such a manner, it seems reasonable to look for similar abilities with regard to the animals.

Known-Aged Animals

The most valuable and widely accepted method for determining the age of elephants is that of Laws (1966) [other methods include those of Sikes (1971), Krumrey and Buss (1968), based on Johnson and Buss (1965)]. The determination of a relationship between the assigned age classes and actual ages depends largely on correlations with known-aged animals. Few such animals were available at the time of the original work, and as Laws et al. (1975) maintained, only further work could provide the information needed to calibrate the methods.

Recent data from the Basal Zoo indicates that known-aged captive elephants grow quickly, mature relatively early, and exhibit molars that move forward faster than at previously accepted rates. The five elephants from Basel lost all of each set of molars within a few months of each other. The first set were lost during the second year, the second at the end of the fourth and beginning of the fifth years, the third at the end of the tenth year, and the fourth at the twentieth year.

Two of the elephants had to be destroyed. Their teeth as well as all of the lost molars previously were aged using Laws' (1966) criteria. Three other elephants of known and estimated ages are described in Laws et. al. (1975). Jaws from four known-aged elephants from the New York Zoo were examined in the American Museum of Natural History. These data are plotted in Figure 1. All except the elephant named "Moonflower" were captive. Roman numerals in Figure 1 indicate the age at loss of each of molars I to IV of elephants at Basel Zoo, compared with the ages at which laws estimated this should happen in the wild.

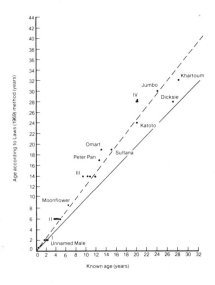

Figure 1 A plot of ages estimated using Laws' (1966) method against the known ages of 28 elephants. (All except Moonflower were captive.) Roman numerals indicate the age at loss of each of molars I to IV of elephants at Basel Zoo, compared with the ages at which Laws estimated this should happen in the wild. Sample sizes by molar category were I = 5, II = 5, III = 5, IV = 2. Names indicate age at death of the named elephant compared with application of the tooth aging method described (Laws, 1966). Solid line is the line of equality ($b = 1$) and dashed line is the regression ($b = 1.25$).

Names identify individual points corresponding to the actual age at death of the named elephant compared with the age estimated through the application of the tooth aging method described in Laws (1966). Some of the data were available at the time of Laws (1966). Others, such as the captive elephants at Basel, and skulls of those of known age that died at the New York Zoo have been the source of more recent data. The total sample size was 28 teeth. Of these five were molar I, five molar II, five were molar III, and two were molar IV. Analysis of covariance showed that the slope of the line fit to these data points (1.25) is significantly different from one ($p < 0.001$). This indicates that the ages assigned by Laws (1966) method are too high and that elephants aged by this method were younger than was previously thought.

Adjustment to Laws' aging technique using the data from kwown-aged animals would thus cause previous estimates of age to be reduced. This suggests that populations wich have been assigned ages using Laws' techniques were actually younger than indicated in previous work. Growth and maturity may therefore occur earlier.

It must be pointed out, however, that there is potential for misinterpretation of this analysis. The differences described above could be the result of more nutritious diet and better conditions for growth as might be found in zoos. If so, variability in tooth erruption with nutrition represents an additional uncertainty in age determination. In either case, the potential clearly exists for earlier maturation than previously recognized. For example, animals thought to mature at age 10 may have matured at an age of 8.

Observed Range of Responses

Douglas-Hamilton (1972) observed rates of increase of elephants in Manyara National Park (Tanzania) of between 3 and 4%. Hanks and McIntosh (1973) concluded from results of a simulation model that a 4% rate of annual increase must be close to the real maximum possible (although slightly below the theoretical maximum). However, a small population in Addo National Park (South Africa), completely protected by fencing, has shown an average annual increase of 6.8% since 1954 (Hall-Martin, 1980). While this may be a somewhat unnatural situation and factors such as age and sex composition need to be considered, it indicates that the realizable rate of increase may be higher than was previously believed possible.

Extrinsic factors clearly act in a number of ways to limit the potential rates of increase. Some of these ways result in observable changes in population parameters. Table 1 summarizes various estimates of population parameters that have been reported for elephants. There is marked variation, both between populations and within populations through time. The numbers of calves born into the population have been found to vary considerably from year to year apparently in reation to conditions characteristic of individual years (e.g., Douglas-Hamilton 1972). Calf mortality, too, can fluctuate annually. Of the calves which were less than 1 year old, 76% died in Amboseli National Park

Table 1 A Summary of Observed Variability in Population Parameters for African Elephants as Found in the Literature (as indicated) and One Recalculation of Previous Work, Corresponding to the Location and Dates Shown

Location	Time	Density (per km²)	Mean age female maturity (years)	Mean age male maturity (years)	Annual calf mort.	Annual adult mort.	Old adult mort. (annual)	MCI[a] (% preg.)	MCI (plac. scars)	Source[b]
Tanzania										
Manyara	1968-1970	4-8	11	—	(0-1 yr) 10%	(2-48 yr) 3.5%	(48 yr) 20%	4-6	—	1
Mkomazi E.	1968	0.8-1.2	12-13	12.5		(10-5 yr) 6%		3.2	3.9	2
			12.2							3
Mkomazi N.	1969	0.8-1.2	12	10.8		2-3%		4.2	6.9	2
			12.2							4
Kenya										
Tsavo	1966	1.2	12	15	(0-1 yr) 36% (1-5 yr) 10%				6.7	4
			14.5						6.8	5
Uganda										
Bunyoro and Toro	1957-1964		7.5					8		6
Murchison and Budongo	1947-1953		8-9					3.8		7
			10-11							8
Murchison N.	1966	1.2-1.5	14	15.6					6-7	4
			16.3						9.1	3
Kabalega (Murchison) N.	1973-1974	3.8-1.3	9.6	14	24%			5.1	4.6	9
Murchison S.	1966	2.3-2.7	18	17.2					8-9	4
									5-6	3

Table 1 (Continued)

Location	Time	Density (per km²)	Mean age female maturity (years)	Mean age male maturity (years)	Annual calf mort.	Annual adult mort.	Old adult mort. (annual)	MCI[a] (% preg.)	MCI (plac. scars)	Source[b]
Kabalega (Murchison) S.	1978-1974	2.7-0.4	9					3.5	5.1	9
Budongo	1966	2.3-3.9	20							5
Forest			23							3
Ruwenzori	1978-1974	1.7-0.6	12.3					4.5	4.2	9
Rhodesia										
Ghona-re-Zhou	1974		12-13							10
Wankie	1972		11					4	3.9	11
Zambia										
Luangwa	1971	1.2-24	14						4	12
South Africa										
Kruger	1954				36%					13
Kruger	1970-1974	0.4-1.1						3.7-4.6	4.5	14

[a]MCI = Mean calving interval estimated as indicated and expressed in years.
[b](1) Douglas-Hamilton (1972), (2) Parker and Archer (1970), (3) Laws and Parker (1969), (4) Laws et al. (1975), (5) Laws (1969a), (6) Smith and Buss (1970), (7) Perry (1953), (8) Recalculated — Laws method, (9) Malpas (1977), (10) Sherry (1975), (11) Williams (1976), (12) Hanks and McKintosh (1975), (13) Pienaar et al. (1966), (14) Smuts (1977).

(Kenya) in 1977 whereas only a handful of the 101 calves born in 1979 died during 1980 (C. Moss, personal communication). Mean ages of female maturity (as defined by Laws, 1966, 1969a) range from 9 to 20 years. The mean ages of maturity of the same populations (Kabalega, formerly Murchison Falls National Park, Uganda) appear to have changed from 14 (north of the Nile) and 19 (south of the Nile) in 1966 to 9.6 (N) and 9 (S) in 1973/1974 (Laws and Parker, 1968; Malpas, 1977). The populations in 1973/1974 were at considerably lower densities than those of 1966, which supports Laws' observation that age at first maturity may be density dependent. Three captive African elephants at the Basel Zoo were first mated at the known ages of 10 years (unpublished data). And, females as young as 8 years were found pregnant in samples from Kruger National Park (South Africa) (Smuts, 1977).

Mean calving intervals (MCI) have been observed to vary from 3.2 to 8 years (Parker and Archer, 1970; Smith and Buss, 1970, using the percent pregnant; Perry, 1953) and from 3.9 to 9.1 (Parker and Archer, 1970; Laws et al. 1975, using placental scars of aged animals; Laws 1969a). The maximum ranges are those quoted by Laws et al. (1975): 2.75 to 13 years. In Kabalega, mean calving intervals also appear to have varied through time, from 3.8 in 1947 to 1951 (Perry, 1953) through 9.1 (N) and 5.6 (S) (Laws et al., 1975) to 5.1 (N) and 3.6 (S) in 1973/1974 (Malpas, 1977).

The MCI, of course, is only one indication of population growth which also depends on concurrent mortality rates and age and sex composition. The ranges of MCI observed for African elephant populations, show that, within periods measured in decades or less, the MCIs and the growth rates, insofar as they depend on birth rates, are capable of showing marked fluctuation.

SIMULATIONS

Methods

The Leslie matrix (Leslie, 1945) was chosen as the appropriate analytical device for the study of elephant populations through the synthesis of information we have presented above. On the one hand, it is not as abstract as the logistic model. It makes use of age-specific events, which we believe to be extremely important in controlling the growth of populations in semiarid environments. On the other hand, it does not demand precise knowledge of transitional probabilities at every state in an animal's life such as that required by the model of Botkin et al. (1977, Chapter 19). Such models, however, will undoubtedly lead to a more intimate knowledge of elephant population dynamics and in time will have important mangagement applications. Given the rate which elephant populations and their habitats are changing and the demands from policy makers for ecological advice, time is the crucial constraint. For our purposes, the matrix approach appears to capture the salient realities of population dynamics as exemplified in Fowler and Barmore (1979). The variable projection matrix methods described by Fowler and Smith (1973) were used in producing the analysis being presented here.

An initial age structure for the population (60 year classes) was developed from those found by Leuthold (1977). A mortality regime from Laws' (1969a) Tsavo data (Figure 2) expressed as the probability of survival during one year was applied. Age-specific fecundity, expressed as the probability of giving birth to a female calf in any one year, was borrowed from that cited by Smuts (1977) (Figure 2). Only a female population was considered; the model was used for projecting over a period of 25 years to examine dynamics exhibited within the simulated population.

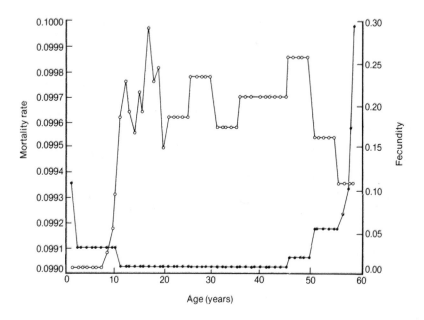

Figure 2 Age-specific mortality (●) (Laws, 1969) and age-specific fecundity (○) (Smuts, 1977) used in the projection model.

Under the initial constant regimes of mortality and fecundity, the age structure stabilized very quickly, indicating that the initial age structure was close to that of the eigenvector of the matrix. The age structure after five years of simulation was nearly identical to that after 25 years. The age structure at 25 years (Figure 3) was therefore taken as the starting age structure from which to develop subsequent analyses.

Since we are concerned here with the hypothesis that elephant populations respond quickly to events involving the habitat, we generated a cyclical habitat event with a period of about 15 years by using a trigonometric function. This representation of periodicity is arbitrary, but matches the real-life fact that the

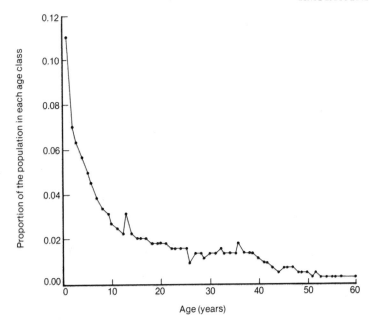

Figure 3   The age structure produced after 25 years of constant fecundity and survival.

1977 floods were last matched in intensity in 1962. The actual period is not of interest here nor even the cyclicity but rather the potential for response to high frequency (relative to the length of life of individual elephants) habitat extremes.

Two vital rates in the simulated population were made functions of the cyclical "habitat factors": the survival of the 1- to 4-year-old age classes, and the fecundity of the 8- to 16-year-olds (Figure 4). The former was chosen as having been found by Hanks and Mackintosh (1973) to be of particular importance in affecting the growth of elephant populations. In addition, a wide range of estimates for juvenile survival has been observed (Table 1). The latter represents the range of the age of first conceptions (e.g., Smuts, 1977; Laws and parker, 1968). The amplitudes of these functions were chosen to be within limits as expected on the basis of variation shown in Table 1. Annual survival of those under 5 years of age were made to vary between about 80 and 95% (e.g., Hanks and Mackintosh, 1973). The average fecundity of the young females was quite likely on the conservative side, dropping to zero as a minimum and peaking at only 0.15 female per female per year. It was, however, deliberately kept low in an attempt to simulate the uncertainties of first conceptions.

It should be noted that this model makes the assumption that there is no density-dependent response in the population at the levels involved. The habitat factors were incorporated in the model so as to involve the kind that vary from year to year (drought versus no drought, for example) to examine regulatory effects which are independent of density.

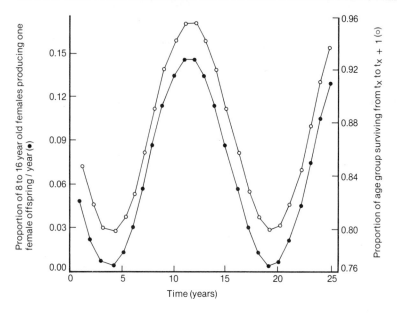

Figure 4   The range and temporal distribution of changes in the survival rate of 1- to 4-year-olds (○) and for the fecundity of 8- to 16-year-olds (●) which were produced as a function of the cyclical "habitat factors."

Simulation Results

Changes in the size and growth rate of the simulated population over 25 years, and the final age structure, are presented here as the results of four simulated situations. (1) In the first, the average annual survival for each age class between 1 and 4 years of age varied between 80 and 95% as dependent on the "habitat factor," while fecundity was held constant with onset of puberty at a "medium" age of 11 years (Figures 5 and 6). (2) In the second, the average fecundity of the 8- to 16-year-olds was varied between 0 and 15% as dependent on the "habitat factor," while survival was held constant (Figures 5 and 6). (3) In the third, a combination of variability in survival and fecundity were made dependent on the "habitat factor" (Figure 5 and 7). (4) Finally, both survival and mortality were held constant with medium age at the onset of puberty (Figures 5, 7, and 8).

   The change in population growth rate as a function of 1- to 4-year-old survival (case 1) and 6- to 8-year-old fecundity (case 2) is shown in Figure 9 (r = 0.98) and Figure 10 (r = 0.98), respectively. In each of the first three cases, growth rates, calculated as proportional change in the population from the previous year, closely tracks the cyclic "habitat factor." The timing of the periods in the growth rates is virtually that of the 1- to 4-year-old survival rates (Figure 4) and the 8- to 16-year-old fecundity rates (Figure 4), with a lag of not greater than a year in each case. The maxima and minima of the population growth rates, as well as the other parameters used in the simulation are shown in Table 2 (on page 312).

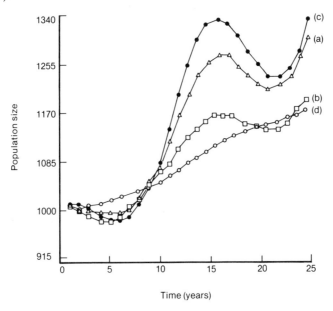

Figure 5   Change in population size over 25 years: in case *(1)* the survival of 1- to 4-year-olds varies (Δ), in case *(2)* the fecundity of 8- to 16-year-olds varies (□), in case *(3)* the survival of 1- to 4-year-olds and the fecundity of 8- to 16-year-olds vary (●), and in case *(4)* the survival and the fecundity are held constant (○).

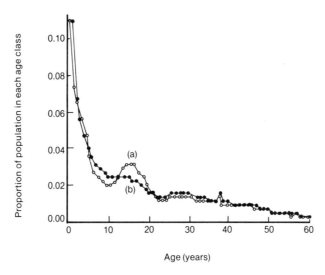

Figure 6   Population age structure after 25 years: in case *(1)* the survival of 1- to 4-year-olds varies (○), and in case *(2)* the fecundity of 8- to 16-year-olds varies (●).

of semiarid elephant supporting ecosystems is that they must turn their attention to planning to accommodate elephants and their habitat in a man-dominated world.

The habitat destruction by populations at high densities may not be a real ecological threat in situations where changes in composition may have, to an extent at least, compensated for the abundance of elephants. This seems to be the case for situations in which we have waited long enough to observe the changes. Policy makers may now address matters of what might be called economic density, and ask such questions as: What are the minimum numbers of elephants necessary to satisfy a particular utilization objective, be it consumptive or nonconsumptive? At what population densities do the benefits outweigh the costs of maintaining elephants? In other words, how few elephants are needed to achieve a particular management objective? What will happen to the composition of the habitat at low elephant densities?

The next step, then, in the modeling exercise, backed where possible with field verification, is to predict the theoretical lower limits at which populations, such as those of the elephant, can safely exist. A related question would be: what is the minimum area requirement (MAR) for elephant populations? It is regrettable that the time has come when we have to look for lower limits.

Even in answering such depressing questions, population ecology still has an important role to play. We must know how much time there is left to make decisions (to shoot or not to shoot, for example). Knowledge is needed to help manage populations that will be pushed to the very limits of their ranges and resources by man. We need to know the probabilities of survival and reproduction at each and every stage in an elephant's life cycle if our recommendations are to help conserve elephants.

LITERATURE CITED

Agnew, A. 1968. Observations on the changing vegetation of Tsavo National Park (East). E. Afr. Wildl. J. 6:75-80.

Beard, P. 1965. The End of the Game. Viking Press, New York.

Botkin, D. B., R.S. Miller, L. S. Wu, and J. P. Muratore. 1977. Long-lived species in time-varying environments: The African elephant as an example (unpublished manuscript).

Cobb, S. 1976. The distribution and abundance of the large herbivore community in Tsavo National Park. Ph. D. Thesis, University of Oxford.

Corfield, T. 1973. Elephant mortality in Tsavo National Park, Kenya. E. Afr. Wildl. J. 11:339-368.

Croze, H. 1972. A modified photogrammetric technique for assessing age structures of elephants populations and its use in Kidepo National Park. E. Afr. Wildl. J. 10:91-115.

Croze, H. (1980) Aerial surveys undertaken by the Kenya Wildlife Mangement Project: Methodologies and results. FAO/UNDP Project Working Document No. 16.

Croze, H., H. Mbai, P. M. Mumiyuka, and M. L. Owaga. (1980). Limits to primary productivity in the Athi-Kapiti ecosystem, Kajiado District, Kenya. FAO/UNDP Kenya Wildlife Management Project Working Document No. 17.

Dunne, T. (1980) Intensity and controls of soil erosion in Kajiado District. FAO/UNDP Project Working Document.

Douglas-Hamilton, I. 1972. On the ecology and behaviour of the African elephant. Ph. D. thesis, University fo Oxford.

Douglas-Hamilton, I. 1978. Report on aerial census of the Tsavo ecosystem April 1978. IUCN Elephant Survey and Conservation Programme.

Eltringham, S. K., and R. C. Malpas. 1976. Elephant numbers in the Ruwenzori and Kabalega National Parks, Uganda. *in:* September 1976 Report to the Director and Trustees Uganda National Parks.

Fowler, C. W., and W. J. Barmore. 1979. A population model of the northern Yellowstone elk herd. Pp. 427-434 *in:* R. M. Linn (ed.). Proceedings of the 1st Conference on Scientific Research in the National Parks, Vol. 1. National Park Service Transactions and Proceedings Series, No. 5.

Fowler, C. W., and T. Smith. 1973. Characterising stable populations: An application to the African elephant population. J. Wildl. Manage. 37:513-523.

Glover, J. 1963. The elephant problem at Tsavo. E. Afr. Wildl. J. 1:30-39.

Glover, J., and D. L. W. Sheldrick. 1964. An urgent research problem on the elephant and rhinoceros populations of the Tsavo National Park in Kenya. Bull. Epizoot. Dis. Afr. 12:33-38.

Hall-Martin, A. 1980. Elephant survivors. Oryx 15:355-362.

Hanks, J. 1972. Growth of the African elephant *(Loxodonta africana)* E. Afr. Wildl. J. 10:251-272.

Hanks, J., and J. A. E. Mackintosh. 1973. Populations dynamics of the African elephant *(Loxodonta africana)* J. Zool. Lond. 169:29-38.

Johnson, O. W., and I. O. Buss. 1965. Molariform teeth of male African elephants in relation to age, body dimensions and growth. J. Mammal. 46:373-384.

Krumrey, W. A. and I. O. Buss. 1968. Age determination, growth and relationships between body dimensions of the female African elephant. J. Mammal. 49:22-31.

Laws, R. M. 1966. Age criteria for the African elephant. E. Afr. Wildl. J. 4:1-37.

Laws, R. M. 1969a. Aspects of reproduction in the African elephant *(Loxodonta africana).* J. Reprod. Fert. Suppl. 6:193-217.

Laws, R. M. 1969b. The Tsavo Research Project. J. Reprod. Fert. Suppl. 6:495-531.

Laws, R. M. 1970. Elephants as agents of habitat and landscape change in Africa. Oikos 21:1-15.

Laws, R. M., I. S. C. Parker. 1968. Recent studies on elephant populations in East Africa. Symp. Zool. Soc. London 21:319-359.

Laws, R. M., I. S. C. Parker and R. C. B. Johnstone. 1975. Elephants and their Habitats: Oxford, London.

Leslie, R. J. 1945. The use of matrices in certain population mathematics Biometrica. 35:215-245.

Leuthold, W. 1976. Age structure of elephants in Tsavo National Park Kenya. J. Appl. Ecol. 13:435-444.

Malpas, R. C. 1978. The ecology of the African elephant in Uganda. Ph.D. Thesis, University of Cambridge.

McNaughton, S. J. 1976. Serengeti migratory wildebeest: Facilitation of energy flow by grazing. Science 191:92-94.

Norton-Griffiths, M., J. Bunning, and F. Kurji. 1973. Woodland changes in the Serengeti ecosystem. (Unpublished manuscript).

Parker, I. S. C. and A. L. Archer. 1970. The status of elephant and other wildlife and cattle in Mkomazi Game Reserve with management recommendations. Wildlife Services Report to the Tanzania Government.

Parker, I. S. C., and I. Douglas-Hamilton. 1976. The decline of elephants in Uganda's Ruwenzori and Kabalega National Parks. Report to the Uganda Ministry of Tourism and Wildlife.

Perry, J. S. 1953. The reproduction of the African elephant *(Loxodonta africana)* Philos. Trans. R. Soc. Ser. B 237:93-149.

Petrides, G. A., and W. G. Swank. 1966. Estimating the productivity and energy relations of an African elephant population. Int. Grassland Congr. 9:831-842.

Pienaar, U., and J. W. van Nierkerk. 1963. Elephant control in National Parks. Oryx 7:35-38.

Sherry, B. Y. 1975. Reproduction of elephants in Gona-re-zhou, south-eastern Rhodesia. Arnoldia (Rhod.) 29:1-13.

Sikes, W. K. 1971. The Natural History of the African Elephant. Weidenfield and Nicholson, London.

Sinclair, A. R. E. 1973. Regulation and population models for a tropical ruminant. E. Afr. Wildl. J. 11:307-361.

Slobodkin, R. B. 1961. Growth and Regulation of Animal Populations. Holt, Rinehart and Winston, New York.

Smith, T. D. 1973. Variable population projection matrices: Theory and application to the evaluation of harvesting strategy. Ph.D. Thesis, University of Washington, Seattle.

Smith, N. S., and I. O. Buss. 1973. Reproductive ecology of the female Arican elephant. J. Wildl. Manage. 37:524-534.

Smuts, G. L. 1977. Reproduction and population characteristics of elephants in the Kruger National Park. J. Afr. Wildl. Manage. Assoc. 5(1):1-10

Western, D. 1979. The relationship of growth rates to body sizes. Afr. J. Ecol. 13:185-204.

Williams, B. R. 1976. Reproduction in the female African elephant in the Wankie National Park, Rhodesia. S. Afr. Wildl. Res. 6:89-93.

SIXTEEN

Simulation and Optimization Models for a Wolf-Ungulate System

CARL J. WALTERS

MAX STOCKER

GORDON C. HABER

INTRODUCTION

Management of large ungulates in North America has been largely based on the assumptions that population numbers are determined primarily by hunting and by habitat factors such as food supply and climate. Natural predators are often assumed to take only weak, ill, and aged individuals and thus to have a beneficial impact on the ungulate population. A few authors (Bergerud, 1974; Pimlott, 1967; Haber, 1977) have argued that this viewpoint is too simplistic, and there is evidence that predators may cause or exaggerate population declines when combined with adverse habitat changes or overhunting (Gasaway et al., 1977; Mech and Karns, 1977; Rausch and Hinman, 1977). These experiences suggest that ungulate population models should take into account at least the functional and numerical responses of predators such as wolves. Also it is time to take a closer look at optimal policies for coupling predator control with ungulate harvest management.

This chapter has two objectives. First, it reviews a simulation model of wolf-ungulate dynamics developed by Haber et al. (1976) from an eight-year field study conducted by Haber (1977) in Denali (Mt. McKinley) National Park, Alaska. This model suggests that wolf predation can become an increasingly important (depensatory) mortality agent on ungulate populations that are reduced from natural levels by hunting. Second, we present optimal moose harvesting and wolf control policies estimated by Stocker (1977) using dynamic programming, for a simplified stochastic model devised by "compressing" the original simulation formulation. These optimal policies are of a feedback character: for any combination of wolf and moose population densities resulting from past effects of harvesting and environmental variation, they specify the next harvest that should be taken. We are not suggesting that harvesting be allowed in Denali National Park, but merely using the data to study management policies for wolf-ungulate systems in general.

THE WOLF-UNGULATE SIMULATION

Haber (1977) studied the "Savage" and "Toklat" wolf packs, whose territories cover more than half of Denali National Park. These wolves and their associated prey form a natural, largely unhunted system. The main prey are moose and Dall sheep; migratory caribou are seasonally important, but will here be considered a nonvarying factor. Haber made detailed, seasonal observations on prey distributions, movement patterns, and behavior of wolves, and birth-death rates in wolves, moose, and sheep.

From this wealth of field data, Haber et al. (1976) produced a simulation model intended to examine long-term (30 to 50 year) dynamics of prey and pack size in a single wolf territory. This section discusses the functional relationships uncovered during the development of that model; these relationships form the basis for the simplified optimization model to be reviewed in later sections.

The detailed simulation is a discrete-time stochastic model with four major submodels: environmental conditions (snowfall), prey dynamics, wolf pack dynamics, and wolf-prey interactions. Effects of food supply and other regulatory mechanisms in the ungulate populations are not modeled explicitly, but are considered to generate density-related changes in birth or death rates. Seasonal phenomena such as reproduction and changes in prey availability due to snow depth are considered. In the following review of the four submodels, we mention only those relationships that were expected to have the most significant impact on long-term predictions.

Environmental Variation

Total winter snowfall is considered to be the best overall index of environmental variation in the Denali area. The historical record suggests a nonrandom pattern of snowfalls (Table 1), involving runs of mild and harsh winters. In the simulation, the snowfall for each year was chosen randomly with probabilities for mild to harsh depending on snowfall the previous simulated year. We hypothesized that population decline and/or recovery phenomena would be dramatically exaggerated by the resulting tendency for bad (or good) years to follow one another.

Table 1 Snowfall probabilities—year $t \rightarrow t + 1$ (from Haber, 1977)[a]

Cumulative Snowfall in Winter $t+1$	Cumulative Snowfall in Winter t				
	1	2	3	4	5
1	0.28	0.20	0.07	0	0
2	0.28	0.40	0.50	0.25	0.67
3	0.28	0.25	0.36	0.25	0.33
4	0	0.15	0	0.25	0
5	0.16	0	0.07	0.25	0

[a]Classes 1-5 represent ranges of snowfall from 1 = very low to 5 = very high.

The model assumes several connections between snowfall and biology of the system. First, the birthrates of moose and sheep are inversely related to snowfall (Figure 1). Second, death rate in the sheep is related to a habitat use index consisting of the product of population size and snowfall (Figure 2). The idea behind this index is that snowfall and population size act jointly to reduce the usable winter "habitat" per sheep, and the relationship in Figure 2 results in sheep population buildups and crashes in basic agreement with historical records (Figure 3). Finally, snowfall influences the winter success of wolves attacking moose, as indicated by the percentage of prey tests (pursuits by wolves) that result in kills (Figure 4).

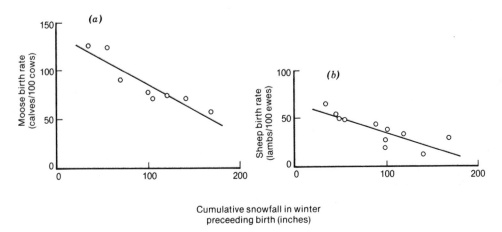

Figure 1   The impact of snowfall on (a) moose birthrate ($r = -0.92$, $Y = 138.1 - 0.52 X$), and (b) sheep birthrates ($r = -0.81$, $Y = 65.9 - 0.31 X$). From Haber (1977).

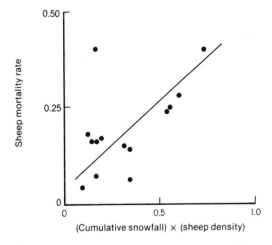

Figure 2   Sheep mortality rate as a function of cumulative snowfall and density. From Haber (1977).

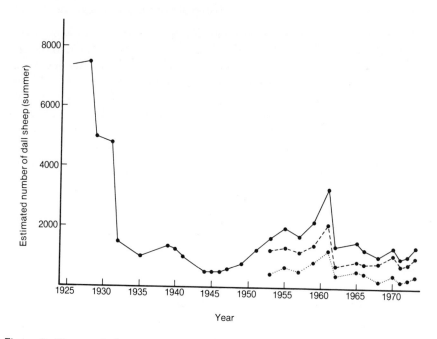

Figure 3 Temporal changes in the size of Denali Dall sheep populations. Dotted line indicates changes west of East Fork River (Toklat range); dashed line east of East Fork River (Savage range); solid line both areas combined. From Haber (1977).

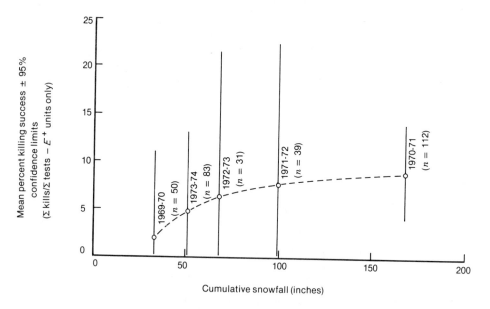

Figure 4 Percent success in killing moose as a function of cumulative snowfall, Savage pack, from winter 1969-1970 to winter 1973-1974. From Haber (1977).

Prey Dynamics

Moose and sheep populations are represented in the simulated population by age class, where each class is assigned a fertility rate, baseline (nonpredator) mortality rate, and vulnerability to wolf predation as measured by pursuit success as explained above. The baseline mortality rates are assumed to generate a pool of carcasses or carrion that are treated as a separate type of prey for wolves. Carcasses are assumed to disappear rapidly (within one month) in summer but to remain usable throughout the winter. The assumption of a carcass pool is critical in the model, since for normal prey populations it can "absorb" a large share of the predation impact that might otherwise be directed to healthy prey individuals.

Age-specific fertility and nonpredator mortality rates were estimated from moose and sheep life tables reported in the literature. These rates followed the general mammalian pattern reported by Caughley (1976): high death rates and low fertility for very young and old animals, and vice versa for animals of intermediate age.

Density-dependent mechanisms not related to predation are assumed to operate in both prey populations. For sheep, there is no evidence of density-dependent changes in fertility. Instead, overall death rates appear to increase with population size, mediated through snowfall (Figure 2); we assumed that the effect is concentrated in young and old animals.

For moose, the picture is less clear because no long-term population records are available for the area. Haber found weak evidence for density-dependent birthrates (Figure 5); similar patterns have been documented for a variety of

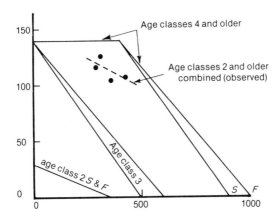

Number of Moose/600 mi² territory

Figure 5 Density-dependent birthrates for moose. Dots are observed rates, corrected for impact of snow in preceding winter using Figure 1a. Solid lines are birthrate responses assumed in model (S = "sharp," F = "flat"). From Haber (1977).

ungulate populations by Gross et al. (1973) and other large mammals by Fowler (Chapter 23). We assumed that moose birthrate follows one or another of the extreme density-dependent patterns shown in Figure 5, modified by snowfall, as indicated in Figure 1. Lacking evidence to the contrary, we assumed that moose death rates are not density dependent.

The net effect of the assumptions concerning the dynamics of prey in the absence of wolf predation is to predict a sheep population with periodic irruptions and crashes, and a relatively stable moose population.

Wolf Pack Dynamics

In these simulation models there are two types of numerical response in the wolf population: changes in pack size and changes in the number of packs. Haber's work indicates that pack size is regulated primarily through density-dependent dispersal or "splitting" of young wolves from the pack (Figure 6). Mech (1977) found that some adult wolves can tolerate long periods of near starvation, though pup production may cease entirely at low prey densities. It appears that younger, socially subordinate individuals often get a smaller ration from each prey killed than do higher-ranking wolves, and that hunger increases their tendency to split off as small hunting groups. However, at large pack sizes young wolves also split off when estimated rates of food consumption indicated little if any stress. Thus, splitting may also be triggered by social factors related solely to pack size. At any rate, more young wolves disappeared as splitting frequency increased; hunting observations indicated that on their own or in small groups, most obtained below-maintenance food rations, and some probably emigrated to other regions.

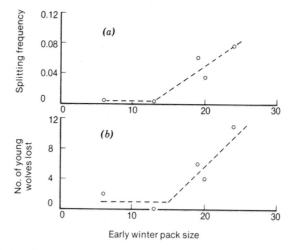

Figure 6 Splitting frequency (*a*), number of subunits recorded per hourly sample period, and overwinter losses of young wolves (*b*), due to emigration or mortality, as a function of early winter pack size, Savage pack, from winter 1969-1970 to winter 1973-1974. From Haber (1977).

For the simulation, it was decided to model effects of food intake on pack size by (1) decreasing pup production, and (2) increasing juvenile (pup and yearling) dispersal-death rates with decreasing food intake. We could find no good quantitative data on either of these relationships, so we used estimates of wolf maintenance ration in winter (about 2 kg food/day) to establish an over-simplified rule: for mean winter ration above maintenance, pup production is set to 6.0 (normally only one dominant female breeds in the pack) and juvenile dispersal depends only on pack size; for ration below maintenance, pup production is proportional to ration while juvenile dispersal increases (to 100% at zero ration).

Changes in the number of wolf packs in an area, in relation to changing prey density, were inferred from a broad examination of wolf territory sizes across North America (Figure 7). It appears that territory sizes are adjusted in the long run so as to prevent depletion of prey within each territory. If a wolf pack is not to deplete its prey (and thus be forced to face the dangers of dispersal), it must take an annual kill K less than the intrinsic rate of prey increase rN per unit area times territory size A:

$$K < rNA$$

This is equivalent to saying that territory size must be

$$A > \frac{K}{rN} = \frac{\alpha}{N}$$

where $\alpha = K/r$. Note that α will tend to be stable across prey types, since both K and r tend to increase as prey body size decreases (e.g., an Alaskan wolf pack will kill fewer moose than a Minnesota pack will kill deer, and moose have a lower r than deer). K will be in the neighborhood of 100 to 300 prey/year (see next section), and r will usually be between 0.1 and 0.4. Thus, the prediction becomes

$$A > \frac{1000}{N}$$

Most of the points (observed A values) in Figure 7 do indeed satisfy this prediction.

If the dotted line in Figure 7 is taken as a target territorial pattern for wolves, we are still left with the problem of estimating how fast the territory size adjustment can occur in the face of rapid changes in prey density. Mech (1977) documented a seven-year Minnesota deer decline in which at least one wolf pack became reduced in size to the dominant mating pair, yet did not give up its territory. Thus it is conservative to assume that territories can be maintained until the adult wolves die of old age (about 10 years); in a statistical sense, the single simulated pack should be able to expand its territory by 10%/year (1 out of every 10 surrounding packs, presumably subject to similar prey conditions,

There has been an independent "test" of the predictions in Figure 11. Gasaway et al. (1977) report low calf survivals and retarded moose recovery following hunting in the Tanana Flats area, Alaska.

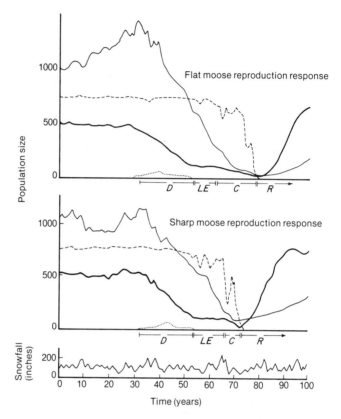

Figure 11 Typical time predictions for the Denali sheep (light-solid line), wolf (dashed line), and moose (heavy solid line) populations, and snowfall, from the simulation model. Moose harvest (dotted line, period *D*) levied over years 30 to 50 is followed by intensified wolf predation (period *LE*), collapse of the wolf pack (period *C*), and recovery of the moose population (period *R*). From Haber (1977).

Figure 12 shows combinations of moose and sheep populations (per 600 mi²) such that immediate recovery of both populations and persistence of wolves are likely to occur if all outside (e.g., hunting) disturbances are removed. Combinations above and to the right of the stippled areas (depending on Figure 5 moose reproduction assumption) will rapidly recover to near the upper right corner of the graph. Starting combinations below the stippled areas are followed by the irruptive syndrome of Figure 11, simulation years 50 to 80. The stippled areas themselves represent combinations where the outcome (recovery or crash) is uncertain due to stochastic effects of snowfall. The areas above and below the

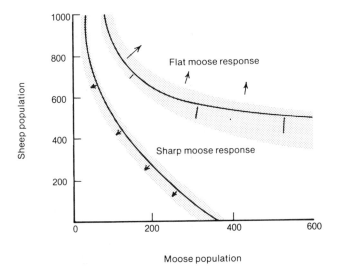

Figure 12 Moose-sheep phase plane. The shaded bands delineate critical population combinations for the "sharp" and "flat" moose reproductive responses assumed in the simulation model. Below each band, collapse of the moose and sheep populations is likely; above each band recovery to normal levels (about 450 moose and 1000 sheep) is likely, without loss of wolves. From Haber (1977).

lines may be considered two "stability regions" in terms of ecological resilience (Holling, 1973); the assumption of a sharp change in moose reproduction leads to a larger stability region for coexistence (moose and/or sheep must be disturbed more radically before the system will collapse).

Much of the computational complexity of the simulation model is due to representation of prey age distributions and related age-specific parameters. It is important to ask whether this complexity (and associated compounding of uncertainties through an increase in the number of parameters to be estimated) has a significant effect on the predictions. Only one such effect was obvious in the many simulations we tried: in scenarious involving moose irruptions (Figure 11), the final few years before the wolf pack collapse are characterized by a moose population consisting largely of progressively older individuals. Heavy wolf predation takes most of the calves but continued calf production helps to sustain the wolf population. Then a large proportion of the moose begin to die of old age, leading to a final sharp population decline followed immediately by the wolf collapse. Had we pooled all moose age classes except calves into a simple population variable with some average adult mortality rate, we would not have predicted the final sharp decline. Instead the simulated adult moose stock would decline more smoothly and over a longer period, though ultimately the collapse of the wolf population would still occur. For scenarios where the moose and sheep populations remain in the upper stability region of Figure 12, the simulated population's age structure remained remarkably stable and could thus be ignored (age-specific rate calculations could be replaced by estimates of average rates across the population).

SIMPLIFICATION FOR OPTIMIZATION

While the simulation can give some feeling for the importance of various dynamic relationships in wolf-ungulate systems, it is too complex to automatically give clearer perceptions of how much systems should be managed in terms of ungulate harvests and wolf control. The remainder of this chapter will attempt to show how optimal (or at least improved) management policies can be estimated by simplifying the simulation to the point where existing optimization techniques such as dynamic programming can be applied to it.

In a recent review of optimization models for ecological management, Walters and Hilborn (1978) noted that existing techniques can produce two types of policies: (1) "open loop" — best actions are specified as a time series into the future, without regard for uncertainties concerning future states of the system; or (2) "closed loop" or "feedback" — best actions are specified as a function of the state of the system at future times, recognizing that it is impossible to predict now what future states will actually occur. Generally, closed-loop policies give considerably better results (see review in Walters and Hilborn, 1978). While optimal open-loop policies can be computed for very complex dynamic models, it is not known how to compute the more powerful feedback policies for models with more than a few state variables. We have the choice of pretending that the world is deterministic and complex, or stochastic and simple. The latter choice was made for this study.

If there are only one or two state variables, for example population sizes, the optimal feedback harvest policy can usually be expressed as a simple, time-independent function. For any state that arises in year t as a result of natural forces and past management actions, the function specifies best harvest to be taken in that year; if the same state rises in another year, the same action will be optimal. To compute the function it is necessary to specify (1) an objective function, for example maximize the sum of harvests over time; and (2) a dynamic model which, for any starting state (i.e., population sizes) and action (i.e., harvest), specifies a probability distribution for the state that will result after one year. If there are two state variables, for example, ungulate population U_t and wolf population W_t, the model must specify a probability distribution for U_{t+1}, W_{t+1} given U_t, W_t and any ungulate harvest and/or wolf control actions during year t. The model may involve an exceedingly complex set of calculations for changes between t and $t + 1$ (for example, seasonal events), but these calculations must be based only on U_t and W_t; such complications as the age distribution of U_t must be ignored or assumed independent of t.

We chose to make four major simplifications in order to compress the simulation into a format suitable for feedback policy calculations. First, we treated only one ungulate population (moose) as a state variable, by assuming that sheep, caribou, and other prey provide a fixed source of potential food for wolves. This is a conservative management assumption, since it implies that the wolf population will respond less rapidly to reductions in moose abundance than would be the case if all prey are driven down together.

Second, we assumed that the age structure of yearling and older moose is relatively stable and can thus be ignored. More precisely, we assumed that the proportion of moose v_i of age $i > 1$ is independent of time. Then if the number of calves C born per year had been calculated in the simulation as

$$C = \sum_i b_i N_i$$

where b_i = age specific birth rates, N_i = number of age i moose, the existence of fixed v_i implies that

$$C = \sum_i b_i v_i U$$

$$= U \sum_i b_i v_i$$

$$= \bar{b} U$$

where U is total moose population and $\bar{b}$ is a weighted mean birthrate per animal. Similar arguments lead to a mean natural mortality rate and mean probability of successful capture by wolves. By defining U as the spring population before calves are born, we allow separate mortality and predation rate calculations within the year for calves, provided these calves are placed into the U_{t+1} pool at the end of the year.

Third, we chose to treat each wolf pack as though it were a single predator, since handling time and pursuit success are largely independent of pack size (Figures 9 and 10). Then W_t becomes the number of wolf packs that occupy the modeled area.

Fourth, the dynamics of W_t are represented as a process of adjustment toward the territory size relationship shown in Figure 7. When W_t times territory size (estimated from dotted line in Figure 7) is less than the total area available, $W_{t+1} = 1.5 W_t$ (0.5 represents our estimate of the maximum intrinsic rate of increase of W_t). When the wolves are overcrowded (W_t times area from Figure 7 exceeds total area), we assume that W_t will decrease by 10%/year. By assuming that W_t begins to adjust immediately to reduced U_t, we are being overly optimistic about the initial response of wolves to declining prey density. However, more complex calculations would require at least one additional "memory" state variable representing accumulated stress on the wolves due to reduced prey availability. In effect the simplified model assumes that (1) the transition periods of decreasing pack size prior to each wolf pack collapse (Figure 11) can be ignored, and (2) prey availability varies between wolf territories such that a few packs are barely holding onto territories (first 10% to drop out) while others will not feel the effects of the overall prey decline for a sustained period.

The simplified model uses the multispecies disc equation as in the simulation to represent the functional response of wolves to prey density. The number of prey categories is reduced to three: moose calves, older moose, and other prey. The disc equation parameters a_i, p_i, and the th_i for other prey are taken to be the same as for sheep.

Stochastic effects are retained in the simplified model by a three-level snowfall variable $S_t =$ low, average, high , and it is assumed that S_{t+1} is independent of S_t. Moose birth rates are assumed to respond to snowfall as in the simulation (Figure 1a), but other snow effects (e.g., pursuit success of wolves) are ignored.

To summarize, the simplified model has two state variables:

$$U_t = \text{moose density}$$

$$W_t = \text{number of wolf packs}$$

U_t is modeled with density- and snowfall-dependent reproduction, density-independent natural mortality (not due to wolf predation), wolf mortality, and harvest. Changes in W_t are modeled as a process of territory size adjustment.

If stochastic effects are ignored, the simplified model leads to two qualitatively different types of predator-prey behavior between wolves and moose (Figure 13). If the wolves are assumed to be relatively ineffective searchers for prey (small "a" parameters in disc equation) and/or the density of alternative prey is low, the model has only one stable equilibrium point as indicated in Figure 13a (isoclines $W_{t+1} = W_t$ and $U_{t+1} = U_t$ cross at equilibrium). If the wolves are efficient and can maintain themselves on other prey when U_t is small (Figure 13b), the system has two stable equilibria: one at a high moose density as in the first case, and one

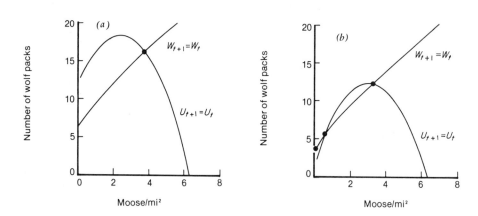

Figure 13   Isoclines of zero moose growth ($U_{t+1} = W_t$) in the wolf pack-moose phase plane. (a) Parameter case 1 of figure 14, with intermediate wolf search efficiency ($A = 550$) and low other prey occurrence ($0 = 3.5$). (b) Parameter case 3 with high wolf search efficiency ($A = 900$) and very low other prey occurrence ($0 = 0.6$). From Stocker (1978).

at a low wolf density but with moose extinct. This second case represents a dangerous management possibility, since it implies that the moose population must be managed to remain within narrow limits unless strong wolf control measures are feasible.

OPTIMIZATION OBJECTIVE FUNCTIONS AND PROCEDURES

State-increment dynamic programming (Larson, 1968) was used to find optimal feedback policies for moose harvesting and wolf control in the simplified three-variable model. It was assumed that the management objective is to maximize the expected value (in a statistical sense) of summed annual returns R_t

$$V_t = \sum_t^T R_t$$

where V_t represents the total returns from t to T, and T is an arbitrary end time planning horizon that can be ignored in practice if it is greater than 10 to 15 years in the future. We considered a simple model for short-term objectives:

$$R_t = u_t - c_w w_t$$

where u_t = moose harvest

w_t = number of wolf packs removed by wolf control

c_w = relative cost factor of wolf control per pack

The cost factor c_w can be considered as a penalty for killing wolves; $c_w = 0$ implies no aversion to control, while c_w large implies a strong conservation lobby against wolf control.

The "principle of optimality" in dynamic programming (Bellman, 1957) states that V_t can be written recursively as

$$V_t = R_t + V_{t+1}$$

But V and R are functions of the system state vector $x_t = (U_t, W_t)$ and of the control actions u_t, w_t:

$$V_t(x_t) = R_t(x_t, u_t, w_t) + V_{t+1}(x_{t+1})$$

Given any starting state x_t and actions u_t and w_t, we can compute R_t and (ignoring stochastic effects for a moment) x_{t+1}. Thus by beginning with some arbitrary end point value $V_T(x_T)$ assigned across states x_T, we can try different choices of u_{T-1} and w_{T-1} at a series of incremental states x_{T-1} that might arise by time $T-1$. After finding the best choice for each x_{T-1}, we will have automatically estimated $V_{T-1}(x_{T-1})$. We then move backward in time to $T-2$, and repeat the search for best u_{T-2} and w_{T-2} choices for each x_{T-2}; the result is V_{T-2}). This "backward recursion" process is repeated for several time steps, and it is usually found that the best (u_t, w_t) choice for each x_t becomes independent of t and, luckily, of V_T. We have then computed the optimal, stationary feedback policy: (u_t, w_t) as a function of x_t. This is done at a series of discrete or incremental states of x_t; thus the name state-increment dynamic programming. We are in essence approximating continuous functions V, u, and w of x_t at a series of discrete values for x_t.

It is very simple to incorporate stochastic effects into the procedure. For any x_t, u_t, and w_t, there are three possible values of x_{t+1}, each with a probability conditional on the S_t (snowfall) element of x_t. We simply take the expectation of $R_t + V_{t+1}$ with respect to these possibilities, by summing probabilities time associated $R + V$ values.

OPTIMIZATION RESULTS

We have computed optimal feedback policies for four different assumptions about other prey and the search efficiency parameters (Figure 14), and for various values of the wolf-control cost parameter c_w. We assumed for all cases that $u_t \leqslant 0.5U_t$, and that $w_t \leqslant 4$ packs/year (equilibrium $W_t = 12$ to 15 packs). The following discussion will concentrate on cases 1 and 3 of Figure 14, since these imply qualitatively different predator-prey associations (Figure 13). The other two cases gave results essentially indentical to case 1 (Stocker, 1978).

Optimal moose hunting rates can be mapped as harvest isopleths on the moose-wolf phase space (Figure 15). For low moose and wolf densities, the optimal policy is no harvesting. Then for higher moose densities, the optimal harvest rate increases steadily.

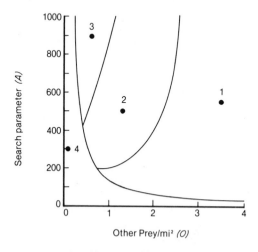

Figure 14   Parameter space of wolf search efficiency (A) and other prey occurrence (0) indicating four regions that exhibit distinct qualitative behaviors of moose and wolf dynamics. Dots indicate parameter combinations for which optimizations were carried out. From Stocker (1978).

For parameter case 1, as defined in Figure 14, the optimal moose harvest is remarkably insensitive to wolf density (near vertical harvest isoclines in Figure 15a); this conclusion is not changed when c_w is varied so as to make the overall optimal policy include or exclude wolf control. Stochastic simulations of the optimal policy being applied to case 1 indicated that the optimal policy tends to stabilize the moose population at about three-fourths the average natural (unharvested) density.

For parameter case 3, the optimal moose harvest policy given large c_w has a special complication. When moose density is low and wolf density is high, it is apparently optimal to harvest the moose to extinction since they will not recover in any case from wolf predation (Figure 15b). When c_w is zero, the optimal harvest contours are as in case 1 (Figure 15a).

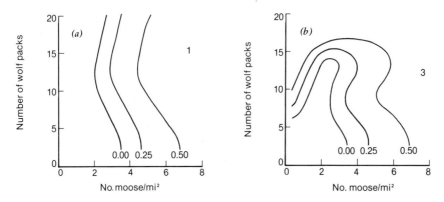

Figure 15 Moose harvest rate isopleths derived from stochastic dynamic programming. (a) For parameter case 1, (b) for parameter case 3. From Stocker (1978).

We found the optimal wolf control policy to be generally and completely insensitive to wolf density in case 1. By optimizing for different values of c_w, we could further isolate threshold values above which wolf control should never be exerted:

Parameter Case (Figure 14)	Critical c_w
1	0.8
2	1.1
3	1.1
4	1.0

Thus, almost independent of predator efficiency, it was best to never exert wolf control when the cost of removing one pack exceeds 1.0 when measured in the same units as moose harvest per square mile. For lower c_w, we found that it was optimal to exert wolf control only when W_t was smaller than some critical value of W^*; for W_t W^*, the optimal $w_t = 0$. We found that W^* increased as c_w decreased, while W^* approached zero as c_w approached 1.0. These results were surprising, since we expected that it would be best to exert wolf control only when the *moose* density was low.

The optimal policies can be summarized very simply: harvest the moose so as to leave a fixed "escapement" (same every year) while essentially ignoring the wolf population; meanwhile, leave the wolves alone or try to remove them entirely if the control cost is low and if they are not abundant in the first place. These results are not intuitively appealing, and we suspect that the problem is with our choice of a simplistic objective function. Had we assumed that wolf control costs increase as the wolf population decreases, the optimal policy would at least involve leaving some wolves around. Likewise, we could doubtless find an objective function such that it would be best to harvest some moose even at very low moose densities.

CONCLUSION

The simulation model has allowed us to examine the dynamics of wolf-ungulate interactions with some precision. We have attempted to capture these dynamics in relatively simple optimization models. For some biological issues that remain unresolved, we have constructed and optimized alternative dynamic models; usually the optimal feedback policy was found to be relatively insensitive to these uncertainties.

The critical problem now is to define optimization objective functions that measure the varied interests and constraints that management agencies face. We could be even more simple-minded about the biology of the problem, but we will not produce credible and useful optimization results until we can deal much more precisely with the human side of the problem.

LITERATURE CITED

Bellman, R. 1957. Dynamic Programming. Princeton University Press, Princeton, New Jersey.

Bergerud, A. T. 1974. Decline of caribou in North America following settlement. J. Wildl. Manage. 38:757-770.

Carbyn, L. N. 1974. Wolf predation and behavioural interactions with elk and other ungulates in an area of high prey diversity. Ph.D. Dissertation, University of Toronto, Toronto, Canada.

Caughley, G. 1976. Wildlife management and the dynamics of ungulate populations. Pp. 183-246 *in*: T. H. Coaker (ed.). Applied Biology, Volume I.

Charnov, E. L. 1973. Optimal foraging: Some theoretical explorations. Ph.D. Dissertation, University of Washington, Seattle. 95 pp.

Clark, K. R. F. 1971. Food habits and behavior of the tundra wolf on central Baffin Island. Ph.D. Dissertation, University of Toronto, Toronto, Canada. 223 pp.

Gasaway, W. C., D. Haggstrom, and O. E. Burris. 1977. Preliminary observations on the timing and causes of calf mortality in an interior Alaskan moose population. Trans. 13th N. Am. Moose Conf. Workshop, Jasper, Alberta, pp. 54-70.

Gross, J. E., J. E. Roelle, and G. L. Williams. 1973. Program ONEPOP and information processor: A systems modeling and communications project. Progress Report. Colorado Cooperative Wildlife Research Unit, Colorado State University, Fort Collins. 327 pp.

Haber, G. C. 1977. Socio-ecological dynamics of wolves and prey in a subarctic ecosystem. Ph.D. Dissertation, University of British Columbia, Vancouver, Canada. 786 pp.

Haber, G. C., C. J. Walters, and I. Mct. Cowan. 1976. Stability properties of a wolf-ungulate system in Alaska and management implications. Institute of Animal Resource Ecology, Research Report R-5-R, University of British Columbia, Vancouver. 104 pp.

Holling, C. S. 1959. The components of predation as revealed by a study of small mammal predation of the European pine sawfly. Can. Ent. 91:293-320.

Holling, C. S. 1973. Resilience and stability of ecological systems. Annu. Rev. Ecol. Syst. 4:1-23.

Kolenosky, G. B. 1972. Wolf predation on wintering deer in east-central Ontario. J. Wildl. Manag. 36:357-369.

Larson, R. E. 1968. State Increment Dynamic Programming. American Elsevier, New York.

Mech, L. D. 1977. Population trend and winter deer consumption in a Minnesota wolf pack. Pp. 55-83 in: R. L. Phillips and C. Jonkel, (eds.). Predator Symposium Proceedings 1975, Montana Forest and Conservation Experiment Station, University of Montana, Missoula.

Mech, L. D., and P. D. Karns. 1977. Role of the wolf in a deer decline in the Superior National Forest. North Central Forest Experiment Station, Forest Service, U.S. Department of Agriculture, St. Paul, Minnesota. 23 pp.

Murdoch, W. W. 1969. Switching in general predators: Experiments on predator specificity and stability of prey populations. Ecol. Monogr. 39:335-354.

Murdoch, W. W. 1973. The functional response of predators. J. Appl. Ecol. 10:335-342.

Oaten, A., and W. W. Murdoch. 1975. Switching, functional response, and stability in predator-prey systems. Am. Nat. 109:299-318.

Peterson, R. O. 1974. Wolf ecology and prey relationships on Isle Royale. Ph.D. Dissertation, Purdue University, Lafayette, Indiana. 368 pp.

Pimlott, D. H. 1967. Wolf predation and ungulate populations. Am. Zool. 7:267-278.

Pimlott, D. H., J. A. Shannon, and G. B. Kolenosky. 1969. Ecology of the timber wolf in Algonquin Provincial Park. Ontario Department of Lands and Forests Research Report (Wildlife). No. 87, 92 pp.

Rausch, R. A., and R. A. Hinman. 1977. Wolf management in Alaska—an exercise in futility. Pp. 147-176 in: R. L. Phillips and C. Jonkel (eds.). Predator Symposium Proceedings 1975, Montana Forest and Conservation Experiment Station, University of Montana, Missoula.

Stephenson, R. O. 1978. Unit 13 Wolf Studies. Vol. I, Project Progress Report, Federal Aid in Wildlife Restoration, Project W-17-8, Jobs 14.8R, 14.9R, 1410.R. Alaska Department of Fish and Game. 75 pp.

Stocker, M. 1978. Optimal harvesting strategies for ungulate populations in relation to population parameters and environmental variability. Ph.D. Dissertation, University of British Columbia, Vancouver.

Van Ballenberghe, V., A. W. Erickson, and D. Byman. 1975. Ecology of the timber wolf in northeastern Minnesota. Wildl. Monogr. 43. 43 pp.

Walters, C. J. and R. Hilborn. 1978. Ecological optimization and adaptive management. Annu. Rev. Ecol. Syst. 9:157-188.

SEVENTEEN

Management and Models of Marine Multispecies Complexes

J. W. HORWOOD

INTRODUCTION

All species interact with others to a greater or lesser degree. Such interactions are usually grouped under simple categories such as predation, competition, or mutualism. The clarity or rigidity of the definitions of these categories has grown with the importance and popularity of theoretical population biology. Predation has been defined as an interaction between two species that is to the benefit of one and the detriment of the other. Competition gives rise to a mutually detrimental effect and mutualism to a beneficial relationship. However, many of the apparent interactions among species are far from direct, and subtle changes in numbers and ecosystem structure may give rise to statistical associations that cannot be defined as simple biological competition or predation. Nevertheless, the existence of such interactions may not be the most important factor for the management of the biotic components of an ecosystem. It is usually the magnitude or effect of such interactions that determines whether they need to be considered in developing management strategies.

Traditionally biological management in marine systems has been based on single-species considerations due to a combination of circumstances especially experience in the field of fisheries. Fisheries have been known to persist for centuries, giving rise to the optimism that the stocks could be fished harder with no serious adverse effects. The early harvesting of such species was not accompanied by correspondingly immediate changes in other species. As a result, tacit acceptance of ecosystems as fluid entities developed. This was typified by the diverse food web of the herring as illustrated by Hardy (1924). The early exploitation of fisheries at low levels (levels not high enough to cause a stock to collapse) gave rise to the yield models that may be used to determine the maximum yield, by weight, from a given recruitment (e.g., the simple model of Beverton and Holt, 1957). This avoided what has been termed "growth-overfishing" (Cushing, 1975) wherein the average age of fish, in the fishery, is regulated and losses in biomass from natural mortality and a decrease in growth rate with age are minimized. Earlier single stock and recruitment models had been developed (Graham, 1935; Schaefer, 1954) but not used extensively. As catches of fish increased with demand the problem of "recruitment-overfishing" grew in importance, especially in fisheries for pelagic species (Cushing, 1975) where some stocks were reduced to levels that could not provide sustainable recruitment. Along with major change in fish abundance, changes in the species composition over a range of trophic levels has been observed, and more attention is now being given to study and management at the level of the community (Cushing, 1975). The setting of overall quotas for higher trophic levels [less than the sum of individual "total allowable catches" (TACs)] in the International Commission for North Atlantic Fisheries (ICNAF) is a manifestation of this new concern with management at the level of the community (Anon., 1974).

To provide insights into resolving marine mammal problems, it is often more useful to study terrestrial mammalian systems than marine fisheries and this is frequently done. However there are major analogies between the management of marine mammals and that of marine fish that may profitably be explored to the benefit of both disciplines. By definition, they live in the same environment and share many of the same food sources. An understanding of what regulates food supply in time and space and how exploitation of the upper trophic levels affects the food supply is necessary to the management of both. The managers share many of the same problems. The animals cover vast areas and cannot be counted directly. Statistics and inference have to be relied upon to a great degree. The nature of fishing and its associated economics means that interactions at the level of species are forced to the attention of the managers by the harvesting strategies used. Finally, it is frequently the same scientists who are involved in reaching decisions concerning the management of both fish and marine mammals, and their decision must to some degree reflect their specific biological training. In this study a range of multispecies interactions is presented relating to both marine mammal stocks and fisheries, along with some of the models of multispecies interactions that assist in the understanding and management of such complexes.

INTERSPECIFIC RELATIONSHIPS IN EXPLOITED MARINE SYSTEMS

Socioeconomically Induced Relationships

Socioeconomic considerations usually determine what species are harvested and preferred. However, in unregulated fisheries or in those where advice has been inadequate or unheeded, overexploitation has frequently occurred and the demands of the market have forced the trade to look for alternative species.

Perhaps the classic example of exploitation on successive species is the baleen whale fishery of the Antarctic. The catches of the four main species are shown in Figure 1 for the period 1909 to 1977. Gambell (1973) records that the blue whale *(Balaenoptera musculus),* being the largest animal, were preferentially hunted, although some fin whales *(B. physalus)* have always been taken. From Figure 1 it can be seen that the peak blue whale catches were taken in about the 1930s after which there was a steady decline only interrupted by World War II. The decline in the availability of blue whales caused increased attention to be given to the fin

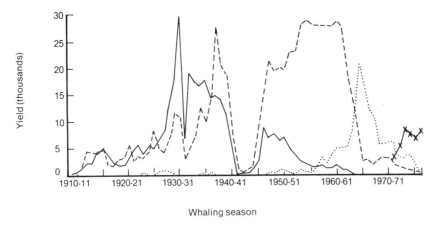

Figure 1 Catches of blue whale (solid line), fin whale (dashed line), sei whale (dotted line), and Minke whale (*X*) for the period 1909 to 1977 in the Antarctic, showing a pattern of overexploitation and switching.

whales, the catches of which remained very high until the early 1960s. Gambell (1973) states that, as the numbers of blue and fin whales declined, the industry had to turn its attention to the smaller sei whale *(B. borealis),* the catches of which reached a peak in 1966. Overexploitation occurred on this latter species as well, but the introduction of the International Whaling Commission's (IWC) New Management Procedure has enabled the sei whales to be held at a much higher percentage of its early stock size than the other whales. These restrictions on the larger whales again forced the industry to look for alternative resources, and during the 1970s the even smaller minke whale *(B. acutorostrata)* has been caught. This illustrates what will occur if demand for a product remains constant or increases above a sustainable level and the industry is unregulated so that it is allowed only to consider short-term objectives. Longer-term objectives could have been better realized by the early acceptance of a mixed fishery.

In the North Pacific, early pelagic whaling for sperm whales *(Physeter catodon)* occurred primarily above 50°N around the Aleutian Islands where the largest male whales spent their summers. The latitudinal movement of the pelagic fleets has been described by Ohsumi (1977) and is illustrated here in Figure 2 for the period 1954 to 1976. Ohsumi states that the change in the sperm whaling ground is closely associated with the change in catches of the baleen whales, since the sperm whale fishery was of lesser importance. Before the mid-1960s the fin whale was the main object of pelagic whaling: this whale was caught north of 50°N. However, the sei whale succeeded the fin whale in being of prime interest until about 1970. The sei whales are usually between 40°N and 50°N in summer. The Bryde's whale *(B. edeni)* succeeded the sei whale being caught between 20°N and 40°N. The Soviet whalers were more interested in the sperm whales than in baleen whales and stayed farther north for a longer period.

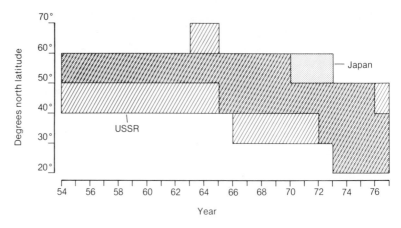

Figure 2   The distribution of Japanese and Soviet whaling effort showing the southward trend associated with different catches of baleen whales.

However, in 1972 the Soviet whalers changed operations and started to process large amounts of baleen meat; a change which may be associated with the increase in Soviet effort in the southern latitudes (Ivashin and Borodin, 1977). Associated with the southerly movement of the fleets, the age composition of the sperm whale catch was altered, with sperm whales being taken at a younger age (Allen, 1977). Clearly, the relative economic importance of target species may have a considerable effect on the nature of exploitation on other species.

Similar patterns of exploitation to those seen in Antarctic whaling are found in fisheries. Figure 3 presents data for six species from the Northwest Atlantic from ICNAF statistics (Anon., 1977a). Each species has been standardized to its maximum catch (= 1.0), and the successive interest shown in the haddock *(Melanogrammus aeglefinus)*, then the cod *(Gadus morhua)*, and then the herring *(Clupea harengus)*, mackerel *(Scomber scombrus)*, squid (various species), and capelin *(Mallotus villosus)* can be observed.

Another example of a complex in which overdemand for one species has consequences is seen in the bowhead whale problem. The Alaskan bowhead whale *(Balaena mysticetus)* is considered to number about 1300 animals (Breiwick and Chapman, 1977) and is regarded as endangered (Anon., 1977b). Along with other species the Alaskan eskimos have traditionally hunted the bowhead, taking between 15 and 20 over the past five years but striking more. If the eskimos were prohibited from taking this whale, hunting would have to switch to alternative species such as various seals, the gray whale *(Rhochianectes glaucus)*, the beluga whale *(Delphinopterus leucas)*, or terrestrial animals. The most promising alternative may have been the Northwest Arctic herd of caribou *(Rangifer tarandus)* which formerly was over a quarter of a million in number is now estimated to be between 60,000 and 75,000 (Anon., 1977c). This stock has a total quota of 3000 males per year with a limit of one per person. A switch from bowhead to caribou could not take place as the caribou is fully exploited. Consequently the Alaskan harvests are being considered with respect to the local ecosystem. It is worth emphasizing that the multispecies relationships are only affected through man's interventions.

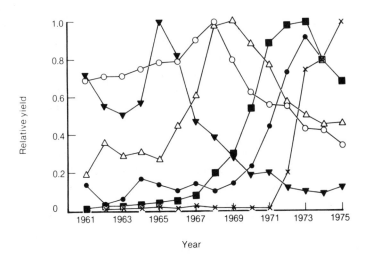

Figure 3   Relative yield of cod (○), squid (●), haddock (▼), herring (Δ), mackerel (■), and capelin (×) from the Northwest Atlantic from ICNAF statistics (Anon., 1977a) with an exploitation pattern similar to that of Figure 1.

The By-Catch Problem

When fishing for a target species of fish other fish are frequently caught, depending on the locality and gear used. This incidental catch is called the by-catch. It is a major feature of many fisheries and was extensively discussed by Beverton and Holt (1957), particularly in relation to the North Sea fishery for plaice *(Pleuronectes platessa)*. Bannister (1977) has explained that the plaice could probably be maximally harvested, for weight, by leaving them until about age 13, at which age they would all be harvested. This cannot be done; some fishing on age groups less than the critical age has to be accepted. The age at first capture can be regulated through mesh size but the mesh size in the North Sea is as much determined by the fisheries for whiting *(Merlangius melangus)* and haddock as by the plaice. As a result plaice show an age of recruitment of 4 years. If the total yield is the main criterion for setting mesh sizes, then there may be some species caught at too early an age and to a degree that is too great to enable the stock to sustain itself. Such a situation has been described by Holden (1974, 1977) for elasmobranchs. Catch per unit of effort has dropped consistently to a third of its former level over the past 20 years in the Irish Sea and Bristol Channel, and Holden considers that over this period the stocks of rays had not been replacing themselves. The stock and recruitment relationship in these species appears linear. Brander (1977) has said that in order to maintain a sustainable fishery for skates and rays in the Irish Sea, or even possibly to avoid extermination, effort must be reduced by some 30%. This is not a trivial problem.

The issue of mesh size for the plaice fishery is further complicated by the different growth and mortality rates of the two sexes. The mortality rate of the males is almost twice as high as that of the females. Males should be fished harder and at a younger age than females. Thus, even for the same species, compromises need to be reached for optimizing total return. A similar problem may be seen in whaling where management and exploitation has been considered for the separate sexes in a stock (see Chapters 11, 23). This occurs in the sperm whale fisheries where the social structure of the whale has allowed an exploitation pattern that leaves a high proportion of females and removes a large proportion of the nonbreeding males compared with baleen exploitation (e.g., Best, 1976). Quotas for males and females in a given divison may be quite different; one sex may be protected and the other not. Clearly if identification of sex is not obvious at the time of the kill, then in an area where one sex is protected, exploitation on both may have to cease. If this is impossible, some other compromise should be adopted.

The major example of marine mammals occurring in by-catches is in the tuna-porpoise fishery in the Pacific. The dolphins associate with yellowfin tuna *(Thunnus albacares)*, and fishermen use their presence to locate tuna whereupon nets are set around both. Some of the dolphins are entangled in the nets and drowned. Norman (1976) recorded that during the 1960s about 400,000 dolphins a year were killed in this way. Estimates of incidental mortality have been calculated for the period 1971 to 1975 by species and are reproduced in Table 1 (Anon., 1976b).

Table 1   Provisional Estimates of Incidental Mortality of Dolphins and Whales in the Eastern Pacific Tuna Fishery. (Anon., 1976b.)

Species/stock	1971	1972	1973	1974	1975
Spotted	185000	273000	120000	75000	106000
Eastern spinner	130000	65000	74000	22000	26000
Whitebelly spinner				18000	39000
Common	4000	9000	23000	4000	8000
Striped	339	848	232	403	1768
Bottlenosed	46	114	31	54	238
Fraser's	2	6	2	3	13
Roughtoothed	2	5	1	2	11
Risso's	3	8	2	4	16
Pygmy killer					
Pilot whale					
Melon-head	0	6	7	6	7

Such a large kill has troubled U.S. authorities, and quotas have been put on the numbers that can be killed before fishing must cease. Observers have been placed on the tuna boats, and gear has been altered. Stock assessments of the dolphins and small whales have been undertaken, and quotas set by stocks (Anon., 1977d). The 1976 quota of a total of 78,000 was reduced to 59,050 for 1977, but with a zero quota for the eastern spinner dolphin which, at very roughly half of its original stock size, is considered to be "overexploited." Norman reports that this restriction on the eastern spinner alone may cost the U.S. tuna fleet 68,000 tons of fish valued at about $45 million. Full statistics are presented in Anon. (1976b).

Interference

Interference is the process in which one species or group adversely affects another without having any effect on its own group. Two examples of interference are presented here. The first illustrates a substantial management problem and can be observed among the plant communities, seals, and birds on a group of islands off the northeast coast of England. One of the breeding sites of the gray seal *(Halichoerus grypus)* is the Farne Islands. The seals' presence has had a major effect on the island habitat (Bonner and Hickling, 1974; Bonner, 1975). The population has increased steadily from the 1920s (Summers et al., in preparation), and early evidence suggests that the seals' impact on the islands' soil and vegetation was small. Damage to the vegetation is caused by the puffins *(Fratercula artica)* and gulls. Natural growth of the vegetation can cope with this problem, but large numbers of seals have prevented regrowth, leaving the thin soil exposed for erosion. This erosion has altered the floral composition of the islands, has robbed the puffins of breeding sites, and has driven off the rosate tern *(Sterna dougalli)*. The seals, of course, are not affected by this process. The interests of the National Trust in the whole island ecosystem have necessitated the introduction of a management plan to reduce or restrict the seals in the area. This has been welcomed by the salmon fishermen, as explained later.

The second example is taken from Jenkins (1921) and involves a curious marine association. It is claimed, but by no means proved, that the whale fishery interfered with the herring fishery. The Scottish herring fisherman in about 1905 objected to the presence of whalers on the fishing grounds and tried to get them banned. The fishermen claimed that the whales' spouting indicated the presence of the herring and hence wanted the whales to remain. Furthermore, it was claimed that the whaling ships disturbed the fish. Whatever the truth of the claims, the herring fishermen believed interference had occurred, adversely affecting them but with no impact on the whales.

Competition

Competitive relationships and effects can be extremely difficult to define in any precise form. Often it is clear that a predator-prey interaction does not exist even though an increase on one species may occur at the same time as a decrease in another. These complementary changes are then frequently regarded as being effected by competitive links. Good examples of such interrelationships can be seen in fisheries and have been described by Cushing (1975).

In the western English Channel between 1925 and 1940 some dramatic changes occurred. From 1927 onward the herring recruitment declined to reach negligible levels about 1933. Starting in 1935 there was a large increase in the number of pilchards *(Sardina pilchardus)*. The interpretation of these events is far from simple, and Cushing has postulated that the winter pilchards, about 6 months old, had so altered the balance of the lower trophic levels that the winter and spring spawned herring could not survive. Other examples of temporal replacement may be seen in (1) the alternation of the Norwegian and Swedish herring stocks, (2) the replacement by the anchoveta *(Engraulis mordax)* of the sardine *(S. caernuleus)* off California about 1955, (3) the replacement by the sardine *(S. sagax)* (to a much smaller degree) of the Peruvian anchoveta *(E. ringens)* after about 1974, and (4) the replacement by the pilchard *(S. ocellata)* of the anchoveta *(Engraulis spp.)* off South Africa about 1960. The interpretation of these changes frequently ignores the changes in ecosystem structure which can be induced by increased or decreased numbers of predators. Many interpretations rely directly on assumptions of environmentally caused changes. Only in extreme examples can the environmental argument be shown to be wrong, and the explanation must frequently lie in a combination of both. This situation is not without parallel in the understanding of marine mammal fluctuations, as will be explained later.

Nevertheless competition has been recognized as a major feature of the Southern Ocean whale biology. Mackintosh (1942) noted an increase in pregnancy rates in the Antarctic blue and fin whale, and this was also noted by Laws (1961) who suggested that the changes were associated with the extra food available to the remaining whales (see Chapter 2). Gambell (1973, 1975) illustrated the change in pregnancy rate over time for the blue, fin, and sei whales. The case of the sei whales is particularly interesting, since Gambell describes a steady increase in apparent pregnancy rate from about 25% post-World War II to about 60% in 1970. However, it will be recalled that significant whaling did not begin on the sei whales until about 1955 or even 1960. Consequently interspecific effects may have occurred. Further information has been obtained through the investigation of the growth layers in the ear plugs of these whales. The age at maturity can be obtained by observing the age of the transition layer. Lockyer (1972, 1974) has recorded the age of maturity of both sexes of fin and sei whales from 1910 in the former and 1925 in the latter. All data sets show a decline in the age of maturity with time, implying that interspecific effects played a substantial part.

However, little insight is gained by merely noting statistical relationships. More information is needed about the ways these population parameter responses were induced. Kawamura (1978) has discussed the food and distribution of the Southern Ocean baleen species and concluded that direct competition for food, in the case of the sei whale, would have been with the right whale *(Eubalaena glacialis)* rather than the blue, fin, or humpback *(Megaptera nodosa).* More recently, though, the sei whale has been observed farther south where it could only have consumed krill *(Euphasia superba).* Evidence that the observed changes in vital rates of the sei whale are real rather than sampling artifacts comes from sighting data off Durban and Donkergat (Anon., 1977e), which show an increasing stock size over the period 1954 to 1965. Further effects of a reduction in the large whales are described by Laws (1977) and Payne (1977). Laws illustrates the range of food types of the whales and seals and shows that, as in the whale stocks, the age at maturity of the crabeater seal *(Lobodon carcinopagus)* has declined noticeably since about 1940. Increases in four penguin species are noted, and Payne has described how the South Georgian fur seal *(Arctocephalus gazella)* has increased from a few hundred in number in the 1930s to over 300,000 today. Laws (1977, Chapter 2) concludes that it is reasonable to assume that the reduced baleen whale density has allowed whales and other animals to grow and breed faster and possibly improve survival.

Predator-Prey Associations

All larger animals feed and are fed upon. This section describes some examples of predator-prey associations that have been identified as having, or likely to have, an obvious effect on man's desires for resource utilization. The process of categorizing these interrelationships is artificial. Man's or nature's interference with balances in predator-prey relationships may cause disturbances in other parts of the ecosystem, but only the most direct associations are considered here.

As Laws (1977) has pointed out, one of the current problems in the management of marine resources is the problem posed by the potential krill harvesting in the Antarctic. Krill dominate the zooplankton of the higher latitudes of the Southern Ocean, and the reduction of the baleen whales as predators from an early level of about 43 million tons to a present level of about 7 million has led to the suggestion that there may be "surplus" krill in the Southern Ocean that can be harvested by man. Horwood (1978a) considered the estimates of potential production from a krill fishery and its implications for whale management. Estimates of the standing stock and krill are between 800 and 5000 million tons and calculations in the early 1970s have suggested that an annual yield of the order of 100 million tons might be taken from the krill. These figures have initiated considerable interest in developing such a fishery, especially when it is noted that the current world catch of fish is between 60 and 70 million tons (Chapter 2). Holt (Horwood, 1978a), however, has reconsidered these estimates in relation to the management of whales and seals but still arrived at figures of the order of tens of millions of tons. Various countries have studied the problems of a krill fishery, and Sahrhage reported (in a lecture given to the Fisheries Laboratory, Lowestoft, England) that data from 209 trawl hauls yielded average catch rates of 10 to 12 tons of krill per hour. The highest recorded value was 35 tons caught in 8 minutes. Detailed results of the Federal Republic of Germany Antarctic expeditions in 1975/1976 are given in a research report (Anon., 1977f). The Southern Ocean complex depends to a large degree on this one species, and exploition of it will affect the Antarctic whales, seals, and birds.

The exploitation of krill *per se* is not the only problem of importance. Of more concern is the amount of exploitation and its effect. At present (1978) neither the processing technology nor the marketing is adequate to cope with the potential catches of krill. It has been estimated that a harvest of 10 million tons would require some 300 ships. International fishing is presently being limited, and there are not 300 ships that could go to the Antarctic. The development of such a fleet would take time, even if the marketing and economic problems can be resolved. Laws (personal communication) suggested that there might be grounds for optimism since concern has been expressed before a fishery has had a chance to develop. Research is currently being carried out to better understand the ecology of the krill.

The Southern Ocean illustrates a complex in which both predator and prey might be exploited. However, a more common problem is one in which man wishes to exploit a prey species and the predator is a competitor with man. Seals take this role. Parrish and Shearer (1977) described three ways in which the seals affect fish catches in Scotland. Seals damage both the gear and fish caught in nets, particularly salmon *(Salmo salar)*, which can be seen mutilated in the nets. These losses are financially small but clearly observable to the fishermen. The major loss is through direct predation of fish by the seals. Estimates of fish eaten, primarily by the gray seal, are about 130 thousand tons annually. If half this amount could be taken by fishermen, then a value of between 15 and 20 million pounds would be realized. Other more subtle conflicts are realized through the carrying of fish parasites, such as the nematode *(Phocanema decipiens)*, of which the gray seal is the final host. The presence of these nematodes reduces the marketability of fish and increases processing time. Other marine mammals are the final hosts of other fish parasites. The management implications are discussed by Summers and Harwood (1978).

Direct predator-prey systems are not so obvious in fisheries, particularly in the shallow sea areas where diversity is high. Anderson and Ursin (1977a,b) have interpreted the increase in North Atlantic gadoids over the past decade and the decline of the pelagic stocks as a multispecies predator-prey assemblage. This latter interpretation has not been without its critics. Hempel (in a lecture given at the Fisheries Laboratory, Lowestoft) described the change in zooplankton and phytoplankton which occurred over the same period and suggested that the change in species composition may have caused a change in the fish abundance. This argument is extended by Garrod and Colebrook (1978). Steele (1980), however, has attributed increased catches of the gadoids to an increase in variance of recruitment and rejects a direct causal relationship between the herring and gadoids. The statistically significant relationship between the two groups, however, has led him to suggest that the stocks of herring and haddock cannot be treated separately. Although the Antarctic ecosystem may be simpler, it is clear that answers are not obvious in the much better sampled areas of the North Sea.

MODELS OF SPECIES INTERACTIONS

Models provide an explicit description of one's conceptualization of a process or system, and their use in science and biology is not new. As evidenced by other chapters in this volume, mathematical models provide perhaps the best techniques for exploring the consequences of conceptualizations. These models can be either heuristic or predictive and can be either qualitative or quantitative. There is a vast literature on single-species models both in theory and practice. As the number of species increases, the abundance of studies declines both in theory and application. The use and understanding of population models can only be fully realized through the background of theoretical models. These models lose immediate application in order to elucidate principles of biological systems. Such single-species models have been reviewed by May (1976a,b). The variance of single and complex systems have been considered by May (1973) and May et al. (1979). Spatial effects have been considered by Steele (1974a) and Levin (1978). May (1973, 1976c,d) has discussed the behavior of multispecies systems. Knowledge from these models facilitates more informed interpretation of results from applied models.

Applied Marine Models of Species Interactions

Technology-Induced Interactions

The mixture of fish species frequently caught in the same area, their basic similarities, and the lack of knowledge of the biological interrelationships of these species has led to the development of models of total biomass for management. Such a model and assessment has been undertaken by Pinhorn (1976) for the groundfish off the east coast of Canada. Brander (1975, 1977) has discussed such models for demersal production for the Irish Sea. These models have some clear advantages particularly in including unspecified implicit interrelationships, and Pope (1976) and Larkin (1963) have concluded that yields from mixed assessments will be less than the sum of individual assessments. However the models falter if fishing can be directed disproportionately onto some of the stocks. Horwood (1976a) considered the model

$$\frac{1}{p}\frac{dp}{dt} = a_1 - b_1 p - \frac{Fp}{p+r}$$

$$\frac{1}{r}\frac{dr}{dt} = a_2 - b_2 r - \frac{Fr}{p+r}$$

The two species p and r are biologically independent but are connected through a proportionate fishing effort. In this simple example the fitting of logistic models to a total stock $(p + r)$ will give biases depending on the range of F. The strategy of fishing will not necessarily take the fishery through the total maximum sustainable yield.

Gatto et al. (1975) have developed a simple model which might be considered in the section under predator-prey systems but as Man is the predator I consider it here. The prey species is an exploited stock and the predator is the fleet. The recruitment relationship used for the stock is that of Ricker (1954), and the two components are related by a catch equation. Through such simple models, and more refined models being developed throughout the world, it is being realized that stock management must also take account of the development of the fleets and try to advise on strategies for the growth and investment involved. With a different recruitment function it is directly applicable to industries taking marine mammals.

Competitive Interrelationships

Simple models of fisheries for competing species have been considered by Beverton and Holt (1957), Larkin (1963), and Pope (1976). Pope (1976) and Pope and Harris (1975) applied a model to two species, the South African anchovy and pilchard, regulated through a logistic relationship and linear competition. The model of Pope and Harris is primarily heuristic, whereas the fisheries model developed by Lett and Kohler (1976) purports to be of greater interest to managers. The model depicts the exploitation on the Gulf of St. Lawrence mackerel and herring. Conclusions based on this study lead to an extreme strategy for managing the two species. However, it is only one conceptualization of the system and the environmental trends are not considered. The existence of statistical correlation does not imply a biological association. The situation is closely analogous to the marine mammal complex in the Southern Ocean described by Horwood (1979).

The marine mammals of the Southern Ocean have undergone biological changes over the past 50 years (e.g., Laws, 1977; Chapter 2) and, as in Lett and Kohler's example from fisheries, competition has been cited as the cause of such changes. Assuming this was indeed the cause, Horwood (1978b) has investigated the effect of interspecific competition on the population dynamics of the sei whale. Only states of equilibrium were considered and the two density- dependent functions used were the linear increase in apparent pregnancy rate with total baleen whale biomass and a linear decline in the age of maturity with total baleen whale biomass. The biology of the sei whale has subsequently been further defined (Anon., 1977e) and the model has been altered accordingly. The true pregnancy rate differs from the observed rate as lactating females are protected; consequently the true pregnancy rate P is calculated from the apparent rate (P') by the relationship

$$P = \frac{P'}{1 + 0.6P'}$$

The value 0.6 accounts for female whales that can be caught after weaning. The age at first capture was also altered and held constant at 7 years for both sexes. The new yields at equilibrium for sei whales, for the total Antarctic, are illustrated in Figure 4 for a range of other baleen whale biomass corresponding to the early and present stages of exploitation. The figure shows that equilibrium yield may have increased from 2000 to 25,000 as the other baleen whales declined and the fishing intensity would have to have increased from about 1% to 4½% to take a maximum sustainable yield.

The model described above illustrates only the logical outcome of one very specific idea of how the sei whales are regulated. Even if it is accepted that competition is occurring, the balance between inter- and intraspecific effects has not seriously been considered. In this model the other whale biomass is given an equal weighting to the sei whale biomass, but it could be argued that a greater weighting be given to sei whale biomass, and so reduce the apparent difference seen in Figure 4. An aspect that can be seen explicitly is that early sustainable yields of sei whales were small and that the early take of a few hundred may have had as significant an impact as taking a thousand later. One use of such models is to highlight pitfalls for managers and to be more explicit about biological ideas. The Scientific Committee of the IWC (Anon., 1977b) has used interspecific models with a multispecies regression on pregnancy data. On the basis of statistical considerations, a model was suggested using regressions of

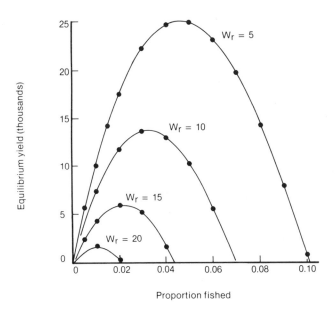

Figure 4 Equilibrium yield of Antarctic sei whales simulated for a range of other baleen whale biomass (W_r in tons $\times 10^6$) corresponding to the early and present stages of exploitation. The model assumes that pregnancy rate and age of maturity is affected by total baleen biomass.

pregnancy rate against blue and sei whale density. The inclusion of fin whales in the analysis did not improve the statistics. Why blue whales might effect the sei whale pregnancy rates and not the fin whales is not discussed (see Chapman, Chapter 14, for further discussion). For management purposes, however, a formal model or conceptualization has to be developed in spite of inevitable reservations.

Predator-Prey Systems

Predator-prey systems have largely been ignored in fisheries, although ble gadoid-herring complex is a notable exception (Anderson and Ursin, 1977a,b). There can be no doubt, however, that the baleen whales and the krill exhibit such a system. A simple model of this complex has been described by Horwood (1976b). Basic details are known about the population biology of the large whales and density-dependent responses may be observed. However we do not understand how the specific responses of changes in fecundity, growth rate, and possible mortality are affected. Obviously their food supply is involved but exactly how has not yet been determined. The dynamics of the krill are even less known. Consequently the model of this whale-krill complex is speculative, certainly in any quantitative sense. Basically it is assumed that the less krill there are the worse off the whales will be. The model is defined as:

$$\frac{dz}{dt} = rz\left(1 - \frac{z}{k}\right) - awz - bz - F_1 z$$

$$\frac{dw}{dt} = r_1 w\left(1 - \frac{w}{cz}\right) - F_2 w$$

where z = amount of krill in millions of tons

w = amount of whales in million of tons

r = instantaneous reproductive rate of the krill at low z

r_1 = instantaneous reproductive rate of the whales at low w

a = predation rate by the whales on the krill

b = predation rate of the other competitor (seals, etc.)

F_1 = fishing rate on the krill

F_2 = fishing rate on the whales

The coefficient c provides a proportionality constant for the "carrying capacity of the whales." This form of the model, wherein the carrying capacity is proportional to prey abundance, has been considered in the theoretical literature (see, e.g., May, 1973). The value of the coefficients are given by Horwood

(1976c). Equilibrium levels of krill and whales for different levels of exploitation on them are shown in Figure 5. Increased exploitation on either of the groups results in lower abundances, but the angle of the isopleths reflects the degree of interaction between the two groups, and the obvious result is that the fewer whales there are the higher the standing stock of krill. Figure 6 shows the total "economic" yield from the krill and whales then the whale meat, per ton, is valued at various levels from 1 to 100 times that of krill. Yields from krill are much higher than from whales, such that, if value is about the same, the exploitation pattern will be dominated by the krill. Up to the level of mortality considered (30%), krill yields increase and are proportionately more the harder the whales are harvested. Maximum whale yields occur at about $F_2 = 0.025$ and accounts for the minima in the yield isopeths at that stage. As the value of whales increases well above that of krill (Figure 6c) the picture changes, and it is very important to insure that whales are harvested optimally, even though greater returns are obtained the harder the krill are fished. Consequently if the two fisheries are harvested and managed independently, conflicts will occur, if a joint management regime is used, economics must play a major role. It must be noted that quantitatively these results are likely to be very wrong. They are presented to illustrate the qualitative arguments involving management of the whale-krill-man complex. A further aspect, not covered in this study, is the dynamics of such a system when not at equilibrium; with the generation times of populations of the two groups almost an order of magnitude different, this poses its own problems.

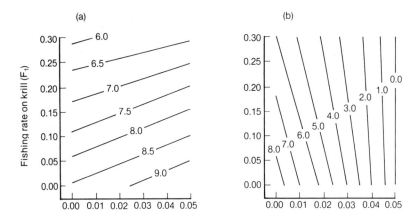

Figure 5   Equilibrium levels for krill (tons $\times 10^9$) (a) and whales (tons $\times 10^8$) (b) for different levels of exploitation for krill (F_1) and whales (F_2) from a predator-prey model. The slope of the lines reflects the degree of interaction.

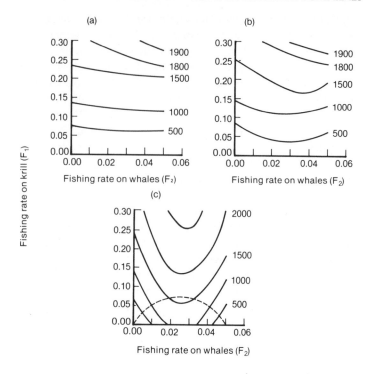

Figure 6   Total equilibrium yield (arbitrary units) from krill and whales when the whale products are valued at 1, 10 and 100 times that of krill by weight. The dashed line (c) is where whales provide 50% of the total "economic" yield.

Marine Ecosystem Models

There is a considerable quantity of published literature concerning the marine ecosystem and attempts to model it. The oceanic systems have been described by Vinogradov and Menshutkin (1977). Temperate-sea systems have been describ- ed by Riley (1946), Horwood (1976c), and Kremer and Nixon (1978). Structural models have been developed by Steele (1974b) and Denman and Platt (1976). Models involving spatial heterogeneity have been developed by Steele and Henderson (1976b, 1977) and Fasham (1978). The conclusions reached in this work are that some understanding of the structure of the systems has been achieved but important detail is still lacking. Prediction is not being attempted by those engaged in serious study of ecosystem dynamics (see also Chapter 14).

   Elton (1927) recommended the high Arctic for ecosystem level study since the interrelationships of such an impoverished fauna could be more completely characterized. Summerhayes and Elton (1923) described these interrelationships along with the energy flow for Bear Island, south of Spitsbergen. A contem- porary study involving a dynamic model of the Antarctic marine system has been undertaken by Green (1975, 1980) for the Ross Sea. As with all models that broach a substantial topic, it has provided a useful frame of reference for ideas as to how the Antarctic marine system might work. It shows clearly our lack of knowledge (or the unevenness of it) over the whole system and has provided a means whereby the biologists can talk to the systems ecologists. None of these are trivial results of models. The model further provides a system of accounting for the energy flow of the Ross Sea that helps us consider different components in perspective. However, there are aspects of the model that are subject to question. One is the use of linear equations to describe any ecosystem. Citing Patten's (1975) concepts of how terrestrial systems might work, Green has adopted a linear approach. At equilibrium Green is able to illustrate that early this century the predation pressure by the whales was not substantially larger than from the rest of the predators. Such perspectives are necessary. Even so the linear approach to the *dynamics* of this system is probably wrong. May (1976b) concludes that "nonlinear systems are surely the rule, not the exception, outside the physical sciences." (Indeed, the traditional, and not illusionary, problem of communication between the natural and physical sciences may be based on this difference in emphasis.) Green's linear model, if all the compartments were well described, may provide limited insights concerning the exploitation of krill and its consequences. These consequences should, however, be interpreted in view of the linear nature of the model.

The energy budgets of the great whales are reasonably well known (e.g., Lockyer 1976, 1978). The cost of maintenance for whales is high, and at puberty it is some 16 to 17 times that for growth. The increase in growth rate observed over the past 40 years could have been effected through a small increase in food supply. The cost of pregnancy is low but that of lactation is extremely high. If it can be assumed that (1) a whale will fulfill its maintenance ration first (since if it is a "bad" year it is better off surviving until the next year rather than risking itself by having to feed a calf; see Chapter 22); (2) growth is to some degree genetically determined and, as such, growth takes second priority in the allocation of the ration; and (3) pregnancy and lactation receive third priority, then if some krill are removed, it would be the lactation that would suffer most and result in high calf mortality (see Chapter 3). This would continue to occur until the younger whales, fewer in number and smaller, had replaced the older, larger whales, thus involving a time lag. The system would take some time to settle down to a new equilibrium, and over the time scales of interest the responses would be highly nonlinear even if the end result appears linear. Consequently the use of linear models to describe dynamics may be unsuitable. It is, however, an important part of model development and provides an initial understanding of the energy flows.

DISCUSSION

The general problems involving the interactions, management, and models of multispecies have been discussed. There remains to be described only the hierarchy of uses of the models and a basic fundamental problem. This is the problem of the balance between environmentally and biologically caused effects.

Single species models have been used for some time now, and as other chapters in this volume show, are extremely useful and will probably remain our main tool for some years to come. However some problems are concerned essentially with multispecies systems and cannot be avoided (see Chapters 14-16, 18, 19). Some in which the two species are intimately linked (such as parasite-host or plant-pest systems,) have been usefully resolved and have given rise to a better understanding of population dynamics (Conway, 1977). Other systems may be linked by a somewhat more one-way relationship. Some fish populations are probably sufficiently large to satisfy seals, but damage to the economics of the fishery by seals can be significant. The predation on the cockles *(Cerastoderma edule)* by the oystercatcher *(Haematopus ostralegus)* has a major effect on the cockles, but the range of other feeding sites available to the whole oystercatcher population means that there is little impact from the cockle fishery on the birds (Horwood and Goss-Custard, 1977). Total multispecies systems have not received a great deal of attention for management, and there are still many problems associated with our understanding of them. Very simple models help to account for the major energy flows, but at present only very special situations have lent themselves to adequate dynamical studies. Many more studies are needed, both to describe biological complexes and to understand how to best use models of them. With better descriptions of the whole complex better models will be formulated.

Many of the models, both single and multispecies, are deterministic. This is largely due to the mathematical training of both biologists and mathematicians. Nevertheless, there are also a range of stochastic models and there are combinations of the two (see Chapters 6, 19). Frequently, however, the stochastic effects, usually designed to depict environmental changes, are imposed theoretically with little concept of how they are generated. In marine systems major changes have been interpreted by some as climatically induced and by others as being caused by changes in the higher trophic levels. In marine mammal management this can be seen in the interpretation of changes in the abundance of North Atlantic whales. Elton (1927) maintained that, historically, the Arctic seas swarmed with whales. With Dutch and British whaling, over a period of 150 years, he claimed that they nearly all disappeared. Undoubtedly the numbers killed were extreme, but is this the whole story? Koch (1945) has described the change in ice cover from 860 to 1940. During the period of intense exploitation the area covered by ice increased dramatically around Iceland. Vibe (1967) differs from Elton and concludes that this mass advance of the ice forced the whales off the shelf areas into regions of poor food supply and caused reduced fertility of the whales. He considered that this was a greater cause of the collapse of stocks than whaling. We clearly do not know. It is likely that a combination of the two causes is involved.

In the case of the Antarctic, changes in the mammalian reproductive parameters have been discussed here only in relation to changes in the density of the great whales. Piggot (1977) has shown that rainfall at South Georgia has steadily increased by 33% since 1910 and that the air temperature has increased by 2 °C; a remarkable change. Similar trends, but to a lesser extent, have been recorded over other parts of the Antarctic (Anon., 1978; Horwood, 1979). Again we do not know what effects these changes had on the marine ecosystem. For this reason the interpretation of models of such systems needs to be cautious and biological criticism more rigorously encouraged. The apparent problems associated with modeling the lower trophic levels and the apparent successes of the higher trophic level models lead to several questions. Do we need to understand the lower trophic levels for management? Is the lack of success the result of inadequate data or is the apparent success of the higher trophic level models the result of lack of data? These questions of course have yet to be answered and reflect the moot point of whether the marine ecosystems survive in spite of, or because of, management.

LITERATURE CITED

Allen, K. R. 1977. Size distribution of male sperm whales in the pelagic catches. Rep. Int. Whaling Comm. Sperm Whales: Special Issue: 51-56.

Anderson, K. P., and E. Ursin. 1977a. A multispecies extension to the Beverton and Holt theory of fishing, with accounts of the phosphorus circulation and primary production. Meddr. Danm. Fisk. Havunders N.S. 7:319-435.

Anderson, K. P., and E. Ursin. 1977b. The partitioning of natural mortality in a multispecies model of fishing and predation. Pp. 87-98 in: J. H. Steele (ed.). Fisheries Mathematics. Academic Press, London.

Anon. 1974. Redbook. Int. Comm. Northwest Atlantic Fish. 154 pp.

Anon. 1976a. Int. Whaling Statistics 77:1-50.

Anon. 1976b. Report of the workshop on stock assessment of porpoises involved in the Eastern Pacific yellow fin tuna fishery. La Jolla, California, 1976. Southwest Fisheries Center. Admin. Report LJ-76-29.

Anon. 1977a. Statistical bulletin. Int. Comm. Northwest Atlantic Fish. 25:1-234.

Anon. 1977b. Report of the Scientific Committee. Rep. Int. Whaling Comm. Vol. 28.

Anon. 1977c. Final environmental impact statement. International Whaling Commission's deletion of native exemption for the subsistence harvest of bowhead whales. Vol. 1. U.S. Dept. Commerce, N.O.A.A. 245 pp.

Anon. 1977d. Federal Register 42 (40) Pt. 5. U.S. Dept. Commerce, N.O.A.A.

Anon. 1977e. Report of the Scientific Committee, Rep. Int. Whaling Comm. 28:335-341.

Anon. 1977f. Research and Exploration of the Resources of Krill and Food Fish in the Antarctic Report of the 1975/76 Antarctic Expedition of the Federal Republic of Germany. Federal Research Center for Fisheries, Hamburg. Hamburg, April 1977. 54 pp.

Anon. 1978. Annual report 1976-77. British Antarctic Survey, Maddingley Road, Cambridge.

Bannister, R. C. A. 1977. North Sea plaice. Pp. 243-282 in: J. A. Gulland (ed.). Fish Population Dynamics. John Wiley and Sons, London and New York.

Best, P. B. 1976. A review of world sperm whale stocks. FAO ACMRR/MM/8. Paper presented at meeting on Scientific Consultation on Marine Mammals. Bergen, Norway, 31 August-9 September, 1976.

Beverton, R. J. H., and S. J. Holt. 1957. On the dynamics of exploited fish populations. Fish. Invest. London Ser. 2, 19.

Bonner, W. N. 1975. Population increase of gray seals at the Farne Islands. Rapp. P. V. Cons. Int. Explor. Mer 169:366-370.

Bonner, W. N., and G. Hickling. 1974. Seals of the Farne Islands, 1971-1973. Trans. Nat. Hist. Soc. Northumberland 42:65-84.

Brander, K. M. 1977. The management of Irish Sea fisheries. Lab. Leafl. MAFF Direct. Fish. Res. Lowestoft 36:1-40.

Breiwick, J. M., and D. G. Chapman. 1977. Population analysis of the Alaska bowhead whale stock (Unpublished report). Doc. 13. I.W.C. Special meeting on North Pacific sperm whales, Cronulla, 1977.

Conway, G. R. 1977. Mathematical models in applied ecology. Nature (London) 269:291-296.

Cushing, D. H. 1975. Marine Ecology and Fisheries. Cambridge University Press, London.

Denman, K. L., and T. Platt. 1976. The variance spectrum of phytoplankton in a turbulent ocean. J. Mar. Res. 34:593-601.

Elton, C. S. 1927. Animal Ecology. Sidgwick and Jackson, London.

Fasham, M. J. R. 1978. The application of some stochastic processes to the study of plankton patchiness. Pp. 131-156 in: J. H. Steele (ed.). Spatial Pattern in Plankton Communities. Plenum Press, New York and London.

Gambell, R. 1973. Some effects of exploitation on the reproduction in whales. J. Reprod. Fert. Suppl. 19:533-553.

Gambell, R. 1975. Variations in reproductive parameters associated with whale stock sizes. Rep. Int. Whaling Comm. 25:182-189.

Garrod, D. J., and J. M. Colebrook. 1978. Biological effects of variability in the North Atlantic Ocean. Proc. Joint Oceanographic Congress, Edinburgh, 1976.

Gatto, M., S. Rinaldi, and C. Walters. 1975. A predator-prey model for discrete time commerical fisheries. IIASA Res. Rep. 38 pp.

Graham, G. M. 1935. Modern theory of exploiting a fishery and application to North Sea trawling. J. Cons. Int. Explor. Mer 10:263-74.

Green, K. A. 1975. Simulation of the pelagic ecosystem of the Ross Sea, Antarctica: A time varying compartmental model. Ph.D. Thesis, Texas A&M University, College Station.

Green, K. A. 1980. Modelling Antarctic ecosystems. In: S. J. El-Sayed (ed.). Biological Investigations of Marine Antarctic Systems and Stocks (BIOMASS), Vol. 2. Scientific Committee on Antarctic Research.

Hardy, A. C. 1924. The herring in relation to its animate environment. I. The food and feeding habits of the herring with special reference to the East coast of England. Fish. Invest. London Ser. 2,7:1-53.

Holden, M. J. 1974. Problems in the rational exploitation of elasmobranch populations and some suggested solutions. Pp. 117-137 in: F. R. Harden Jones (ed.). Sea Fisheries Research. Elek. Science, London.

Holden, M. J. 1977. Elasmobranchs. Pp. 187-214. in: J. A. Gulland (ed.). Fish Population Dynamics. John Wiley and Sons, London.

Horwood, J. W. 1976a. Interactive fisheries: A two species Schaefer model. Int. Comm. Northwest Fish. Select. Pap. 1:151-155.

Horwood, J. W. 1976b. On the joint exploitation of krill and whales. FAO ACMRR/MM/SC. Paper presented at meeting on Scientific Consultation on Marine Mammals. Bergen Norway, 31 August-9 September, 1976.

Horwood, J. W. 1976c. A model of primary and secondary production in Loch Striven, and its stability. Pp. 297-307 *in:* G. Persoone and E. Jaspers (eds.). Proc. 10th European Symposium on Marine Biology, Vol. 2, Ostende, 1975.

Horwood, J. W. 1978a. Whale management and the potential fishery for krill. Rep. Int. Whaling Comm. 28:187-190.

Horwood, J. W. 1978b. The effect of interspecific and intraspecific competition on the population dynamics of the sei whale *(Balaenoptera borealis)*. Rep. Int. Whaling Comm. 28:401-404.

Horwood, J. W. 1979. Competition in the Antarctic? Rep. Int. Whaling Comm. (Special meeting on sei whales, Cambridge, 1979).

Horwood, J. W., and J. Goss-Custard. 1977. Predation by the oystercatcher *(Haematopus ostralegus)* in relation to cockle fishery in the Burry Inlet, South Wales, J. Appl. Ecol. 14:139-158.

Ivashin, M. V., and R. G. Borodin. 1977. A note on catch per unit effort trends in USSR North Pacific sperm whale operations. Doc. 16. Special meeting on North Pacific sperm whales, Cronulla, 1977.

Jenkins, J. T. 1921. A History of the Whale Fisheries. H. F. and G. Whitherby, 326 High Holborn, W. C.

Kawamura, A. 1978. An interim consideration on a possible interspecific relation in southern baleen whales from the viewpoint of their food habits. Rep. Int. Whaling Comm. 28:411-419.

Koch, L. 1945. The East Greenland ice. Meddr. OM Gronland 130, 3:1-374.

Kremer, J. N., and S. W. Nixon. 1978. A Coastal Marine Ecosystem. Springer-Verlag, Berlin, London, New York.

Larkin, P. A. 1963. Interspecific competition and exploitation. J. Fish. Res. Board Can. 20:647-678.

Laws, R. M. 1961. Reproduction, growth and age of southern fin whales. Discovery Rep. 31:327-486.

Laws, R. M. 1977. Seals and whales of the Southern Ocean. Philos. Trans. R. Soc. London Ser. 279:81-96.

Lett, P. F., and A. C. Kohler. 1976. Recruitment: A problem of multispecies interaction and environmental perturbation, with special reference to Gulf of St. Lawrence Atlantic herring *(Clupea harengus harengus)*. J. Fish. Res. Board Can. 33:1353-1371.

Levin, S. A. 1976. Population dynamic models in heterogeneous environments. Annu. Rev. Ecol. Syst. 7:287-310.

Levin, S. A. 1978. Population models and community structure in heterogeneous environments. *In:* S. A. Levin (ed.). Mathematical Association of America Study in Mathematical Biology, Vol. 2, Population and Communities.

Lockyer, C. 1972. The age of sexual maturity of the Southern fin whale *(Balaenoptera physalus)* using annual layer counts in the ear plug. J. Cons. Int. Explor. Mer. 34:276-94.

Lockyer, C. 1974. Investigation of the ear plug of the southern sei whale *(Balaenoptera borealis)* as a valid means of age determination. J. Cons. Int. Explor. Mer 36:71-81.

Lockyer, C. 1976. Growth and energy budgets of large baleen whales from the southern hemisphere. FAO, ACMRR/MM/SC 41. Paper presented at meeting on Scientific Consultation on marine mammals. Bergen, Norway, 31 August-9 September, 1976.

Mackintosh, N. A. 1942. The southern stocks of whalebone whales. Discovery Rep. 22:197-300.

May, R. M. 1973. Stability and complexity in model ecosystems. Princeton University Press, Princeton, New Jersey.

May, R. M. 1976a. Models for single populations. Pp. 4-25 *in:* R. M. May (ed.). Theoretical Ecology. Blackwell Scientific Publications, Oxford.

May, R. M. 1976b. Simple mathematical models with very complicated dynamics. Nature (London) 261:459-467.

May, R. M. 1976c. Models for two interacting populations. Pp. 49-70 *in:* R. M. May (ed.). Theoretical Ecology. Blackwell Scientific Publications, Oxford.

May, R. M., J. R. Beddington, J. W. Horwood, and J. G. Shepherd. 1979. Exploiting natural populations in an uncertain world. Math. Biosci. 42:219-252.

Norman, C. 1976. Dolphin dissonance. Nature (London) 264:598-600.

Ohsumi, S. 1977. Review of Japanese fishing effort on sperm whales in the North Pacific. Doc. 2. Special meeting on North Pacific sperm whales, Cronulla, 1977.

Parrish, B. B., and W. M. Shearer. 1977. Effects of seals on fisheries. ICES CM 1977/M:14, 6 pp. (mimeo).

Patten, B. C. 1975. Ecosystem linearization: An evolutionary design problem Am. Nat. 109:529-539.

Payne, M. R. 1977. Growth of a fur seal population. Philos. Trans. R. Soc. London Ser. B. 279:67-80.

Piggott, W. R. 1977. The importance of the Antarctic in atmospheric sciences. Philos. Trans. R. Soc. London Ser. B. 279:275-286.

Pinhorn, A. T. 1976. Catch and effort relationships of the groundfish in ICNAF subareas 2 and 3. Int. Comm. Northwest Atlantic Fish. Select. Pap. 1:107-116.

Pope, J. G. 1976. The effect of biological interactions on the theory of mixed fisheries. Int. Comm. Northwest Atlantic Fish. Select. Pap. 1:157-162.

Pope, J. G. and O. C. Harris. 1975. The South African pilchard and anchovy stock complex, an example of the effects of biological interactions between species on management strategy. Spec. Meet. Int. Comm. Northwest Atlantic Fish., 1975. Res. Doc. 113 (mimeo).

Ricker, W. E. 1954. Stock and recruitment. J. Fish. Res. Board Can. 11:559-623.

Riley, G. 1946. Factors controlling phytoplankton populations on Georges Bank. J. Mar. Res. 6:54-73.

Schaefer, M. B. 1954. Some aspects of the dynamics of populations important to the management of the commercial fish populations. Int. Trop. Tuna Comm. Bull. 1,2:27-56.

Steele, J. H. 1974a. Spatial heterogeneity and population stability. Nature (London) 248:1-83.

Steele, J. H. 1974b. The Structure of Marine Ecosystems. Blackwell Publications, Oxford.

Steele, J. H. 1980. Some problems in the management of marine resources. *In:* T. H. Coaker (ed.). Applied Biology.

Steele, J. H., and E. W. Henderson. 1976. Simulation of vertical structure in a planktonic ecosystem. Scot. Fish. Res. Rep. 5. 27 pp.

Steele, J. H., and E. W. Henderson. 1977. Plankton patches in the Northern North Sea. Pp. 1-19 *in:* J. H. Steele (ed.). Fisheries Mathematics. Academic Press, London. 198 pp.

Summerhayes, V. S., and C. S. Elton. 1923. Contributions to the ecology of Spitsbergen and Bear Island. J. Ecol. 11.

Summers, C. F., and J. Harwood. 1978. The grey seal problem in the Outer Hebrides. *In:* J. M. Boyd (ed.). The Natural History of the Outer Hebrides. Proc. R. Soc. Edinburgh Ser. B.

Summers, C. F., W. N. Bonner, and J. Van Haaften. 1980. Changes in the seal populations of the North Sea (in preparation).

Vibe, C. 1967. Arctic animals in relation to climatic fluctuations. Middel. OM Gronland 170:1-227.

Vinogradov, M. E., and V. V. Menshutkin. 1977. The modelling of open-sea ecosystems. Pp. 891-922. *in:* E. D. Goldberg, I. N. McCave, J. J. O'Brien, and J. H. Steele (eds.). The Sea, Vol. 6. Marine Modelling. Wiley-Interscience, New York.

EIGHTEEN

What We Do Not Know About
the Dynamics of Large Mammals

GRAEME CAUGHLEY

INTRODUCTION

To begin, a conclusion of this chapter is stated as an axiom: until we understand the dynamics of the most simple of natural ecological systems that include animals—a simple herbivore population feeding on a simple supply of vegetation—we are unlikely to get far with a more complicated system that includes predators and competing herbivores. And yet we know very little about the simple system.

This chapter is limited, therefore, to the simple plant-herbivore system. It examines why our progress in understanding it is, to date, less than spectacular.

Models of ecological systems are constructed for a number of purposes but only two need to be discussed here. One purpose is to predict; the other is to aid understanding of the system. But there is overlap. A predictive model may provide incidental biological insight, although that is not a necessity. Its constants need not correspond with biological processes and its form need not duplicate the structure of the system. Its function is primarily to predict. Extreme examples are provided by Ptolemy's cosmology and by the polynomial regression model. In contrast, a consonant model must include the component parts of the system linked in the same way as they are linked in the system that is modeled. Not only its outcome but its structure must agree with reality. The construction of a consonant model begins with a study of the component processes of the system. These are assembled into a coherent whole that models the outcome of interaction between the components. Precise biological meaning can be ascribed to each of its constants. Consonant modeling was introduced to ecology comparatively recently, largely by Holling (1959). Entomology still holds a monopoly on it, the best examples being in that field (see particularly the work of Gilbert et al., 1976).

Both predictive and consonant models are prescient, their predictive power being a test of their utility. Tight prediction is necessary and sufficient to validate a predictive model, necessary but insufficient to validate a consonant model. Predictive ability does not of itself demonstrate that the model is consonant with the system, but that is no drawback if the model is used only to predict. The trouble starts when we assume that a predictive model or a contracted model (see next section) is a comprehensive picture of reality.

361

SINGLE-SPECIES MODELS

There are many single-species models of animals, but there are no animals in single-species systems. An animal must eat plants or other animals, and consequently its ecological relationships at their most simple are those of an animal within a two-species system. Single-species models of animal populations are not consonant with real systems, although they may be close enough for practical purposes when the animal does not affect the level of resources available to its next generation. Alternatively, they may be discordant with biological processes but yet serve well as predictive models. But they seldom provide a useful conceptual framework for understanding how a population works, and they rarely generate the questions that would guide our search for understanding.

The essence of a single-species model is the assumption that the rate of increase of a population can be described mathematically entirely in terms of its density. The commonly used models duplicate our feeling about how a population should operate: it should grow vigorously at low density and sluggishly, or not at all, at high density. Most of the models are logistic, or minor modifications of the logistic, or discrete analogues of the logistic. May and Oster (1976) list most of the forms in current use, and Eberhardt (1977) demonstrated that both the Beverton and Holt and the Ricker stock-recruitment models are essentially logistic.

Where such a model fails to track the data, a better fit can usually be forced by introducing an arbitrary time lag into the model. Rate of increase is then viewed not as a function of density at the time both are measured but as a function of density a constant interval of time previously.

Fisheries Models

Almost all fisheries models are of the single-species type. In a recent book on fish population dynamics Ricker (1977) warned that "we omit questions of fish foods and food supply, early development, environmental effects, and many aspects of population and community interactions." That is another way of stating that single-species models will be used throughout. Quoting Ricker (1977) again: "An early analysis by Baranov (1926) sought to bring the fish's food supply into the equation explicitly, but the results proved unsatisfactory."

Single-species modeling is ingrained in fisheries dynamics. If that approach leads ultimately to a satisfactory explanation of how a population works, then fisheries research should be near that stage by now. Progress certainly has been made, but somehow clear insight into how fish populations operate seems to have eluded us. The models have predictive power when it comes to estimating sustained yield but they have not added greatly to our understanding of population biology. That is a controvertible conclusion with which many would disagree. Unfortunately, it is not easily debatable because our personal judgments on what represents a satisfactory synthesis are dictated by the level of resolution that we seek and by the scale of approximation that we are willing to accept.

It may be that our imperfect knowledge of fish populations reflects only the conditions under which fisheries biologists operate. A second possibility is that, in opting early for a single-species approach, they chose a subset of the real world too small to yield a clear picture. I suspect the answer lies midway. Fisheries biologists are faced by an inability to see the populations on which they work, by nearly impossible problems of measurement, by a necessary reduction of questions to those conceivably answerable, and by a resultant contraction of possible models to those structured on very simplified assumptions.

What has all this to do with large mammals? The single-species model that fisheries investigations produce need not be adopted as a paradigm by those of us lazy enough to demand a fully visible population. Single-species models may be the only answer to estimating fish yields, but they are not forced on us by circumstances, nor dictated by observed processes, when we study the dynamics of large terrestrial mammals.

Contracted Models

Single-species models need not be as naive as they appear. They can be used as a mathematical summary of a complex process represented in the model by constants no longer interpretable in explicit biological terms. Caughley (1976a) showed that the delayed logistic could be used to summarize the growth of a herbivore population that interacted without lag with a population of plants. The "delay" term took care of the fact that present plant density was a function of the number of animals that had been eating it over an extended period. May (1976) provided another contracted model whereby the rate of increase of a prey population was described by a single-species equation incorporating predator effects into a delay term.

Such contracted models may be useful where explicit data are not forthcoming from one or more components of a multilevel system. They may predict reasonably well, but they hide biological processes rather than exposing them. They do not aid our understanding of how the system works.

INTERACTIVE MODELS

The alternative to single-species models is embodied in interactive models. The level of resources available to the population appears explicitly in this type of model rather than being indexed by the population's present or past density. This is a considerable advantage if the resources the animals seek are plants which have their own dynamics.

In the simplest case a plant-herbivore system can be constructed of three blocks: (1) the growth response of plants as a function of edible plant biomass, (2) the numerical response of herbivores to edible plant biomass, and (3) the link between the two provided by the functional response of herbivores to edible plant biomass.

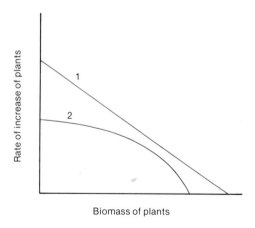

Figure 1   Two possible trends for the rate of increase of plant biomass, per biomass unit, against standing crop of plants. Curve 1 leads to the parabolic relationship between growth increment and standing crop characteristic of the logistic model. Curve 2 implies a similar relationship but with the peak displaced to the right.

Plant Growth

Most of our knowledge on how fast plants grow and reproduce at different densities or levels of biomass comes from studies of artificial forests or pastures (Harper, 1961, 1977; Donald, 1963; Donald and Hamblin, 1976; Sagar and Mortimer, 1976). Growth is usually logistic or nearly so. Hence the absolute increment to the standing crop is maximized when plants are held by grazing at a biomass well below the ungrazed level. McNaughton's (1976) findings from the natural pastures of the Serengeti Plain show this to be the case.

The close approach of plant growth to the logistic model is to be expected because plants are one of the few groups of organisms that come close to validating strict logistic assumptions (Caughley, 1976b). Their resources (water and sunlight) are renewed at a rate independent of the population's standing crop. The logistic, which, at best, is a predictive or contracted model of population growth for animals, comes close to being a consonant model for plants.

Figure 1 diagrams two possible trends of rate of increase of plant biomass, per unit of biomass, against standing crop of plants. The first trend, a straight line, leads to the parabolic relationship between growth increment and standing crop characteristic of the logistic model. The second trend implies a similar relationship but with the peak displaced to the right. The examples of growth responses collected from the literature by Noy-Meir (1975) all fit one or the other of these patterns. Absolute productivity is maximized when plants are held at a biomass of one-half, or a little above one-half, the biomass at ungrazed equilibrium.

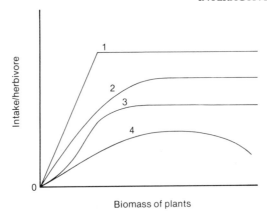

Figure 2 Functional response of herbivores to plant biomass level. Curves 1, 2, and 3 correspond to Holling's three types, while curve 4 represents the situation when grazing declines as the grass becomes so dense as to inhibit consumption.

Functional Response of Herbivores

For clues on how a herbivore's rate of intake changes as plant biomass changes we must go again to the agricultural literature. Most reported trends (see Noy-Meir, 1975, for examples) are similar to the type 2 response of Holling (1959) diagramed as the second curve of Figure 2, but often the data near the origin are too few to differentiate between this curve and a sigmoidal trend (Holling's type 3). The two forms have different implications for stability of a grazing system. Holling (1959) argued that the sigmoidal functional response would be most common among vertebrates eating more than one kind of food, but Hassell et al. (1977) showed that invertebrates too can respond in this way to prey density even when food is of one kind. The type 3 functional response may, therefore, be expected from grazing systems, but it has not yet been reported.

Functional responses may be simple or complex depending on the structure of the vegetation. A good example of the first is provided by Arnold's (1975) study of corriedale ewes grazing a pasture. When height of the vegetation averaged 25 mm, each sheep ate around 250 g/day. At 175 mm the daily intake was 1000 g. Between these extremes the trend of intake against height arched as a type 2 response.

Intake may, however, be influenced by age of the animals (Langland, 1968) and by the ratio of leaf to stem (Stobbs, 1975). Chacon and Stobbs (1976) reported that cows select leaves from the upper layers of vegetation. As the quantity of leaf available is reduced the animals take smaller bites, increase time spent in grazing, and increase the rate of biting.

The work of Bell (1971) in the Serengeti and Stobbs (1975) in northern Australia suggests that the intake of some grazing species may decline when grass is dense. This possibility is diagramed as curve 4 of Figure 2.

Numerical Response of Herbivores

The trend of rate of increase of a population against biomass of edible vegetation has not, to my knowledge, been established for any vertebrate herbivore. One would expect it to parallel the functional response because the better an animal eats the more likely it is to survive and reproduce. The numerical responses of Figure 3 have been drawn accordingly, but they are displaced to intersect the abscissa at points removed from the origin. The intercept on the abscissa represents the minimum standing crop of plant biomass needed by an animal to ensure its replacement in the next generation.

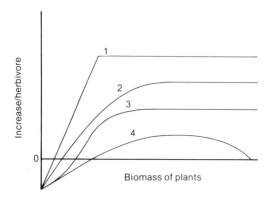

Figure 3   Numerical responses of herbivore to plant biomass level. Curves 1, 2, and 3 correspond to Holling's three types, while curve 4 represents the situation when grazing declines as the grass becomes dense. See text for details.

Stability of Interactive Systems

The question that the single-species framework forces — "How does a population regulate *itself?*" becomes a species of question begging when an interactive framework is adopted. The appropriate question in the latter case is: "By what processes does a plant-herbivore system attain an equilibrium?" It is a subtly different question. Plants and herbivores are accorded equal stress in an interactive model. In the single-species models plants are declared inert with respect to the herbivores; the blame being placed squarely on the herbivores for any tendancy toward establishment of an equilibrium.

When the system is viewed as plants and animals interacting, the principal stabilizing feature is seen as the negative slope of the plant growth response in Figure 1. The positive slopes of the herbivore's functional and numerical responses (Figures 2 and 3) pull against this and can, according to their forms, reduce the stability of the system (May, 1976). But there is a larger generality within this context. May (1972, 1975) showed by way of the Poincare-Bendixson theorem that, regardless of the details of the functional and numerical responses, essentially all two-layered interactive systems lead either to equilibrium or to a stable limit cycle.

Theory has outstripped empiricism. If we had information on the three response functions of a plant-herbivore system, we could, by elegant analysis or electronic brute force, calculate a fair approximation to the position of the equilibrium, or the stable limit cycle, for any system. Furthermore, we could gauge the strength of the equilibrium—whether it is easily displaced and tardy in its return or whether it snaps back promptly. Noy-Meir (1978) had done that for the 16 combinations of four forms of the growth response of plants and four of the herbivores' functional response.

Taking as an example the first response function of Figure 1 and the second of Figures 2 and 3, stability is enhanced when:

1   The intrinsic rate of increase of the vegetation is high.
2   The standing crop of plants when ungrazed is low.
3   The maximum intake per herbivore is low.
4   The herbivore is not skilled in finding food when food is scarce.
5   The herbivore's intrinsic rate of increase is low.
6   The herbivore is unable to maintain density when food is scarce.

SECOND-ORDER EFFECTS

The generalizations outlined above follow when the interactive system is stripped to its bones. They allow reasoned guesses as to the behavior of a plant-herbivore system, but these guesses can be converted into estimates only when the model is fleshed out by the second-order influences on stability. The term "second-order" is used in the sense that these factors would logically be fitted onto the model after its framework had been constructed from the three response functions. Their effects may, in extreme cases, override totally the stability properties of the skeletal model.

Second-order effects on plant-herbivore systems include vegetational complexity, succession, time lags, age distributions, seasonality, year-to-year variation in weather, dispersal, territoriality, sociality, and spatial heterogeneity. Only a sample of these can be considered in this chapter (see also Chapter 23).

Age Distributions

Although a herbivore population's schedules of age-specific fecundity and survival will usually respond within a year to a change in food supply, its rate of increase is less reactive (see Chapter 15). The rate of increase implied by the current schedules of survival and fecundity is not precisely the one allowed by its current age distribution unless those schedules have remained constant for some time. Age distributions introduce a reactive lag into a plant-herbivore system and this lag reduces stability. This influence on stability is probably seldom dominant because the stable age distribution is a robust configuration and convergence toward it is rapid.

Instability of age distribution becomes a more important influence if it results from surges in recruitment rather than from trends in fecundity and survival brought about by the population progressively depleting the standing crop of plants. Newsome (1977) showed that the highly unbalanced age distribution of red kangaroos *(Macropus rufus)* in central Australia during the early 1960s was dominated by cohorts released by surges of plant growth in the late 1940s, 1953-1955, and 1958. Such "boom and bust" recruitment could obliterate the stability properties of a simple interactive model.

See Chapter 23 for further considerations of the effect of age structure.

Territoriality

There are theoretical reasons for suspecting that territoriality would increase the stability of a plant-herbivore system, but we lack a demonstration of this from the field. Owen-Smith (1977) listed 41 species of ungulates known to be territorial and 32 that are not. Some genera contain both territorial and nonterritorial species. A comparison of the stability of populations of pairs of closely related species, one territorial the other not, would be of great interest. Such a comparison could, for example, involve the territorial roan antelope *(Hippotragus equinus)* living in the same area as the nonterritorial sable antelope *(Hippotragus niger)*.

See Chapters 6 and 16 for further considerations of territoriality at the level of predators.

Spatial Heterogeneity

Spatial heterogeneity would likewise be expected to increase stability or persistence (Birch, 1971; Maynard Smith, 1974). To date, however, there is little evidence of this from large mammals. Hilborn (1975) showed by simulation that it need not necessarily increase the stability of an interactive system.

Plant Succession

Following the pioneering work of Clements (1916), succession was seen as the inexorable process by which a whole community of plants reconstituted itself after a catastrophe. More recently succession has been viewed as a resultant of individual species reacting to environment in their own individual ways (Drury and Nisbet, 1973; Horn, 1974). As Horn (1976) puts it: "The only sweeping generalization that can safely be made about succession is that it shows a bewildering variety of patterns." He holds that "...the general pattern of a succession is largely determined by biologically interpretable properties of the individual species that take part in the succession."

If the vegetation of a large area were at a single subclimax stage, and successional advance were unhindered by the herbivores of the area, a plant-herbivore equilibrium would be impossible until the climax was attained. That situation is rare. More commonly the average size of successional cells is considerably smaller than the home range of a large mammalian herbivore, and these cells are at different stages of succession. As seen through the eye of a herbivore population, such an area comprises a mosaic of habitat types, some of which are being created while others are being destroyed. The overall trend, at the scale appropriate to the ecology of a large mammal, often approaches zero.

But more important, the herbivore population itself can accelerate, reverse, or halt trends in plant composition. Herbivores can hold vegetation stable in composition at a "successional stage" that would not survive in the absence of grazing—the so-called grazing subclimax. That stage could be given a better name. A large and often dominant influence on the structure and composition of vegetation is the regime of defoliation to which it is subjected. Yet climax is often defined as an outcome in the absence of defoliation, as if grazing and browsing were pathological influences having no rightful place in a decently organized botanical world. This climax is often an aberrant plant community found nowhere outside exclosure plots. For most vegetation the natural end point of succession is some equilibrium between plant composition and grazing pressure.

The mosaic of successional cells provides habitats that may be utilized selectively. Martinka (1976), for example, showed that in Glacier National Park mule deer *(Odocoileus hemionus)* favored the shrub fields of the early postfire succession, elk *(Cervus elaphus)* were more closely tied to stands of young conifers, and whitetail *(Odocoileus virginianus)* mostly inhabited forests.

I suspect that plant succession is very much a second-order influence on plant-herbivore equilibria. However, the propensity for the composition of vegetation to change when grazing pressure is relaxed, introduces the interesting possibility that some plant-herbivore systems may have more than one stable state. This idea has been explored in a broader context by Holling (1973) and specifically with respect to herbivory by Noy-Meir (1975) and May (1977).

DISCUSSION

Some simple models are reasonably successful in predicting the effect on a population of changing its density, even though they are poor blueprints of the actual mechanisms producing the outcome. These models are useful in management because they operate on relatively few data to provide an acceptable approximation to outcome, particularly when intrinsic rates of increase are low. They are less successful in revealing how the system actually works, and they may even fail in their primary purpose if they are too simple or if intrinsic rates of increase are high.

Plant-herbivore systems tend to be interactive when the herbivore is a large mammal. The herbivores increase at a rate influenced by the availability of food, and the food plants increase at a rate influenced by the number of animals eating them. Mammalian herbivores strongly affect the composition and standing crop of the vegetation, and since plant productivity is itself a function of that standing crop, the animals indirectly control the vegetation's productivity.

The relationship is too complex to be usefully summarized, even as a metaphrase, by a density-dependent single-species model. The alternative is an interactive model that mirrors the actual relationships between the animals and the vegetation. No grazing system containing a large herbivore has yet been studied in a way that allows the dynamics of the plants, the dynamics of the animals, and the interaction between the two to be integrated within a realistic model. The closest approach is probably Sinclair's (1977) study of the African buffalo.

A major concern is the lack of detailed studies of the growth of plants. To learn more about the dynamics of large herbivores we must learn much more about the reaction of plants to grazing and about the reaction of an animal's grazing behavior to changes in plant density. We must also discover at what level of vegetational complexity a single growth response must be replaced in a consonant model by a suite of growth responses.

After progressing along these lines it would be interesting to investigate how a plant-herbivore system reacts to the introduction of a predator (see Chapter 16). The effect can be modeled consonantly by adding to the model of the plant-herbivore system the functional and numerical responses of the predator. The system now has three tiers instead of the two of most prey-predator models. The latter usually seal off the system from below by ascribing logistic growth to the prey, thereby denying direct interaction between herbivores and plants, and indirect interaction between those and the predators. The three-tier model, in contrast, allows explicitly for such obviously important interactions, and it may point out more clearly what data are required to unravel the workings of a plant-herbivore-predator system.

LITERATURE CITED

Arnold, G. W. 1975. Herbage intake and grazing behaviour in ewes of four breeds at different physiological states. Aust. J. Agric. Res. 26:1017-1024.

Bell, R. H. V. 1971. A grazing ecosystem in the Serengeti. Sci. Am. 225:86-93.

Birch, L. C. 1971. The role of environmental heterogeneity and genetical heterogeneity in determining distribution and abundance. Proceedings Advanced Study Institute Dynamics Numbers Population. Oosterbeck, 1970, pp. 109-128.

Caughley, G. 1976a. Wildlife management and the dynamics of ungulate populations. Pp. 183-246 in: T. H. Coaker (ed.). Applied Biology, Vol. 1. Academic Press, London.

Caughley, G. 1976b. Plant-herbivore systems. Pp. 94-113 in: R. M. May (ed.). Theoretical Ecology: Principles and Applications. Blackwell Scientific Publications, London.

Chacon, E., and T. H. Stobbs. 1976. Influence of progressive defoliation of a grass sward on the eating behaviour of cattle. Aust. J. Agric. Res. 27:709-727.

Clements, F. E. 1916. Plant succession: An analysis of the development of vegetation. Carnegie Institute of Washington Publ. 242. 512. pp.

Donald, C. M. 1963. Competition among crop and pasture plants. Adv. Agron. 15:1-118.

Donald, C. M., and J. Hamblin. 1976. The biological yield and harvest index of cereals as agronomic and plant breeding criteria. Adv. Agron. 28:361-404.

Drury, W. H., and I. C. T. Nisbet. 1973. Succession. J. Arnold Arboretum Harvard Univ. 54:331-368.

Eberhardt, L. L. 1977. Relationship between two stock-recruitment curves. J. Fish. Res. Board Can. 34:425-428.

Gilbert, N., A. P. Gutierrez, B. D. Frazer, and R. E. Jones. 1976. Ecological Relationships. W. H. Freeman and Company, Reading, Massachusetts.

Harper, J. L. 1961. Approaches to plant competition. Symp. Soc. Exp. Biol. 15:1-39.

Harper, J. L. 1977. Population Biology of Plants. Academic Press, London.

Hassell, M. P., J. H. Lawton, and J. R. Beddington. 1976. Sigmoidal functional responses by invertebrate predators and parasitoids. J. Anim. Ecol. 46:249-262.

Hilborn, R. 1975. The effect of spatial heterogeneity on the persistence of predator-prey interactions. Theor. Pop. Biol. 8:346-355.

Holling, C. S. 1959. The components of predation as revealed by a study of small mammal predation of the European pine sawfly. Can. Entomol. 91:293-320.

Holling, C. S. 1973. Resilience and stability of ecological systems. Annu. Rev. Ecol. Syst. 4:1-23.

Horn, H. S. 1974. The ecology of secondary succession. Annu. Rev. Ecol. Syst. 5:25-37.

Horn, H. S. 1976. Succession. Pp. 187-204 in: R. M. May (ed.). Theoretical Ecology: Principles and Applications. Blackwell Scientific Publications, London.

Langland, J. P. 1968. The feed intake of grazing sheep differing in age, breed, previous nutrition and liveweight. J. Agric. Sci. 71:167-172.

Martinka, C. J. 1976. Fire and elk in Glacier National Park. Proc. Tall Timbers Fire Ecol. Conf. 14:377-389.

May, R. M. 1972. Limit cycles in predator-prey communities. Science 177:900-902.

May, R. M. 1975. Stability and complexity in model ecosystems, (2nd ed.). Princeton University Press, Princeton, New Jersey.

May, R. M. (ed.). 1976. Theoretical Ecology: Principles and Applications. Blackwell Scientific Publications, London.

May, R. M. 1977. Thresholds and breakpoints in ecosystems with a multiplicity of stable states. Nature 269:471-477.

May, R. M., and G. F. Oster. 1976. Bifurcations and dynamic complexity in simple ecological models. Am. Nat. 110:573-599.

Maynard Smith, J. 1974. Models in Ecology. Cambridge University Press, Cambridge.

McNaughton, S. J. 1976. Serengeti migratory wildebeest: Facilitation of energy flow by grazing. Science 191:92-94.

Newsome, A. E. 1977. Imbalance in the sex ratio and age structure of the red kangaroo, *Macropus rufus,* in central Australia. Pp. 221-233 in: B. Stonehouse and D. Gilmore (eds.). The Biology of Marsupials. MacMillan Press, London.

Noy-Meir, I. 1975. Stability of grazing systems: An application of predator-prey graphs. J. Ecol. 63:459-481.

Noy-Meir, I. 1978. Stability in simple grazing systems: Effect of explicit functions. J. Theor. Biol. 71:347-380.

Owen-Smith, N. 1977. On territoriality in ungulates and an evolutionary model. Q. Rev. Biol. 52:1-38.

Ricker, W. E. 1977. The historical development. Pp. 1-26 in: J. A. Gulland (ed.). Fish Population Dynamics. John Wiley and Sons, London.

Sagar, G. R., and A. M. Mortimer. 1976. An approach to the study of the population dynamics of plants with special reference to weeds. Pp. 1-47 *in*: T. H. Coaker (ed.). Applied Biology, Vol. 1. Academic Press, London.

Sinclair, A. R. E. 1977. The African Buffalo. A Study of Resource Limitation of Populations. The University of Chicago Press, Chicago.

Stobbs, T. H. 1975. The effect of plant structure on the intake of tropical pastures. III. Influence of fertilizer nitrogen on the size of bite harvested by jersey cows grazing *Setaria anceps* cv Kazungula swards. Aust. J. Agric. Res. 26:997-1007.

NINETEEN

How Ecosystem Processes are Linked to Large Mammal Population Dynamics

DANIEL B. BOTKIN

JERRY M. MELLILO

LILIAN S.-Y. WU

INTRODUCTION

One of the problems in much of the existing population theory for large mammals is its failure to view the animals within an ecosystem context. Standard approaches to population theory, including the logistic, Lotka-Volterra, and Leslie matrix models are deterministic and abstract, in that populations are treated independently of most of the variability of the environment and biographical forces. Such models are obviously justified only under certain circumstances (i.e., for certain objectives) but in many cases are necessary because of limitations in analytical techniques or the data that are available. In this paper we will discuss how ecosystem processes are linked to large mammal population dynamics in an attempt to compensate for some of the deficiencies in other approaches.

In recent years, population theory for large mammals has made considerable progress, particularly in regard to human populations. Stochastic models of conception and birth have been developed by Sheps and others (for an overview, see Sheps and Menken, 1973; Mode, 1975), and stochastic population models by Pollard (1966), Sykes (1969), and Namkoong (1972), among others. These provide a beginning for the consideration of the relation between a population of large mammals and a stochastic environment. The major purpose of this chapter is to demonstrate that to understand population regulation of large mammals one must consider their ecosystems explicitly (see also Chapters 2, 16-18).

The importance of ecosystems becomes obvious through the analysis of data on 5000 elephants taken as part of management control programs in Africa. These and other data suggest that age structures are nonstationary, and that there is a relationship between annual or seasonal rainfall and processes of birth and death (Laws et al., 1975; Wu and Botkin, 1980). The inapplicability of standard models to these data forced us to an approach that involves more detail, especially as related to the life history of the species involved. The model used in this approach considers the life history *states* open to an individual and the probability of transition between states. (We postpone discussion of the model until a later section.) Such an analysis forced us to consider the more general problem of what regulates large animal abundance and what role rainfall plays in this regulation.

The importance of ecosystems as the context for populations becomes clear when one reconsiders the observation that the abundance of herbivores in the east and southern African grasslands and savannahs greatly exceeds that found almost anywhere else in the world. The abundance of large mammals in the African plains and savannahs reaches 280 kg/ha, nearly 10 times that estimated to characterize the North America prairies prior to European colonization (Table 1). The abundance is even more striking when compared to boreal and north temperate forests. For example, in Isle Royale National Park, the abundance of moose (7 kg/ha), the island's sole large herbivore, is one-fortieth that of Murchison Falls in Uganda, although the density at Isle Royale is the highest known anywhere in the world for moose.

WHAT REGULATES HERBIVORE ABUNDANCE IN EAST AFRICA?

One common explanation for the regulation of abundance of animals in East Africa, as summarized by Coe et al. (1976), involves the observation that net primary production and animal abundance seem to be linearly related to rainfall. It is tempting to conclude that rainfall patterns determine animal abundance. This explanation can lead to the assumption that there is a causal linkage between rainfall, the total amount of net primary production, and herbivore abundance as illustrated in Figure 1a. In other words, in explaining abundance, it might be concluded that certain aspects of population dynamics can be ignored since rainfall seems so important.

Clearly such broad-scale correlations for average values can only be applied for large areas and over long periods of time and cannot be used to make estimates for specific areas at specific times. This is because

1   The current state of the habitat cannot be ignored. Indeed, the abundance of animals at Tsavo National Park, Kenya, immediately before and after the great drought of 1970-1971 would be quite different even for two years with the same rainfall period.

2   Herbivores are affected by the quality of food, that is, the concentration of essential nutrients. This has been well established for domestic livestock and some wild herbivores. For example, nutrient limitations have been reported for the African elephant, even though this species is one of the most generalized feeders of all mammals, utilizing more than 100 species of vegetation (Bax and Sheldrick, 1963).

The diet of elephants varies considerably with habitat. Their preferred diet appears to be approximately three-quarters browse (leaves and twigs of woody

Table 1 Comparison of Mammalian Herbivore Biomass in East African Woodlands, Savannas, and Grasslands with that in North American Woodlands and Grasslands, Showing the Ratio of Biomass to that of Isle Royale

Location	Habitat	No. of Species	Biomass	Ratio (kg/ha)	Ref.
Murchison Falls	Riverine	7	280	40	Laws et al. (1975)
Wankie	Savannah	17	268	38	Dasmann (1964)
Serengeti	Grassland-bush	10-12	53-175	8-18	Talbot and Talbot (1963)
Tsavo	Grassland-bush	9	45	6	Leuthold and Leuthold (1976)
East Africa	Cattle on Savannah	1	37-56	5-8	Henderson (1950), Ledger et al. (1961)
Oklahoma	Cattle on Prairie	1	37	5	Petriedes (1956)
South Dakota	Prairie	4	36	5	Calahane (1952), Petriedes (1956)
Michigan	Woodland	1	6-10	0.9-1.4	Van Etten (1951), O'Roke and Hammerstrom
Isle Royale	Woodland	1	7	1	Botkin et al. (1973)

plants) and one-quarter herbs. In some areas grasses compose most of their diet, while in others, elephants survive on underground roots and rhizomes. In spite of this adaptability, elephants appear to be subject to nutritional stress in dry seasons (McCullagh, 1969). Indeed, it is reported that elephants that died during the famous Tsavo drought were observed to have stomachs full of vegetation that was extremely low in protein.

The lack of sufficient protein suggests that nitrogen can be a limiting nutrient for elephants. Other evidence suggests nitrogen is limiting to other herbivores during the long dry season in the Serengeti (Sinclair, 1974). A rapid decline in the quantity of high quality vegetation occurs at the beginning of the dry season. Superficially, a large standing crop of food appears to remain, but chemical analyses show that crude protein in the grasses drops from 8% to less than 3%. At least 4 to 5% crude protein is required by ungulates to maintain their body weight (Sinclair, 1974). Sinclair's finding underscores the problems of using measurements of the quality of food that have been taken as mean values covering a whole year.

In addition to nitrogen there may be other nutrients that may be limiting for elephants and other herbivores. These include sodium (Weir, 1972; Dougall et al., 1964), potassium, calcium, and phosphorus (Harris, 1972). A simple relationship between rainfall and herbivore abundance could be observed, under these circumstances, if the availability of nutrients in vegetation and the production of vegetation were both directly correlated with rainfall, regardless of other climatic or edaphic factors (Figure 1a). But the system is probably not that simple. The quality of vegetation is, among other things, a function of precipitation, the nutrients in the soil, and the nutrients in and biomass of vegetation. The biomass of vegetation is, in turn, a function of herbivore activity. As McNaughton (1979) has shown, net primary production (aboveground) in the Serengeti grasslands is stimulated by moderate herbivory and reduced by extreme herbivory. Moderate grazing improves vegetation quality by converting a senescent plant community to a productive one.

The abundance of animals and quality of the vegetation are linked by a complex set of interactions mediated by the hydrological cycle. Among other things, rainfall influences the level of free-standing water, net primary production, and soil nutrients. Soil nutrients affect vegetation quantity and quality, both of which are also affected by free-standing water and the quality and quantity of the vegetation. Soil nutrients are also affected by the vegetation and animal populations (Figure 1b). It is through interrelationships such as this that animal population dynamics can be linked to the dynamics of ecosystems.

RELATING ECOSYSTEM NUTRIENT STATUS AND ANIMAL ABUNDANCE

In some ecosystems, dead organic matter and inorganic clays provide sites for storage of large quantities of various elements. However, in contrast to many

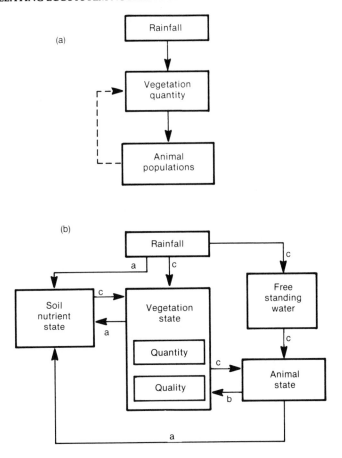

Figure 1 Relationship between rainfall and animal population dynamics. (*a*) Availability of nutrients in vegetation and vegetation products directly correlated with rainfall, regardless of other climatic or edaphic factors. (*b*) Animal abundance and the quality of vegetation linked by a complex set of interactions mediated by the hydrological cycle.

temperate zone ecosystems, those of east and southern Africa typically have little dead organic matter. In particular, some of the African habitats with the highest abundance of animals exhibit small amounts of clay in the soil and very little dead organic matter. It thus becomes of interest to determine how nutrients are retained in these ecosystems. In eastern and southern African ecosystems, two mechanisms appear potentially important. At low levels of rainfall, purely physical processes may prevent the loss of essential elements. All precipitation is returned to the atmosphere through evaporation, and the process of evaporation leaves the dissolved elements in the soil surface (Steward et al., 1967; Power et al., 1973; Rogler and Lorenz, 1974; Kilmer, 1974; Viets, 1975). At higher levels of rainfall, water moves out of the soil into streams and through subsurface runoff. In an area devoid of life, these hydrological processes would lead to soil infertility. There is evidence that in these areas, specific biological mechanisms retain nutrients (Bormann and Likens, 1967; Cole et al., 1968; Vitousek and Reiners, 1975).

The work of Scott (1962) on soils of east Africa supports the argument that the mechanisms of nutrient retention are rainfall dependent (Figure 2a). At very low levels of rainfall (< 500 mm/year), saturation is maintained regardless of type of soil. With increasing rainfall the levels of nutrients decrease exponentially, from 100% of saturation levels at approximately 500 mm/year, to a minimum of about 15% saturation at 740 mm/year. A discontinuity occurs at this minimum. The levels of nutrients increase with rainfall up to a maximum of 55% of saturation levels at about 1200 mm/year, and declines for higher levels of precipitation.

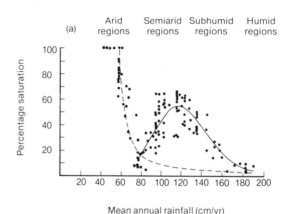

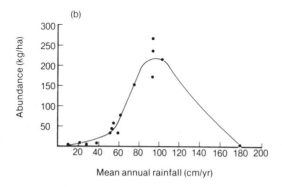

Figure 2   (a) Percent soil base saturation as a function of mean annual rainfall ranging from arid to humid conditions. Fitted curves describe relationship at increasing rainfall levels, with an apparent discontinuity at approximately 74 cm/year. (b) Animal abundance as a function of mean annual rainfall. See text for details.

It is interesting that a graph of large mammal biomass versus rainfall shows a similar pattern, increasing with rainfall to a peak at near 1000 mm/year, decreasing to much lower values at very high rainfall, in tropical rain forests (Figure 2b). The statistical relationship observed by Coe et al. (1976) between herbivore productivity, biomass, and rainfall is based primarily on values that lie within a restricted range wherein precipitation and the abundance of animals do seem to show a simple relationship, as shown in Figure 2b.

Scott (1962) interprets the increase in nutrient levels between 740 and 1200 mm/year to be due to the establishment of tight intrasystem chemical cycling between plants and soils. As a somewhat more general hypothesis, we suggest that animals act together with plants so that nutrients are retained in the ecosystem in spite of the forces of hydrological and eolian erosion. Quantitatively this biotic effect on ecosystem nutrient retention is the difference between the upper solid line in Figure 2a and the lower dashed line.

Some of the major parks and preserves in Africa have precipitation at levels that place them near the minimum of the scale of concentration of nutrients shown in Figure 2a. For example, Kenya's Masai-Mara National Park, in the northern Serengeti, is characterized by mean precipitation high enough to suggest that nutrient retention would be at least partially under biotic control. Other areas in the southern Serengeti in Tanzania have rainfall values less than 500 mm/year, indicating that on the average nutrient retention in these areas would be under more abiotic control. With annual variation in precipitation, it is possible that the control of nutrient retention shifts back and forth from biotic to abiotic mechanisms within the same ecosystem.

The dynamics and interrelationships described above suggest that the processes involved in the population dynamics of large mammals in these ecosystems must involve the effects of the large mammals and the vegetation on nutrient cycling. An ecosystem from which the large mammals (or the large mammals and most of the vegetation) were removed for a sufficient length of time could undergo severe loss of nutrients from which recovery might be slow, even much slower than would be predicted from the simpler assumption that rainfall is linearly and directly related to net primary production and animal abundance.

It appears that both higher plants and mammalian herbivores are involved in nutrient retention in savannas and woodlands where rainfall is moderate to high. Higher plants act both directly and indirectly to conserve a system's nutrients. Through rapid uptake of available soil nutrients and their incorporation into plant tissue, plants act directly to conserve the system's nutrients. By exerting control on the system's hydrological cycle and channeling precipitation through transpiration rather than runoff, plants act indirectly to conserve nutrients. Allelopathic control prevents nitrification by East African grasses (Munro, 1966a,b; Stiven, 1952), and if operative (Purchase, 1974), would be another example of plants acting indirectly to retain nutrients. During nitrification hydrogen ions are released into the soil solution, where they displace cations, thus rendering the cations more susceptible to loss via leaching (Nye and Greenland, 1960; Likens et al. 1970).

In East African grasslands, savannas, and woodlands the role of mammalian herbivores in nutrient cycling processes may be quite large compared to the role of mammalian herbivores in temperate zone prairies and forests. This may be clarified by reference to Figure 3, which distinguishes the major pathways of nutrient cycling in terrestrial ecosystems. In temperate zone ecosystem the flux of nutrients through the mammalian herbivores is small, whereas in the ecosystems of East and Southern Africa the flux of nutrients through the large mammals may be substantial, thus providing a major control mechanism for nutrient retention within those ecosystems.

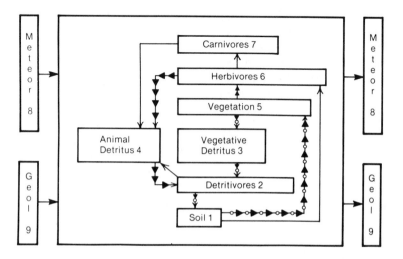

Figure 3   Major storage units (1-9) and major pathways of nutrient cycling in terrestrial ecosystems. Solid lines denote flux pathways. Pathways that are dominant in temperate ecosystems are marked with circles. Those dominant in grasslands, savannas, and woodlands of East Africa are marked with triangles.

In grassland systems Wiegert and Evans (1967) found that herbivore consumption rarely exceeded 10% of the net primary production (NPP) of the vegetation. The only exception appears to be in the East African tropical grasslands which support large ungulate populations. There it was estimated that between 30 and 60% of the primary production was consumed. Working with a more complete data base for the Serengeti Plains, Sinclair (1974) estimated that the NPP consumed by ungulates is between 19% in the tall grasslands and 34% in the short grasslands. Thus, it appears that large mammalian herbivores in African grasslands are a major pathway of material flux. The relationship between this pattern of the flux of chemical elements through the herbivores and the retention of nutrients in the ecosystem is unclear at this time. The cycling of nutrients through the herbivores may help to maintain a pool of nutrients near the surface of the soil, readily available to the vegetation. The herbivores may stimulate the uptake of nutrients through the consumption of vegetation resulting in rapid turnover for the various elements. McNaughton (1979) has shown that the NPP aboveground is stimulated by moderate grazing in the Serengeti. The nutrients required for this increased growth of plants aboveground must come from one of two sources: the nutrients in the soil or nutrients stored in the roots. The uptake of nutrients from the soil and their incorporation into plant tissue would make them less susceptible to leaching. Translocation of nutrients from the roots to aboveground plant parts, followed by consumption of those plant parts by herbivores would result in lower nutrient content of root litter. These processes would result in the maintenance of a nutrient pool near the soil surface. This seems particularly likely for cations, but the situation may be more complex for nitrogen. It may be that the cycling of nitrogen through herbivores increases the potential for nutrient losses via volatilization of NH_3 from animal urine and feces (Woodmansee, personal communication).

Up to this point in our discussion only the roles of higher plants and mammalian herbivores have been stressed insofar as they relate to nutrient retention in ecosystems. The activity of arthropods, in particular ants and termites, has often been suggested as important in mineral cycling. Fire also can play an important role. Fire and grazing have certain similar effects on vegetation. Both remove aboveground stems and tend to convert elements from comparatively immobile forms to comparatively mobile ones. Both can decrease the amount of matter in the form of dead vegetation. A complete explanation of the causes of nutrient retention in ecosystems would take into account, among other things, fire, soil microorganisms, arthropods, large mammals, and vegetation.

At this point we are faced with several conjectural explanations regarding the role of the biota in nutrient cycling, and the importance of these roles in determining the abundance of the large mammals. These explanations are:

1   Herbivores in African ecosystems are limited, at least under some conditions, by the nutritional quality, not the quantity of vegetation.
2   The nutritional quality of vegetation depends, over long time periods, on the retention of essential elements by the ecosystems.
3   The retention of nutrients in ecosystems involves mechanisms that are biotic at moderate to high rainfall and abiotic at low rainfall.
4   Removal of vegetation at moderate to high levels of precipitation (between 74 and 160 cm/year) will have a rapid and severe effect on the nutrient retention in these ecosystems.
5   At moderate to high levels of precipitation, removal of the herbivores will lead to a decrease in the retention of nutrients in these ecosystems followed by decrease in production by the vegetation.

Any realistic theory for the population dynamics of large herbivores must account for the cycling of nutrients at the level of the ecosystem, since individual large mammals live long enough to have profound effects on nutrient retention and to respond during their lifetime to changes in the nutrients within their ecosystem.

A MODEL FOR LONG-LIVED SPECIES

If such a large number of factors must be taken into account, it would appear that the study of populations of large mammals must depend, in part, on complex models. Such models might appear, at first consideration, to be unwieldy, difficult to analyze mathematically, and highly abstracted from data. This need not be the case. As explained below, we have developed a theoretical model for the population dynamics of long-lived species that permits consideration of many features of their environment and life history while remaining conceptually simple and closely tied to field observations. It characterizes a population as groups of individuals each in a particular stage in their life history.

The model was developed from data made available to us for the African elephant, but it is general in form and may be useful for any species whose individuals live through two or more breeding periods and whose breeding may not be synchronized. In this approach various optional life history states are considered for the individuals in a population along with the probability of transition between states. Figure 4, for example, illustrates the states and transition probabilities concerning the birth process for female elephants. A female is characterized by a state with three components (i, b, e) where i is her age category ($i = 0,1, \ldots ,60$), b denotes her biological states ($b = 1,\ldots5$), and e the environment ($e = 1, \ldots ,n$). Specifically, the biological states are immature,

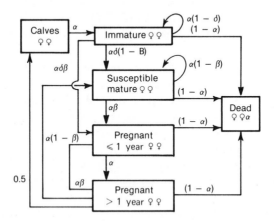

Figure 4   A representation of the life history of a female elephant, including the states (boxes) and transition probabilities between states (lines and arrows). a is the annual probability of survival, β is that of conception, and d is that of maturation. Each is a vector, with one element for each age class. Each may be a function of biotic and abiotic factors. An even sex ratio at birth is assumed, so that probability of a female calf being born is 0.5.

mature but not pregnant, pregnant for less than one year, pregnant for more than one year, and death. The probability of transition from one state to another depends on the probability of survival(α), maturation(δ), and pregnancy(β). Each may be a vector with as many elements as there are age classes, and each may be a function of environmental and biological factors (such as population density). Where data are lacking or indicate that a simpler characterization is sufficient, or if the question is simple enough, α, δ, and β may be scalars or may be the same for many of the age classes.

This model allows for much of the flexibility that is available in other approaches (e.g., see Chapters 5-10, 13-16, 20-23). By focusing on individuals, however, the model is based on parameters that can be determined directly. For example, elephants have overlapping pregnancies, because gestation lasts for 22 months and breeding occurs annually. This means the probability of a conception depends on whether an elephant is already pregnant, and this probability can be estimated from observations of the number of conceptions among those individuals susceptible to pregnancy. In the formulation embodied in a projection matrix, for example, age-specific fertility rates, adjusted for mortality, in combination with the number of females in each age group, are used to determine the number of live young after one time period. However, in populations with overlapping calving intervals, difficulties often arise in the estimation of the average number of young produced per female that remain alive at the junction of the time intervals involved. These difficulties exist for the Leslie model for any time unit (see Leslie, 1945; Keyfitz, 1968, on how aggregation of states is usually treated). Such difficulties may be overcome by determining the composition of the female population with respect to reproductive state.

It should be emphasized that, although the life history model is more flexible than the Leslie matrix formalisms, it requires no more precision in the estimation of parameters. And, if data are sparse, it can be utilized with no more parameters than the Leslie matrix. Mathematical properties and results obtained with this model have been discussed elsewhere (Wu and Botkin, 1980). These include a method for calculating the average number of female offspring produced in the lifetime of one female, the probability of extinction of a population and the composition of a population. The conditions under which a population will increase in size and those for reaching a stable age distribution may also be determined. It is useful for choosing an appropriate time unit for measurement and analysis of a particular process. It can also be used to investigate time-varying aspects of long-lived populations even when samples are obtained from a single point in time. The model produces simulated elephant populations with an age structure resembling that of observed data (e.g., Laws et al., 1975).

RAINFALL AS A SIMPLE EFFECT ON POPULATION PROCESSES

We have used the model described above to consider whether it is sufficient to link rainfall in a very simple way to population processes. Although the data we have analyzed in combination with those reported by Laws suggest that conception rates are related to rainfall, others have suggested that precipitation is more important in its effect on infant mortality. Since birth and juvenile mortality are both linked to early recruitment it is tempting to conclude that a distinction between the effects of precipitation on the two is unnecessary. The records concerning precipitation for the areas near the parks where the elephant data were collected, show no serial correlation (Fishman, personal communication). Thus, rainfall may be treated as a random variable with a distribution within any single year that is not dependent on previous years. With this information, we used the model to consider two cases, identical in initial conditions and all parameters except that in the first it is assumed that precipitation affects conception linearly, while in the second, first-year survivorship is linearly dependent on precipitation. These cases lead to markedly different population projections (Figure 5). Population growth is much lower where mortality is involved compared to that in which conception is affected. This is because gestation lasts 22 months; when a female loses a calf, she has lost two opportunities for conception.

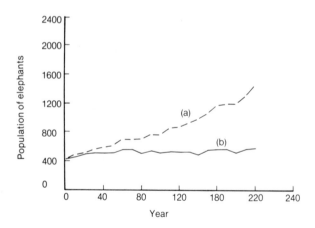

Figure 5   Simulated total population size as a function of time when rainfall affects the probability of conception (*a*) or the probability of infant mortality (*b*).

These results show that we must know whether rainfall acts on conception, infant mortality, or both. It is important to know the extent of each effect as well. These results also suggest that an understanding of the mechanisms that link rainfall to the population processes is necessary. Since these mechanisms may involve mineral cycles within the ecosystem, it is necessary to consider the population dynamics of elephants within the context of this ecosystem. Obviously the same will hold for most long-lived large herbivores in terrestrial ecosystems.

CONCLUSIONS

Previous analyses have suggested that total animal biomass is correlated with total net primary production, and total net primary production with rainfall. Therefore, it has been concluded (elsewhere) that a sufficient estimate of the change in a specific mammal population from one year to the next can be based simply on changes in total rainfall. However, several studies have shown that large mammalian herbivores are limited by the quality of their food rather than the quantity. We have accumulated evidence that suggests that the population dynamics of elephants and other large mammalian herbivores in Africa are linked to ecosystem dynamics through mineral cycling. We believe the birth and death rates are affected by these mineral cycles, and in turn the storage and flux of nutrients in East and Southern Africa is influenced by the living vegetation and animals. In this chapter we suggest mechanisms linking rainfall, mineral cycling, and biotic processes, and argue that the study of such mechanisms is essential for further advances in the development of population theory for large mammals as well as the development of theories for the dynamics of terrestrial ecosystems.

Previous analyses of the population dynamics of large mammals have been primarily based on two standard models used in classical population theory: (1) the deterministic, continuous, differential equation model, typified by the logistic equation, in which a population is characterized by its size and its dynamics as a function of size; and (2) the deterministic, discrete time and age dependent model typified by projection matrices. Both kinds assume that any effects of the environment are mediated within the population through the size or structure of the population. This ignores changes in the environment that occur independently of changes in population size. Both assume that a population, if undisturbed, will attain a constant, equilibrium total size and a stationary age structure. It is generally assumed that these models are sufficient descriptors for analyses of certain fundamental and applied questions. Other problems require different approaches.

To better understand the population dynamics of long-lived organisms, we have studied the African elephant, which, with its extreme longevity and being one of the best studied wild mammals, represents an excellent case history. Based on our analysis of data on more than 5000 elephants taken as part of control programs in Africa, we developed a life history model that is a stochastic, discrete-time model allowing consideration of the effects of the environment, the biological state of an individual, and the effect of the rest of the population on the birth and reproductive processes of individuals.

Our use of this model indicates that the effects of the environment cannot be ignored in studying the dynamics of long-lived organisms. What appear to be insignificant details (such as whether an environmental factor influences conception or infant mortality) can have significant effects on population dynamics, and it is necessary to understand the mechanisms by which the environment affects such population processes. Since the mechanisms may involve such things as mineral cycles at the level of the ecosystem, it is necessary to consider the population dynamics of large herbivores within an ecosystem context.

LITERATURE CITED

Bax, N., and D. L. W. Sheldrick. 1963. Some preliminary observations on the food of elephants in the Tsavo Royal National Park (east) of Kenya. E. Afr. Wildl. J. 1:140-153.

Bormann, F. H., and G. E. Likens. 1967. Nutrient cycling. Science 155:424-429.

Botkin, D. B., P. A. Jordan, A. S. Dominski, H. D. Lowendorf, and G. E. Hutchinson. 1973. Sodium dynamics in a northern terrestrial ecosystem. Proc. NAS 70:2745-2748.

Botkin, D. B., J. M. Melillo, and L. S.-Y. Wu. 1980. Why are there so many animals in Africa? Manuscript in preparation.

Calahane, N. H. 1952. A report of wildlife conditions in 1952. Wildlife Resources of the National Park System. 135 pp. Unpublished manuscript.

Coe, M. J., D. H. Cummings, and J. Phillipson. 1976. Biomass and production of large African herbivores in relation to rainfall and primary production. Oceologia 22:341-354.

Cole, D. W., S. P. Gessel, and S. F. Dice. 1968. Distribution and cycling of nitrogen, phosphorus, potassium and calcium in a second-growth douglas fir ecosystem. In: H. E. Young (ed.). Primary Productivity and Mineral Cycling in Natural Ecosystems. University of Maine Press, Orono.

Dasmann, R. F. 1964. African Game Ranching. Pergamon Press, London.

Dougall, H. W., V. M. Drysdale, and P. E. Glover. 1964. The chemical composition of Kenya browse and pasture herbage. E. Afr. Wildl. J. 2:86-121.

Harris, L. D. 1972. An ecological description of a semi-arid East African ecosystem. Range Science Dept. Science Series No. 11. Colorado State University, Fort Collins. 80 pp.

Henderson, G. R. 1950. Rumeruti grazing trails. Annu. Rev. Dep. of Agriculture, Kenya.

Keyfitz, N. 1968. Introduction to the Mathematics of Populations. Addison-Wesley, Reading, Massachusetts.

Kilmer, V. J. 1974. Nutrient losses from grasslands through leaching and runoff. Pp. 341-362 in: D. A. Mays (ed.). Forage Fertilization. Am. Soc. Agron., Crop Sci. Soc. Am., Soil Sci. Soc. Am., Madison, Wisconsin.

Laws, R. M., I. S. C. Parker, and R. C. B. Johnstone. 1975. Elephants and Their Habitats. Clarendon Press, Oxford.

Leslie, P. H. 1945. The use of matrices in certain population mathematics. Biometrika 33:183-212.

Leuthold, L., and B. M. Leuthold. 1973. Ecological studies of ungulate in Tsavo (east) National Park, Kenya. Tsavo Research Project, Kenya National Parks. 67 pp.

Likens, G. E., F. H. Bormann, N. M. Johnson, D. W. Fisher, and R. S. Pierce. 1970. Effects of forest cutting and herbicide treatment of nutrient budgets in the Hubbard Brook Watershed Ecosystem. Ecol. Monogr. 40:23-47.

Ledger, H. P., W. J. A. Payne, and L. M. Talbot. 1961. A preliminary investigation of the relationship between body composition and productive efficiency of meat producing animals in the dry tropics. Trans. Int. Congr. Anim. Prod. 8 (Hamburg).

McCullagh, K. G. 1969. The growth and nutrition of the African elephant. I. Seasonal variations in the rate of growth and the urinary excretion of hydroxyproline. E. Afr. Wildl. J. 7:85-90.

McNaughton, S. J. 1979. Grazing as an optimization process: Grass-ungulate relationships in the Serengeti. Am. Nat. 113:691-703.

Miller, R. S., and D. B. Botkin. 1974. Endangered species: Models and predictions. Am. Sci. 62:172-181.

Mode, C. J. 1975. Perspectives in stochastic models of human reproduction: A review and analysis. Theor. Pop. Biol. 8:247-291.

Munro, P. E. 1966a. Inhibition of nitrifiers by roots of grass. J. Appl. Ecol. 3:227-229.

Munro, P. E. 1966b. Inhibition of nitrifiers by grass-root extracts. J. Appl. Ecol. 3:231-238.

Namkoong, G. 1972. Persistence of variances for stochastic discrete-time, population growth models. Theor. Pop. Biol. 3:507-518.

Nye, R. H., and D. H. Greenland. 1960. The soil under shifting cultivation. Harpenden England Tech. Bull. No. 151. 156 pp.

O'Roke, E. D., and F. N. Hammerstrom, Jr. 1948. Productivity and yield of the George Reserve Herd. J. Wildl. Manage. 12:78-86.

Petrides, G. 1956. Big game densities and range carrying capacities in East Africa. Trans. N. Am. Wildl. Conf. 21:525-537.

Pollard, J. H. 1966. On the use of direct matrix product in analyzing stochastic population models. Biometrika 53:397-415.

Power, J. F., J. A. Lessi, G. A. Reichman, and D. L. Grunes. 1973. Recovery, residual effects, and fate of nitrogen fertilizer sources in a semi-arid region. Agron. J. 65:765-768.

Purchase, B. S. 1974. The influence of phosphate on nitrification. Plant Soil 41:541-547.

Rogler, G. A., and R. J. Lorenz. 1974. Fertilization of mid-continent range plants. Pp. 231-254 in: D. A. Mays (ed.). Forage Fertilization. Agron. Soc. Am., Crop Sci. Soc., Soil Sci. Soc. Am., Madison, Wisconsin.

Scott, R. M. 1962. Exchangeable bases of mature, well drained soils in relation to rainfall in East Africa. J. Soil Sci. 13:1-9.

Sheps, M., and J. Menken. 1973. Mathematical Models of Conception and Birth. University of Chicago Press, Chicago, Illinois.

Sinclair, A. R. E. 1974. The resource limitation of trophic levels in tropical grassland ecosystems. J. Anim. Ecol. 44:497-520.

Steward, B. A., F. G. Viets, Jr., G. L. Hutchinson, and W. A. Kemper. 1967. Nitrate and other water pollutants under fields and feedlots. Environ. Sci. Technol. 1:763.

Stiven, G. 1952. Production of antibiotic substances of the roots of grass (*Trachypogen plumosus* (H. B. K. Nees) and of *Pentanisia variabilis* (E. Mey) Harv. (Rubiacea). Nature 170:712-713.

Sykes, Z. M. 1969. On discrete stable population theory. Biometrics 2:285-293.

Talbot, L. M., and M. H. Talbot. 1963. The high biomass of wild ungulates on East African savanna. Trans. N. Am. Wildl. Conf. 38:465-476.

Van Etten, R. C. 1955. Deer range ecology in a 647 acre enclosure in northern Michigan. 17th Midwest Wildl. Res. Conf. Mimeo.

Viets, F. G., Jr. 1975. The environmental impact of fertilizers. CRC Critical Rev. Environ. Control 5:423-453.

Vitousek, P. J., and W. A. Reiners. 1975. Ecosystem succession and nutrient retention: A hypothesis. Bio. Sci. 25:376-381.

Weir, J. S. 1972. Spatial distributions of elephants in an African National Park in relation to environmental sodium. Oikos 23:1-13.

Wiegert, R. G., and F. C. Evans. 1967. Investigations of secondary productivity in grasslands. Pp. 499-518 in: K. Petruzewicz (ed.). Secondary Productivity of Terrestrial Ecosystems (Principles and Methods). Pol. Acad. Sci., Warsaw-Krakow.

Wu, L. S.-Y., and D. B. Botkin. 1980. Of elephants and men: A discrete, stochastic model for long-lived species with complex life histories. Am. Nat. 116:831-849.

TWENTY

Incorporation of Density Dependence and Harvest into a General Population Model for Seals

DOUGLAS P. DeMASTER

INTRODUCTION

Population models constructed in an attempt to simulate the behavior of real populations of seals have historically incorporated convenient and sometimes arbitrary mathematical functions that have mathematically derived equilibria. At present, it is impossible to know what mathematical function best describes the form of the density dependence responses that are assumed operative in nature. Various functions have been applied (Chapman, 1961, 1973; Fowler and Smith, 1973; Allen, 1975; Lett and Benjaminsen, 1977; Eberhardt and Siniff, 1977), but until recently (Fowler, 1981; Chapter 23) no clear patterns have emerged. The various functions do seem to form a continuum ranging from conservative linear functions (Allen, 1975), to nonlinear functions (Eberhardt and Siniff, 1977; Fowler, 1981; Chapter 10). Differences are related to the degree of density dependence expressed at any particular density, with linear functions having a gradual and consistent depressing effect on reproduction or survival when compared to nonlinear functions which concentrate changes into certain ranges of population levels. Therefore, simulation with linear and nonlinear functions should produce a range of types of population responses that should bracket the true population response in nature. Also, many population models assume that the age structure of the harvest has little or no effect on the predictions of the model (but see Chapter 23). In addition, many population models ignore the fact that density dependence may be restricted to specific age classes with reproduction and survival of other age classes being essentially density independent (at least over the observed range of densities). If only those age classes beyond the ages in which reproduction and survival are density dependent are harvested, then it is conceivable that harvest mortality may not give rise to strictly compensatory responses (Brownie et al., 1978). Finally, some population models that incorporate density dependence are based on the assumption that the regulatory mechanisms are operative throughout the entire year. Some evidence exists that this is not the case for Weddell seals (DeMaster, 1978). It appears that, for this species at least, regulatory mechanisms are only operative at specific and predictable times of the year. The purpose of this chapter is to investigate the ways several factors influence the dynamics of a population of Weddell seals as represented by a collection of assumptions in the form of a population model. These include (1) changing the density-dependent functions from linear to nonlinear form, (2) changing the age at which survival is density dependent (from pups to adults), (3) changing the age structure of the harvest from a harvest of pups to a harvest of adults, and (4) changing the time of the harvest. The latter allows early density-dependent harvest compensation (to be defined later) to occur in one case but not in another.

A PROJECTION MATRIX

A 25×25 projection matrix (Leslie, 1948) may be used for representing a population of female Weddell seals with 25 age classes. As constants, the entries of the top row (age specific reproduction as females born per female) were: $b_0 = b_1 = b_2 = 0$; $b_3 = 0.10$, $b_4 = 0.25$, $b_5 = 0.30$, $b_6 = 0.35$, and $b_7 = b_8 = \cdots b_{24} = 0.38$, as derived from DeMaster (1978), Siniff et al. (1977), and Stirling (1971). Similarly, the subdiagonal entries (age-specific survival) may all be set equal to 0.85 (DeMaster, 1978). Such a matrix model, with constant parameters, produces a constant growth rate and age structure (Leslie, 1945).

A more realistic approach involves a similar model that incorporates some type of density dependence into the transition matrix (Leslie, 1948; Fowler and Smith, 1973; Fowler and Barmore, 1979). Such density dependence is often assumed to be linear (Allen 1975), even though there is growing evidence to support nonlinear functions for large mammals in general (Fowler, 1981; and Chapter 23). In this analysis the following four density-dependent functions were used:

$$P_0 = A - M_0 X_1 \tag{1}$$
$$P_0 = A\left(1 - e^{-B(X_k - X_1)}\right) \tag{2}$$
$$P_i = A - M_1 X_2 \qquad i = 1, 25 \tag{3}$$
$$P_i = A\left(1 - e^{-B(X_K - X_2)}\right) \qquad i = 1, 25 \tag{4}$$

where P_0 = annual pup survivorship

P_i = annual survivorship of seals i years old

A = maximum annual survivorship

M_0 = constant associated with the number of female pups and their survivorship

M_1 = constant associated with the number of females 4 years of age or older and their survivorship

X_1 = number of female pups

X_2 = number of females 4 years of age or older

B = a constant that describes the shape of the curve in equations (2) and (3)

X_k = the maximum number of female pups in the population

X_K = the maximum number of females 4 years of age or older in the population

The following values were used in the 22 simulations summarized in Table 1: $A = 0.85$; $M_0 = 0.000125$; $M_1 = 0.00003594$; $B = 0.02$; $X_k = 350$; $X_K = 1000$. The estimated maximum annual survivorship A is a value derived from the available empirical data (DeMaster, 1978). X_k and X_K were approximated by the maximum number of pups and adult females that have previously been recorded in the McMurdo Sound study area (DeMaster, 1978). The value of B was arbitrarily set equal to 0.02 (Eberhardt and Siniff 1977) and represents a relatively rapid change in the shape of the nonlinear function (Figure 1). M_0 was determined by simulating the population with the nonlinear form of pup survivorship until an equilibrium was reached, and then using the equilibrium values of P_0 and X_1 to solve for M_0 (Figure 1). A similar approach was used to solve for M_1.

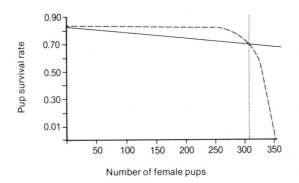

Figure 1   Linear (solid line) and nonlinear (dashed line) relationships between the survival rate for pups and the number of female pups born. The linear form is fit through the initial point (0, 0.86) and the equilibrium point (308, 0.69). Dotted line shows equilibrium level of female pups born.

The simulation was allowed to continue until an equilibrium was reached or until the population went extinct as was done by Fowler and Smith (1973) and Allen (1975):

$$X_{t+1} = L(X_t)X_t - H_t \qquad (5)$$

where X_{t+1} = the 25 × 1 population vector for year $t + 1$

$L(X_t)$ = the 25×25 variable projection matrix expressed as a function of X_t

X_t = the 25×1 population vector for year t

H_t = the 25×1 harvest vector in year t

The original population vector was derived from Stirling (1971) and DeMaster (1978):

Age	Frequency	Age	Frequency
1	232	10	44
2	179	11	37
3	152	12	30
4	130	13	26
5	108	14	22
6	90	15	18
7	75	16	15
8	63	17	12
9	53	18	10

At each iteration of the model, a new population vector is calculated. Appropriate portions of the new population vector are then used to calculate the entries for the projection matrix using various combinations of equations (1)-(4). Thus, the influence of changing the form of the density-dependent function and the age at which survival is density dependent can be investigated. By changing the harvest vector from one containing all pups to a vector of only seals 4 years of age or older (harvest from each age class was proportional to the size of the particular age class), the influence of varying the age structure of the harvest can be investigated.

Equation (5) is representative of a situation where the harvest occurs after the age-specific survival and reproduction have been realized. If the harvest occurs after the rates of survival and reproduction have been determined, there is no chance for further compensation by the nonharvested segment of the population during that unit of time. In other words, there is no increase in the rates of reproduction or survival as a result of the removal of a segment of the population by harvesting. Such reactions, by necessity, can occur only in the next time step. This may or may not be a realistic situation. It is possible that populations may react to the density immediately following harvest (rather than prior to harvest) thus compensating for the harvest in the same time step (a reversal of the situation above). To determine the influence that any such harvest compensation may have on the equilibrium population a simulation that calculated the harvest before the projection was also investigated. This is represented in equation form as:

$$X_{t+1} = L(X_t)(X_t - H_t) \tag{6}$$

In this case, the entries for the projection matrix are calculated from the population vector subsequent to the harvest.

The simulations of female Weddell seal populations, as produced for this study (Table 1), can be broken down into three categories. These are (1) simulations with no harvest (Table 2), (2) simulations in which only pups are harvested (Table 3), and (3) simulations in which only animals 4 years of age and older are harvested (Table 4). Cases 1 and 2 and cases 3 and 4 must necessarily be equivalent in all ways at equilibrium because of the way in which the slope of the linear form of the density dependence was derived. Cases 5 and 6 (Table 2) incorporated both adult and pup survivorship that was density dependent. In these cases, the equilibrium population and number of female pups are not the same. This is because the equilibrium survival rates (P_0/P) are not associated with the same number of female pups and the number of adult females for the linear and nonlinear models.

Table 1   Summary of 22 Simulations

Case	P_0	P	Harvest Compensation	Composition of Harvest
1	Linear	—	No	0
2	Nonlinear	—	No	0
3	—	Linear	No	0
4	—	Nonlinear	No	0
5	Linear	Linear	No	0
6	Nonlinear	Nonlinear	No	0
7	Linear	—	No	Pup
8	Nonlinear	—	No	Pup
9	—	Linear	No	Pup
10	—	Nonlinear	No	Pup
11	Linear	Linear	No	Pup
12	Nonlinear	Nonlinear	No	Pup
13	Linear	—	Yes	Pup
14	—	Nonlinear	Yes	Pup
15	Linear	—	No	Adult
16	Nonlinear	—	No	Adult
17	—	Linear	No	Adult
18	—	Nonlinear	No	Adult
19	Linear	Linear	No	Adult
20	Nonlinear	Nonlinear	No	Adult
21	—	Linear	Yes	Adult
22	—	Nonlinear	Yes	Adult

RESULTS OF SIMULATIONS

The results of simulations without harvest (Table 2) show that different density-dependent mechanisms will produce different equilibrium populations. Dif-

Table 2 Results of Weddell Seal Population Simulation in Which There Is No Harvest[a]

Case (see Table 1)	Population	X_1	P_0/P	Mean Age	Time to Return (20% reduction)
1. Linear P_0	1701	308	0.6937	5.05	250
2. Nonlinear P_0	1701	308	0.6936	5.06	90
3. Linear P	1790	301	0.8202	4.51	230
4. Nonlinear P	1790	302	0.8201	4.51	50
5. Linear P_0 and P	893	156	0.7708/0.8346	4.76	170
6. Nonlinear P_0 and P	1666	290	0.7715/0.8347	4.76	40

[a]See Table 1 and text for details concerning the nature of the model in each case.

ferent levels of pup production will be associated with these equilibrium populations. The largest difference in equilibrium occurred between cases 4 and 5. When both pup and adult survival are linearly density dependent, the number of pups and adults has to be substantially lower to reach equilibrium than when only one of these density-dependent mechanisms are employed. It is interesting that when only the survival of pups (P_0) is density dependent its equilibrium is less than when only the adult survival is density dependent. This is due to the differences in the composition of the resulting populations in the two cases, as explained in Chapter 23.

Changing the age classes subject to density-dependent survival thus has a predictable effect on the mean age of the equilibrium population. The same is true for the time necessary for the population to return to the equilibrium. When the survival of pups is density dependent, the mean age of the equilibrium population is greater than the mean age of the population when the survival of both pups and adults are density dependent. Also, when survival for all age classes is density dependent, the mean age at equilibrium is greater than when the survival for adults alone is density dependent (see Chapter 21).

Estimates of the time necessary for the population to return to equilibrium were generated by reducing each age class of the equilibrium population by 20%, and then finding how long it took for the population to reach a constant mean age (determined by a predetermined and uniform criterion). The nonlinear models were found to return more rapidly than the linear models because the nonlinear models used in this study have essentially maximum rates of growth until they approach the maximum population, while the linear models have growth rates that are constantly declining as the population increases. The most rapid return took 40 years, while the slowest return took 250 years (Table 2). DeMaster (1978) documented a 50% reduction in the breeding population of the Weddell seal in McMurdo Sound between 1975 and 1976. Stirling et al. (1977) documented a 50% reduction in the number of ringed seals *(Phoca hispida)* in the western Canadian Arctic. Returning from such reductions, of course, would require even more time than indicated by these models.

With a pup harvest as part of the population model, the population giving rise to a maximum sustainable yield (MSYP) and the maximum sustained yield itself

Table 3 Results of Weddell Seal Population Simulation in Which There Is a Harvest of Pups[a]

Case (see Table 1)	MSYP[b]	MSYP Equilibrium	H_1	X_1	Yield (%)	Maximum Population	H_1 max
7. Linear P_0	938	0.55	17	170	10	—	—
8. Nonlinear P_0	1396	0.82	43	253	17	—	—
9. Linear P	1018	0.57	14	177	8	—	—
10. Nonlinear P	1481	0.83	43	265	16	—	—
11. Linear P_0 and P	574	0.64	7	101	7	—	—
12. Nonlinear P_0 and P	1373	0.82	40	247	16	—	—
13. Linear P_0 compensation	1011	0.59	20	184	11	—	—
14. Nonlinear P_0 compensation	1665	0.98	52	301	17	1810	40

[a]See Table 1 and text for details concerning the nature of the model in each case.
[b]Population giving rise to a maximum sustainable yield (MSY).

(MSY) were determined by finding the equilibrium population at various levels of harvest. The largest harvest that the population could sustain was considered to be the MSY. By altering the form of the density-dependent function, the age at which it operates, and the time at which it operates, a range of corresponding MSYP were determined. The MSYP, expressed as a percentage of the equilibrium population (no harvest) was found to vary between 55 and 98 (Table 3). In general, if only the form of the density dependence was changed, simulations with nonlinear density dependence had higher MSYP than the models with linear density dependence. Similarly, nonlinear models tended to have higher MSY values when expressed as a percentage of the total pup production. A unique situation arose with nonlinear models in that some harvest rates actually increased the equilibrium populations (Table 3, case 14). In this case, harvesting 40 pups increased the survivorship of the remaining pups to such an extent that the equilibrium population of the model was 109 seals larger than the equilibrium population with no harvest at all.

Table 4   Results of Weddell Seal Population Simulation in Which There Is a Harvest of Adults[a]

Case		MSYP	
(see Table 1)	MSYP	Equilibrium	Harvest
15. Linear P_0	1166	0.69	7
16. Nonlinear P_0	1570	0.92	22
17. Linear P	831	0.46	10
18. Nonlinear P	1437	0.80	24
19. Linear P_0 and P	665	0.74	3
20. Nonlinear P_0 and P	1475	0.89	21
21. Linear P compensation	831	0.46	10
22. Nonlinear P compensation	1486	0.83	25

[a]See Table 1 and text for details concerning the nature of the model in each case.

When the age structure of the harvest is changed from exclusively a pup harvest to a situation in which all seals 4 years of age or older (approximate age of sexual maturity and subsequent return to breeding population: Stirling, 1971; Siniff et al., 1977) are harvested, the MSYP, when expressed as a percentage of the nonharvested equilibrium population, varied between 46 and 92 (Table 4). This is essentially the same as the range of MSYP for the pup harvests. The cases with adult harvests were similar to cases with pup harvests in that those with linear density dependence had lower MSY values and equilibrium populations than those with nonlinear density dependence. The maximum adult harvest occurred in the case where nonlinear density dependence of adult survivorship and harvest compensation were incorporated. In comparing the maximum yields for pup and adult harvests, roughly twice as many pups can be harvest from one age class as adults can be harvested from 21 age classes. This supports the prediction of Eberhardt and Siniff (1977) that marine mammal populations are extremely sensitive to increases in adult mortality.

The simulations that incorporated pup harvests of various types had MSY values that ranged from 7% (Table 3, case 11) to 17% (Table 3, cases 8 and 14) of the pup numbers. Comparable ranges were even greater when harvest of adults is included in the cases being compared. For all harvest strategies that were examined in this work the MSYP, given as a percentage of the nonharvested equilibrium population, varied between 46 and 98. This analysis suggests that reasonable management decisions must incorporate information concerning the regulatory mechanisms, harvest compensation, time at which density dependence is expressed, and the age structure of the harvest. Specifically, models and predictions for one species should not necessarily be accepted as general guidelines for the management of other species.

DISCUSSION

The Model

The main purpose of this chapter is to demonstrate the effects of different regulatory processes, types of harvests, and harvest schedules. Therefore, the particular hypothetical relationships used to model the regulatory processes can be considered only as rough approximations of what may actually happen. The four types of density dependence in equations (1)-(4) represent a wide range of reasonable types of density dependence. By examining the extremes of a reasonable spectrum of possibilities, minimum and maximum estimates can be produced to create an interval within which the true value will occur.

Other modes of regulation need investigating. Eberhardt (1977), for example, suggests that in addition to the survival of pups, age of first reproduction is a very important mechanism in the regulation of populations of marine mammals. For simplicity, age-specific reproduction was assumed constant in this model. This needs to be investigated. Some of the effects of changing this variable are examined by Fowler and Smith (1973), Fowler and Barmore (1977), and Eberhardt and Siniff (1977), and in Chapter 23. Preliminary modeling in this study suggested that manipulating the age of first reproduction may cause cyclic behavior (see May and Oster, 1976).

The particular nonlinear function used in this work is attractive because it comes close to approximating the concept of a threshold density. That is, survival, for example, is essentially density dependent only when density is above the threshold. Chapman's 1973 presentation of data on northern fur seals *(Callorhinus ursinus)* seems to support the threshold concept because recruitment was found to appear independent of density at low population levels. These are dynamics that conform to the general pattern described by Fowler (1981) and in Chapter 23. In this pattern, most density-dependent change is restricted to a range of levels close to the equilibrium for large mammals.

Models Without a Harvest

This analysis indicates that mechanisms that have relatively small effects on the growth rate of a population indeed may regulate those populations. However, the time necessary for this regulation to occur may be longer than the period during which environmental conditions are constant enough to allow for a constant equilibrium population. Because the minimum time necessary for the population to return to the equilibrium (after a 20% reduction in the population) was 40 years in this work, it seems unlikely that seal populations would ever reach a stationary age distribution (age distribution where age structure is constant). If it is assumed that declines can occur much more rapidly than increases (Stirling et al. 1977; DeMaster, 1978; Siniff et al., 1977; Payne, 1977), it may be that populations of pinnipeds will commonly be found to be increasing to a level their resources can maintain. However, periodically populations may rapidly decline, possibly owing to changes in some aspect of the physical environment. This type of growth pattern needs to be considered as far as future methodologies and research (see Fowler, 1981).

Considering the predictable manner in which the mean age of a population responded to various types of density dependence, it may be possible to use the change in the mean age of the population as it grows to infer something about the age class for which density-dependent mechanisms are operative. That is, if the mean age of the population increases as the population increases, it is most likely that density-dependent factors are affecting the younger age classes or the birth rate (see Chapter 21), and vice versa. This approach may not be particularly sensitive because of the relatively wide confidence interval around the mean age that is generated from a distribution of ages. However, such a technique has been used successfully with southern elephant seals (*M. leonina:* Laws, 1960) to determine the number of bulls to be culled. Further studies of this population parameter should be conducted.

Models Incorporating Harvest

The incorporation of age-specific harvests into a general population model seems to be necessary in light of the findings in this study and as shown in Chapter 23. The fact that the maximum sustainable yield of adult female Weddell seals was only 1.7% in the most liberal population model in this study suggests that harvests as low as 20 seals from a population of 1000 adult females may be critical. Since 1964, an average of 30 adult females have been harvested from the McMurdo population of Weddell seals in the Antarctic. This population contains roughly 1000 adult females. The 1977-1978 harvest of this population was roughly 35 adult females (combined take of United States and New Zealand). The gradual decline of this population since 1967 may be, in part, a result of this adult harvest (DeMaster, 1978).

On the other hand, harvesting 17% of the female pup production could be sustained by this same simulated population. It is even conceivable that some harvests may temporarily increase populations. A similar mechanism has been suggested for black bears (Kemp 1974).

Analyses that have incorporated factors similar to those discussed in this chapter have been presented by Stirling et al. (1976), Allen (1975), and Chapman (1961). Stirling et al. (1976) simulated brown bear and polar bear populations with a fixed age-specific rate of reproduction and survival. Although it is not possible to generate MSY values with their model, a type of sensitivity analysis can be generated. The authors found that various population projections were extremely sensitive to manipulations that affected the mortality of adult females.

Chapman (1961) estimates the MSY of 3-year-old male northern fur seals *(Callorhinus ursinus)* with two different types of density-dependent rates of recruitment. Both models are nonlinear, with one model producing an MSY of 45%, when the population of pups is 68% of the maximum population of pups. The other model produces an MSY of 46%, when the population of pups is 72% of the maximum pup population. The estimated MSY of 3-year-old fur seals is much greater than the estimated MSY of pup Weddell seals in this chapter. This is primarily because the intrinsic growth rate for fur seals is much larger than for Weddell seals. However, the population at which the MSY occurs for fur seals is bracketed by the estimates of the MSY population produced in this chapter for Weddell seals.

CONCLUSIONS

It is difficult to generalize from these results because of the hypothetical nature of the model and because each manipulation was associated with a unique MSY and equilibrium population. Estimates of the MSY were found to be higher for nonlinear than linear models, higher for pup harvests than adult harvests. These results are consistent with the general patterns described in Chapter 23. Models that incorporated harvest compensation produced higher MSY than models that did not.

The management of populations of pinnipeds is confounded by difficulties in assessing population levels, regulatory mechanisms, and controlling factors. The recent criticisms of the concept of MSY (Larkin, 1977) and the current use of the concept of optimum sustainable population (OSP) (Marine Mammal Commission, 1975) seem to stem from an awareness of how sensitive current MSY estimates are to changes in the marine community (particularly of fishery stocks) and changes in the age structure of the harvest. In the future, the population models that are used to estimate MSY values must incorporate these factors. Changes in the community structure will necessarily alter the equilibrium population of seals, and adjustment procedures need to be established. Perhaps initial carrying capacity estimates should be compared with current carrying capacity estimates. Perhaps, the effect of varying environments may have to be incorporated in future models by determining MSYs (stated as a percentage) in advance, but then delaying the determination of the quotas until an assessment of the current populations are made. Finally, the idea that one general model will predict maximum harvest rates and equilibrium stocks for many or all seals seems doubtful. For each species, the regulating mechanisms, and controlling factors must be identified before MSYs can be derived. In addition, the age structure of any proposed harvests must be incorporated in management oriented models because this is critical in determining the MSY for any particular population.

LITERATURE CITED

Allen, R. L. 1975. A life table for harp seals in the northwest Atlantic. Rapp. P.V. Reun. Cons. Int. Explor. Mer 169:303-311.

Brownie, C., D. R. Anderson, and K. P. Burnham. 1978. Statistical Inference from Band Recovery Data — A Handbook. USDI, Res. Publ. 131.

Chapman, D. G. 1961. Population dynamics of the Alaska fur seal herd. Trans. N. Am. Wildl. Conf. 26:356-369.

Chapman, D. G. 1973. Spawner-recruit model and estimation of the level of maximum sustainable catch. Rapp. R.V. Cons. Int. Explor. Mer 164:325-332.

Eberhardt, L. L. 1977. Optimal policy for conservation of large mammals, with special reference to marine ecosystems. Environ. Conserv. 4:205-212.

Eberhardt, L. L., and D. B. Siniff. 1977. Population dynamics and marine mammal management policies. J. Fish. Res. Board Can. 34:183-190.

DeMaster, D. P. 1978. Estimation and analysis of factors that control a population of Weddell seals *(Leptonychotes weddelli)* in McMurdo Sound, Antarctica. Ph.D. Thesis, University of Minnesota, Minneapolis. 81 pp.

Fowler, C. W. 1981. Density dependence as related to life history strategy. Ecology (in press).

Fowler, C. W., and W. J. Barmore. 1979. A population model of the northern Yellowstone elk herd. Pp. 427-434 in: R. M. Linn (ed.). Proceedings of the 1st Conf. on Scientific Research in Nat. Parks, Vol. 1. Nat. Park Serv. Trans. and Proc. Series, No. 5.

Fowler, C. W., and T. Smith. 1973. Characterizing stable populations: An application to the African elephant population. J. Wildl. Manage. 37:513-523.

Kemp, G. A. 1976. The dynamics and regulation of black bear, *Ursus americanus*, populations in northern Alberta. *In:* Bears — Their Biology and Management. IUCN Publ. New ser. 40:191-198.

Larkin, P. A. 1977. An epitaph for the concept of maximum sustainable yield. Trans. Am. Fish. Soc. 106:1-11.

Laws, R. M. 1960. The Southern elephant seal *(M. leonina)* at South Georgia. Nor. Hvalfangsttid. 49:466-476; 520-542.

Leslie, P. H. 1945. On the use of matrices in certain population mathematics. Biometrika 33:183-187.

Leslie, P. H. 1948. Some further notes on the use of matrices in certain population mathematics. Biometrika 35:213-243.

Lett, P. F., and T. Benjaminson. 1977. A stochastic model for the management of the northwestern Atlantic harp seal *(Pagophilus groenlandicus)* population. J. Fish. Res. Board Can. 34:1155-1187.

Marine Mammal Commission. 1975. The concept of optimum sustainable populations. (unpublished report) MMC, Washington, D.C.

May, R. M., and G. F. Oster. 1976. Bifurcations and dynamic complexity in simple ecological models. Am. Nat. 110:573-599.

Payne, M. R. 1977. Growth of a fur seal population. Philos. Trans. R. Soc. London. 279:67-79.

Siniff, D. B., D. P. DeMaster, R. J. Hofman, and L. L. Eberhardt. 1977. An analysis of the dynamics of a Weddell seal population. Ecol. Monogr. 47:319-335.

Stirling, I. 1971. Population dynamics of the Weddell seal in McMurdo Sound, Antarctica. *In:* W. Burt (ed.). Antarctic Pinnipedia. Am. Geophys. Union, Washington, D.C. 226 pp.

Stirling, I., W. R. Archibald, and D. P. DeMaster. 1977. Distribution and abundance of seals in the eastern Beaufort Sea. J. Fish. Res. Board Can. 34:976-988.

Stirling, I., A. M. Pearson, and F. L. Bunnell. 1976. Population ecology of polar and grizzly bears in Northern Canada. Trans. N. Am. Wildl. Conf. 41:421-430.

TWENTY-ONE

Observations on a Heavily
Exploited Deer Population

GEORGE E. BURGOYNE, JR.

INTRODUCTION

With the continuing development of the field of population dynamics and systems analysis, modeling work has moved to increasingly complex models. With the exception of introductory courses in modeling, there appears to be a general disregard, perhaps even disdain, of simple models. Complex models appear more realistic since they incorporate more features of the real system and thus provide additional factors for examination within the model structure. Simple models, on the other hand, sacrifice some of this detail, but have the advantage of being more easily understood. They also have the advantage of capturing the salient features of complex systems, an advantage that will be utilized below.

Another important reason for using simple models is that simple models are much more likely to be incorporated into the management process. As Ackoff (1970) points out, and as discussed in Chapter 12, a mathematical model must be adequately understood by the manager for it to be used. Otherwise the manager will mistrust the model's predictions and will return to using his intuition to solve the entire problem rather than using the model to clarify those parts of the problem for which it is suited and then using intuition to solve those parts of the problem which cannot otherwise be solved.

Simple population models can be valuable in the exploration of the population dynamics of natural populations. More complex models may tend to suppress the expression of simple relationships due to the interactive effects of numerous factors. Mertz (1970) points out that this simplicity "will be an advantage rather than a disadvantage when the models are used to clarify certain ecological concepts."

The principle of Occam's razor may be too often ignored in model building. While it can be interesting and challenging to incorporate as many factors as possible into a model, it may be more useful to identify the minimum set of factors that needs to be accounted for in order to produce a model that will adequately mimic reality.

An example is presented here in which an extremely simple population model is useful in clarifying the interpretation of certain field data collected from several large populations of the white-tailed deer *(Odocoileus virginianus)*.

While the data that gave rise to the motivation for this work came from the State of Michigan, in the United States, similar problems involving the same misinterpretations are in evidence from other areas, nationally and internationally.

MICHIGAN PROBLEM

One of the most collected kinds of data from populations of white-tailed deer is age data based on the tooth-wear methods developed by Severinghaus (1949). States and provinces throughout the United States and Canada have been collecting and compiling these data since the early 1950s. However, there is confusion among wildlife managers over the information content of the age-structure data. For example, in Michigan, a considerable amount of effort has gone into the interpretation of such data with the objective of producing projections of populations subjected to harvest (Eberhardt, 1960). However, there has been little sign that it must be recognized that there are differences to be considered in the interpretaton of data coming from heavily exploited populations and those coming from naturally controlled populations. To provide an example of these kinds of problems, as commonly encountered, a typical situation, as found in Michigan is described below.

Several observations have been made by wildlife biologists in Michigan concerning white-tailed deer. First, the proportion of yearling bucks in the harvest of antlered deer has been consistently higher in region II than in the more northerly region I (Figure 1, Table 1; Burgoyne et al., 1979). This was considered to be a result of the higher productivity thought to characterize the population in region II. To those interpreting the data, a more productive herd should obviously have a greater proportion of young animals.

Figure 1   Regions for the management of deer in Michigan.

However, there have been antlerless deer seasons in region II, while there has been little antlerless hunting in region I during the period considered. This produced the second observation: during this period of antlerless hunting, the proportion of older does in the female component appears to have declined. This was interpreted as evidence that the female population had been overharvested and that the older does in particular had received most of the exploitation. Others argued that antlerless harvests could not have resulted in an overharvest of the population of females and that the data concerning numbers by age were erroneous. To complicate things further, the total herd in region II had declined (Burgoyne and Moss, 1977). It was difficult to conceive of a declining herd with an age structure that was gradually becoming younger.

There is a need to reproduce changes of the nature of those described above in a simple model of populations so that something other than unreconcilable opinion might direct the interpretation of data and guide management. This chapter describes such an effort and its results, particularly as applied to Michigan's deer herd.

Table 1   Proportion Yearling Males in Antlered Buck Harvest in Michigan

Year	Region I	Region II
1960	0.30	0.57
1961	0.42	0.69
1962	0.44	0.71
1963	0.50	0.70
1964	0.65	0.76
1965	0.45	0.68
1966	0.34	0.64
1967	0.43	0.72
1968	0.56	0.71
1969	0.43	0.66
1970	0.42	0.63
1971	0.47	0.64
1972	0.44	0.72
1973	0.55	0.73
1974	0.57	0.74
1975	0.48	0.76
1976	0.45	0.68
1977	0.49	0.75

GENERALIZATIONS IN POPULATION DYNAMICS

Before providing more detail about the specific problems faced in Michigan state, it is useful to consider some of the potentially misleading generalizations of importance in population studies. Discussions of the meaning of age class pyramids, for example, are typified by that found in Odum (1971). A rapidly growing population is characterized by an age structure with a large proportion of young animals, while a stable population is characterized by a population with an older age structure. Odum cites a muskrat population in which the youngest age class comprises 52% of the population as an extreme in which the population's reproductive rate was apparently responding to the high exploitation rate.

The basic guidelines for interpretation which Odum used appear to be based on a natural population that was relatively unaffected by direct exploitation by man. However, as will be shown below, a population that is being heavily exploited, and is responding to the exploitation either directly or indirectly, may have a broad range of representation in the younger age classes without indicating that the population is expanding.

In addition, classical approaches to the study of populations seem to produce the belief that populations behave such that both sexes exhibit the same dynamics, since most models either ignore sex or look only at the female portion of the population (but see Chapter 23). The age structure of the male and female portions of the herd responds differently to changes in net reproduction and to changes in the exploitation rate.

Most wildlife biologists making management decisions have had little training in population dynamics other than a superficial exposure to the principles in the current literature or a brief mention in a wildlife management class. Those who have been exposed to this information are, at best, left with only basic generalizations of the type described above. Although important, these concepts, when applied out of context or in their simplicity, may actually be misleading. The interpretation of data collected by people at this level may be based primarily on intuition or "gut feeling" and can easily lead to misinterpretation of the data.

DATA BIAS

Before developing a model of Michigan's deer population, one must consider the system used for collecting age data and the validity of assuming that it represents the real-world population. Understanding the nature of the data involved will help evaluate the results of work based on models.

The state of Michigan uses a system of collecting biological data concerning hunter-killed deer in which information is provided by the hunters on a voluntary basis. This method is a compromise relative to other techniques and is used because of costs and the particular geography of Michigan. Most of Michigan's deer are killed in the northern two-thirds of the state, while most of the hunters live in the southern third of the state. Hence, by setting up checking stations alongside the major north-south highways leading from the areas hunted to the areas of residence, there is the potential for contacting a large portion of the successful deer hunters. The hunters stop at these stations voluntarily.

There are at least three transformations that may affect the data in hand to make it unrepresentative of the real population. The first transformation involves sampling from the herd in the wild to produce the population of harvested animals. At least two things, animal vulnerability and hunter selection, affect this process. There is conflicting evidence regarding the question of differential vulnerability. Hayne and Eberhardt (1952) reported some evidence that yearling bucks might be less vulnerable than older bucks, while McCullough (personal communication) has evidence of their being more vulnerable than older bucks on the George Reserve.

Hunter selection between age classes probably does occur to some extent, even though the ability to select between yearling and older animals has not been successfully demonstrated. The most obvious selection is between the sexes. This is relatively simple in deer since most older bucks have visible antlers. Selection may also take place between does and the fawns that still accompany them in the fall. In this situation the hunter would have two animals of different age class to compare. However, Gill (1953) concludes that in healthy populations of white-tailed deer, the size difference in the fall is often difficult to detect.

The second transformation is from the population of harvested animals to the population of animals that is sampled at the check stations. Since Michigan uses a voluntary data collection system, whether or not any given animal is seen by biologists is affected by the hunter's willingness to have his deer examined. Based on unpublished Michigan data, it appears there is reluctance on the part of many hunters to bring female deer and fawns to checking stations to be examined by biologists. In Michigan, as elsewhere, there is a stigma attached to killing antlerless deer since most deer hunters take more pride in killing a buck. The data collected at Michigan checking stations show that a deer hunter who kills an antlerless deer is much more likely to transport his deer concealed from view than a hunter who has killed an antlered deer. Thus, he may also be less likely to stop to have his deer examined. However, it is reasonable to assume that a hunter who kills a yearling buck is just as likely to have his deer examined as a hunter who kills an older buck, since their size is very similar. The same should apply to those who kill yearling and older does. Then, although there are biases regarding the deer seen at the checking stations, it can be argued that all antlered bucks probably have the same likelihood of being brought in for checking, and that while does have a different chance of showing up at a station than the bucks, all does that are yearlings and older probably share about the same chance of being checked.

The final transformation is from the population examined to the population recorded. The methods of collecting and recording data may introduce further bias in the information. There is strong evidence that such problems do occur. Ryel et al. (1961) showed that the problem of distinguishing between the deer of age $2\frac{1}{2}$ and $3\frac{1}{2}$ years (age classes 2 and 3) significantly affects the resulting proportions of those deer in the sample. However, this study also showed that the accuracy of correctly assigning deer into the categories of fawn, yearling, and adult was quite high.

As a result, it is obvious that there are several possible sources of bias being introduced in the series of steps from the field population to data bank. It can be argued, however, that the sample of deer which are 1 year old and older should be reasonably representative if it is used only for comparison within sex and only for the division of the population into two categories: (1) yearlings, and (2) animals older than 1 year of age.

A MODEL AND OBSERVATIONS

To gain a first impression of the effects of harvest, natural mortality, and reproduction on the structure of a population by age and sex, we can construct a simple model. In order to begin with a minimum of assumptions, the population model described here accounts only for the males and females, each divided into 20 annual age classes. Age-specific reproductive rates are applied to the females. There are two types of annual harvests. One type is modeled after the Michigan buck harvest, taking a proportion of the males in age class 1 and older. The other type of harvest involves the antlerless deer, in which a proportion of male fawns and a portion of all females are harvested. Within the year used in the model, reproduction occurs before the harvested animals are removed.

The model is a discrete time model with a time interval of one year. There is a discrete annual age scale starting at 0 (fawns). The model makes the assumption that within an age interval birth and death rates remain constant (although they may differ from one interval to the next). N_{ij} is the number of animals in annual age class i (0, 1, 2, .. n) in sex j (male $= 1$, female $= 2$). The annual survival rates s_{ij} are age and sex specific. The reproductive rates p_{ij} are dependent on the age of the mother i and the sex of the offspring j.

Given the rates and the initial population in year 1, it is possible to calculate the population in year 2 as follows:

$$N_{0j} = \sum_{i=1}^{n} p_{ij} N_{i2} \qquad j = 1,2$$
$$N_{ij} = s_{ij} N_{ij} \qquad i = 0,1,2,\ldots,n; \quad j = 1,2$$

where N_{ij} is the number of animals in the respective age-sex class at the beginning of the next annual time period.

The effects of a harvest can be simulated by substituting s_{ij}^{*} for s_{ij}, where $s_{ij}^{*} = s_{ij} - h_{ij}$ and $h_{ij} =$ harvest rate on age class i of sex j.

An antlered harvest can then be simulated by setting

$$h_{i1} > 0 \quad \text{for} \quad i = 1,2,3,\ldots,n$$

otherwise $h_{ij} = 0$ and an antlerless harvest can be simulated by setting

$$h_{0j} > 0 \quad \text{for} \quad j = 1, 2$$
$$h_{ij} > 0 \quad \text{for} \quad i = 1, 2, 3, \ldots, n; \quad j = 2$$

otherwise $h_{ij} = 0$.

In experiments with different harvest rates and different reproductive rates in this model, it was discovered that the age structure of the male component of the population responded to changes in rates differently from that of the females. The age structure of the males responded most strongly to the mortality rate for the males, while a change in the birth rate generated only a temporary change in the age structure of the males. Following a change in reproductive rates alone, the age structure of the males would undergo an initial change then return to its original composition. On the other hand, the age structure of the females did not respond to changes in the female harvest (mortality) rate. It did, however, respond strongly to changes in the reproductive rate.

The remainder of this chapter is an explanation for these responses and a discussion of how these facts can be applied to the interpretation of field data.

ANTLERED MALE AGE STRUCTURE

The interpretation of the age structure of the antlered males (age class 1 and older) appears relatively straightforward. Hayne and Eberhardt (1952) offer a graphical method for estimating survival rates from the age distributions based on the methods of Ricker (1948) for examining fish population dynamics. However, Ryel et al. (1961) show that there can be definite problems with the subjective age determinations of the next two age classes ($2\frac{1}{2}$ and $3\frac{1}{2}$ years old). Estimated numbers in these categories can significantly affect the computed survival. The reliability of discrimination between yearlings and older deer appears much more reliable.

Gill (1953) points out that the recruitment into populations of antlered males is essentially independent of the size of the population before recruitment is added. Recruitment to Michigan's white-tailed deer population does not appear to be very sporadic. As a result, dominant age class phenomenon do not appear to be as common as in some fish species. It is therefore reasonable to explore the outcome of assuming the recruitment may be not only independent of the antlered buck population, but also relatively constant. Then, assuming that mortality is nonselective, the age structure can be computed easily.

If there are N animals entering the population with an annual mortality rate of m and an annual survival rate of s ($m + s = 1$), the number of animals surviving to their second year is $s \times N$. Of these $s \times (sN)$ or s^2N will survive to their third year, s^3N to their fourth year, and $s^{n-1}N$ to their n^{th} year. Since the annual recruitment is constant, the total population will consist of $N + sN + s^2N + s^3N + s^4N + ...$ individuals, a geometric series summing to $N/(1 - s)$. The proportion that would be in the youngest age class is computed by

$$\frac{N}{N/(1 - s)} = 1 - s = m$$

In other words, with these simple assumptions, the proportion of the animals that would be expected to be in the youngest age class would be approximately equal to the annual mortality rate.

Since N cancels out of the calculation of this proportion, the value of the recruitment rate is unimportant. The exploitation rate in combination with the natural mortality rate will control the level of this proportion. Hence, the continuing differences in the proportion of yearlings observed in the antlered populations between the two Michigan regions may very well be a result of the mortality rate or exploitation rate rather than any difference in recruitment. Stochastic simulations in which recruitment, and mortality, are implemented as random variables (either individually or in combination) confirm that this proportion is relatively insensitive to the rate of recruitment, but that it is highly dependent on the mortality rate.

If the yearling age class is harvested at a rate different from that to which the rest of the antlered population is subjected, the proportion of animals in the younger age class will be inflated. The extent of this inflation is estimated by $(1 - p)^{-1}$, where p is the ratio of the survival rate for the older age classes to the survival rate for the youngest age class. This, of course, is also the case if the natural survival of the first age class is different from that of the older age classes.

FEMALE AGE STRUCTURE

Gill (1953) pointed out, through the use of numerical examples, that interpreting the female age structure as a consequence of mortality rate was incorrect. He found that nonselective harvest would not produce a change in the female age structure. A change in the reproductive rate, however, was found to produce a very real and significant change. This he attributed to the fact that the recruitment into the female population was dependent on the population that was present before recruitment. In fact, if the age-specific reproductive rates are not altered, recruitment is proportional to the reproductive population prior to recruitment. This is true whether the population is expanding, declining, or stable.

It is easy to show that the age structure of the female component of a population depends to a great extent on the reproductive rate (see Chapter 11). This may be accomplished through an examination of the equations for the stable age distribution from typical life table calculations.

Beginning with either the equations of the form presented in Lotka (1956) or those of Mertz (1970), it is possible to produce the equation for the proportion of the animals in age class x.

$$C_x = \frac{e^{-rx}L_x}{\displaystyle\sum_{y=0}^{\infty} e^{-ry}L_y}$$

where C_x = proportion of animals in age class x

r = the rate of increase

L_x = the proportion of animals alive at age 0 that are alive at age x

e = base of the natural logarithms

y = age class

The value of r can be decomposed into its components, the instantaneous birthrate b and the instantaneous death rate d (Mertz, 1971; Pielou, 1969), $r = b - d$. Using Lotka's definition of L_x under the assumption of a nonselective mortality rate of d

$$L_x = e^{-dx}$$

By substituting both of these in the original equation for the age class proportion we obtain

$$C_x = \frac{e^{-(b - d)x}e^{-dx}}{\sum\limits_{y=0}^{\infty} e^{-(b - d)y}e^{-dy}}$$

and after combining exponents we have our final equation:

$$C_x = \frac{e^{-bx}}{\sum\limits_{y=0}^{\infty} e^{-by}}$$

The mortality rate cancels out of the calculation entirely. Under the assumptions of this simple model, then, it must be concluded that the age structure of the female population at equilibrium (or the stable age structure of a growing or declining population of females) does not depend on a nonselective mortality.

Those more familiar with working in terms of population projection matrices such as in Keyfitz (1968) or Pielou (1969) may find a slightly different line of reasoning easier to follow. Imposing a nonselective mortality across the entire female age structure is equivalent to multiplying the projection matrix by a constant. This constant multiplier will not affect the relative sizes of the computed stable age distribution. It only changes the rate at which the population would grow or decline under conditions of stable age structure.

On the other hand, a uniform change in the birthrate will only affect the top row of a projection matrix. A change in fawn mortality may also involve only the elements of the first row of projection matrices, since such matrices are often constructed to represent populations in which some mortality has been suffered by the youngest age class at the time of projection. As such, these two factors will each affect the stable age structure by changing the net recruitment to the population.

Other changes (changes that violate the assumptions of the model used above) can give rise to changes in the age structure of a female population. One such change involves the transient effect of a change in reproduction. The relative sizes of any adjacent age classes will not change until the cohort reflecting altered net recruitment rate ages to that point. Age-specific changes in mortality will also produce an initial change in the age distribution. But in all cases such transient changes should disappear over time. Year-to-year variability in reproduction, of couse, will create some variability in the age structure of a population. But, as mentioned above, in large mammals such as deer this is to be expected less than in smaller bodied species. On the average the trends discussed here will override most other factors.

It should be noted that nonselective harvest can directly alter the female age structure *if* it brings about changes such as density-dependent changes in the net recruitment rate to the population. (This may explain the younger age structure in the white-tailed deer population in region II of Michigan.) However, the altered age structure is, again, predominantly a result of the change in recruitment rate. If the recruitment rate does not change, of course, a harvested population could be driven to extinction without showing a significant or noticable change in the female age structure.

DISCUSSION

There are two relatively simple management implications of the findings presented above. First, the composition of the harvest of antlered males gives a direct indication of the overall mortality rate affecting this segment of the population. In fact, the proportion of yearling males in the antlered harvest can be an estimate of the proportion lost annually unless there is considerable difference in the vulnerability of the age classes (for which there may be a correction). Second, however, the age structure of the female portion of the herd is almost entirely controlled by the reproductive rates and does not respond directly to nonselective mortality. Thus, it is incorrect to monitor the female age composition for evidence of overharvest, a relatively nonselective mortality factor in white-tailed deer. The age structure of females primarily reflects the reproductive rate. Using information gained from the separate examination of data from each sex can give the manager considerable information about a population.

Work such as this demonstrates that mathematical modeling can clarify problems posed by wildlife managers and respond to these problems in a form that is readily understandable.

LITERATURE CITED

Ackoff, R. L. 1970. A Concept of Corporate Planning. Wiley-Interscience, New York.

Burgoyne, G. E., Jr., and M. L. Moss. 1977. The 1977 deep pellet group surveys. Surv. Stat. Serv. Rep. No. 160. Michigan Dept. Natural Resources, Lansing. 19 pp.

Burgoyne, G. E., Jr., T. Gamble, and J. Meister. 1979. Deer checking station data—1978. Surv. Stat. Serv. Rep. No. 179. Michigan Dept. Natural Resources, Lansing. 29 pp.

Eberhardt, L. 1960. Estimation of vital characteristics of Michigan deer herds. Game Div. Rep. No. 2282. Michigan Dept. of Conserv., Lansing. 192 pp.

Gill, J. 1953. Remarks on the analysis of kill-curves of female deer. 9th Northeast Section Wildlife Conf. Mimeo. 12 pp.

Hayne, D. W., and L. Eberhardt. 1952. Notes on the estimation of survival rates from age distributions of deer. Paper presented at 14th Midwest Wildlife Conf., Des Moines, Iowa, December 17-19.

Keyfitz, N. 1968. Introduction to the Mathematics of Population. Addison-Wesley, Reading, Massachusetts.

Lotka, A. J. 1956. Elements of Mathematical Biology. Dover Publications, New York.

Mertz, D. B. 1970. Notes on the methods used in life-history studies. 15 pp. in: J. H. Connell, D. B. Mertz, and W. W. Murdoch (eds.). Readings in Ecology and Ecological Genetics. Harper & Row, New York.

Mertz, D. B. 1971. Life history phenomena in increasing and decreasing populations. In: G. P. Patil, E. C. Pielou, and W. E. Waters (eds.). Statistical Ecology, Vol. 2. Sampling and Modeling Biological Populations and Population Dynamics. Pennsylvania State University Press, University Park.

Odum, E. P. 1971. Fundamentals of Ecology, 3rd ed. W. B. Saunders, Philadelphia.

Piolou, E. C. 1969. An Introduction to Mathematical Ecology. Wiley-Interscience, New York.

Ricker, W. E. 1948. Methods of estimating vital statistics of fish populations. Indiana Univ. Publ., Science Series. No. 15. 101 pp.

Ryel, L. A., L. D. Fay, and R. C. Van Etten. 1961. Validity of age determination in Michigan deer. Papers of the Michigan Academy of Science, Arts and Letters. Vol. XLVI, 289-316.

Severinghaus, C. W. 1949. Tooth development and wear as criteria of age in white-tailed deer. J. Wildl. Manage. 13:195-216.

TWENTY-TWO

Life History Analysis of Large Mammals

DANIEL GOODMAN _____

INTRODUCTION

An organism's life table, like other biological traits, is molded by natural selection. The specific form of the life table may be viewed as a solution to adaptive problems that dominated the species' history. Accordingly, an understanding of the adaptive significance of various life history characters can provide insight into the natural economy of a population's survival and growth. Analysis can reveal which stages of the life cycle are most critical in their impact on population increase, and which aspects of a population's demography will be most affected by natural environmental change or by intentional control measures.

Because of pervasive biological constraints — and because the ultimate evolutionary objectives of all living systems are identical — most actual life histories fall into a few classes. Among large mammals, for example, life spans are long, adult survival rates are high, fecundity rates are low and reproductive maturity is usually late.

Two matters associated with such a life history pattern will be our concern in this chapter. First, there is the technical mathematical problem of how to develop models that will capture the essential features of population growth under such a life table and, at the same time, ensure, to the extent possible, that demands for data in applying the model are within reach of practical field programs and that the form of the model facilitates insight into how various measured inputs affect the population growth characteristics. Second, there are the questions prompted by an evolutionary perspective: What is adaptive about such a life table, and what might be the maladaptive consequences of changing certain aspects of the population's environment or of modifying the life table itself?

POPULATION MODELS

Population Projection with a Constant Life Table

When an animal's life span is sufficiently long that population growth cannot reasonably be represented in terms of seasonal or annual replacement of the existing population, two factors, the overlap of generations and the distribution of reproductive events over a considerable fraction of an individual's lifetime,

415

combine to obstruct simple mathematical modeling of demographic processes (see Chapter 19).

At any time, individuals whose ages span a significant range may contribute to the total population's birth and death statistics, so that even a minimal model, assuming no future changes in vital rates, requires three kinds of information (the life table, the initial population size, and the initial age profile) to yield even a short-term projection of population size. Furthermore, the appreciable time that elapses between an individual's birth and its period of major reproductive output gives rise to complicated lags in the demographic responses both to direct perturbation of the population size or composition and to environmentally induced changes in the life table.

These processes may adequately be studied, instance by instance, by computer simulation. Matrix notation conveniently summarizes the set of equations that can project a population vector from one time period to the next:

$$n_{t+1} = A_{(n_t, t)} n_t \qquad (1)$$

where n_t is the vector of numbers of individuals, age class by age class, at time t (i.e., element i of n_t is the number of individuals in age class i), and A is the projection matrix, which may be adjusted with respect to population density or time, as appropriate. This sort of formulation lends itself naturally to computer modeling.

One drawback of such an approach is that, despite its self-evident applicability for modeling specific populations in specific circumstances, it is too cumbersome for ready extraction of some desirable generalizations. When the objective is understanding the demographic consequences of certain broad classes of life histories or predicting the responses to qualitatively specified modification of broad classes of life tables, analytical techniques are to be preferred.

Even when dealing with a single specific population, for which considerable data are available, the simple and direct model suggested by equation (1) may not be well suited for direct exploration of optimal management strategies. Optimization techniques often bog down when presented with too many variables (see the review by Walters and Hilborn, 1978). Successful analysis depends on minimal models which capture the essence of the dynamics of the system in question, but which are simplified sufficiently to ensure that the problem of optimization is not mathematically overwhelming and that the additional demands for data are within reason.

Classically, the problem of population projection, where vital rates are assumed constant, was solved by an integral equation due to Lotka [reviewed by Lotka (1939a) and Keyfitz (1968)]. The method begins by expressing the time sequence of birthrates in the population as a function of past birthrates in the population and the life table:

$$B_{(t)} = \int_0^\infty B_{(t-x)}\, l_{(x)}\, m_{(x)} dx = \int_0^\infty B_{(t-2)}\, \Phi_{(x)}\, dx \qquad (2)$$

where $B_{(t)}$ is the birthrate at time t, $l_{(x)}$ is the fraction of newborn surviving to age x, and $m_{(x)}$ is the per capita birthrate to individuals age x. Since the survivorship schedule and fecundity schedule often appear as a product, as in equation (2), we may profitably condense notation by defining the net maternity function, $\Phi_{(x)}$, as this product, the expectation of eventual births at age x per newborn.

This model assumes that the schedules of survivorship and fecundity remain constant over time. It will be most convenient to consider only the female segment of the population, so that the $l_{(x)}$ schedule refers to the survival statistics of females, and only female births are scored in the calculation of $B_{(t)}$ and $m_{(x)}$. The problem of modeling population growth of both sexes has been treated by Goodman (1967) and Yellin and Samuelson (1977).

The solution to equation (2) is the infinite series

$$B_{(t)} = \sum_{j=1}^{\infty} Q_j e^{r_j t} \tag{3}$$

where the terms r_j in the exponents are solutions to the integral equation

$$1 = \int_0^{\infty} e^{-r_j x} \, \Phi_{(x)} \, dx \tag{4}$$

and the terms Q_j are scale factors, associated with each root, r_j. The magnitudes of the scale factors depend on initial condition—that is, on the starting age profile $c_{(x,o)}$, and population size N_0.

A solution for the birth sequence is a sufficient solution for the trajectory of the entire population, as we can reconstruct the entire age profile of a population given only its prior birth sequence.

Equation (4), known as Lotka's equation, has one real root, which we will designate r_1, and this real root is larger than the real part of any of the complex roots. Accordingly, the term $Q_1 e^{r_1 t}$ will, in time, dominate equation (3), for it will be growing exponentially at the greatest rate. Then the birthrate will essentially be growing at the exponential rate r_1, and, as is well known, the population's age distribution will have changed to a characteristic form—the stable age distribution associated with that life table—and the population as a whole will be growing at the exponential rate r_1. But what can be learned of the transient behavior of a perturbed population that is not in stable age distribution?

The scale factors and the roots, except for the first, occur in complex conjugate pairs; and so, the birth sequence may be represented as a sum of terms one of which is an exponential and the remainder of which are sine waves with exponentially growing or diminishing amplitudes. The exponentials are given by the real parts of the roots of Lotka's equation and the frequencies of the oscillatory components are given by the imaginary parts of the complex roots divided by 2π.

If we were to order the roots according to the size of the real parts, largest first, we could obtain an adequate description of the transient as well as the ultimate behavior of the birth sequence by truncating the series of equation (3) after the first few terms, for the dominant terms would be retained, for example,

$$B_{(t)} \cong Q_1 e^{r_1 t} + 2e^{a_2 t} (A_2 \cos b_2 t - B_2 \sin b_2 t) + 2e^{a_3 t} (A_3 \cos b_3 t - B_3 \sin b_3 t) \quad (5)$$

where the complex roots are written as $a \pm ib$ and their associated scale factors are $A \pm iB$.

This treatment has already accomplished a major simplification of the modeling of transients in the dynamics of populations that are not in stable age distribution. It will prove especially illuminating in a later section, where we will discuss simpler approximate methods for obtaining the roots r and the scale factors Q.

A discrete time model, with similar properties to the integral equation model, was developed by Leslie (1945, 1948). This yields a characteristic equation, analogous to equation (4).

$$1 = \sum_{x=1}^{\omega} e^{-r_j x} \Phi_x \quad (6)$$

where ω is the maximum age, and Φ_x is the product of l_x, the fraction of births surviving to age x, and m_x the per capita number of offspring born to individuals of age $x - 1$ to x, during a unit time interval, with the parents censused at the beginning of the interval, and the offspring censused at the end. The model rests on the matrix projection technique illustrated in equation (1), with the restriction that the matrix not vary over time. The matrix itself has the vector of fecundities m_x as its first row, and the vector of survival rates p_x as its principal subdiagonal, and it is zero everywhere else.

Eigenanalysis of Leslie's matrix operator permits a decomposition of the population sequence, through time, as a sum of products of sine waves and exponentials, in the manner of equations (3). The projected population vector is

$$n_t = \sum_{j=1}^{\omega} Q_j e^{r_j t} c_j \quad (7)$$

where the Q_j are scale factors determined by the starting population, the roots r, as obtained from equation (6), are in fact logarithms of the eigenvalues of the matrix (there will be at most ω of them), and the c are the associated eigenvectors of the matrix.

From equation (7) we see that the vectors c represent characteristic components of the age distribution, for their sum, when weighted by $Q_j\, e^{r_j\, t}$ equals the observed age structure. The importance of each of these components of the age distribution changes with time according to the associated exponential $e^{r_j\, t}$. As in the decomposition in equation (3), the root with the largest real part is the real root r_1, and so, with time, the component c_1 comes to dominate the population's age distribution. The eigenvector c_1 is the stable age distribution associated with the discrete time model. The other components decay in relative importance with time (the ones associated with complex roots do so in an oscillatory manner) at rates governed by their respective r_j.

The one glaring difference between the continuous time and the matrix models is that with the matrix method some projected populations fail to converge to a stable age distribution (Sykes, 1969a), but this anomaly will not be encountered in annually based projections for long-lived, repeatedly breeding animals. In seasonally based models, however, where reproduction itself is seasonal, oscillatory behavior will persist indefinitely, as in fact it should (Gourley and Lawrence, 1977).

Approximate Life Tables

For some mathematically special cases of life history schedules, the integral in Lotka's equation (or the summation in the characteristic equation for the Leslie matrix) can be solved analytically. This will yield a much simpler expression relating r to the life table. In some cases, the new expressions even permit algebraic solution for r.

Almost any biologically likely life table can reasonably be approximated by some class of life-table model that permits analytical simplification. Two approaches will be shown here.

For the most extreme sort of iteroparous life table, appropriate for very long-lived vertebrates where brood size does not change much during the adult life of the organism, four parameters can be used to characterize the species' demography. These are the mean fecundity m, the adult annual survival rate p, the age at first reproduction α, and the survivorship until the age of first reproduction, l_α. Assuming that the adult fecundity and survival rates are essentially independent of age results in a geometric series in the characteristic equation, which yields

$$e^{r\alpha} - pe^{r\,(\alpha-1)} - ml_\alpha = 0 \qquad (8)$$

It does not matter that the assumption of constant vital rates is unrealistic as regards the very old adults, for these contribute only negligibly to the important demographic statistics of the population. If, however, the onset of reproductive maturity actually involves a gradual increase in fecundity over a number of years, this abrupt model will not prove accurate in modeling transients and it will be difficult to calibrate it to the value for α that reliably yields the correct r in a growing or declining population.

The model of equation (8) is especially convenient for examining the sensitivity of the growth rate to marginal changes in the vital rates. Simple differentiation with respect to p or m or l_α or α, keeping all other parameters constant, will reveal the importance of that parameter to the value of r. For parameter values in the range encountered in large mammals, it turns out that the growth rate will invariably be more sensitive to small changes in adult survival rates than to changes in fecundity rates. The sensitivity to immature survivorship depends principally on fecundity. If m is low the growth rate will also be less sensitive to l_α than to p (Goodman, 1980a).

Interestingly, the sensitivity to changes in α, the age at first reproduction, will be zero or essentially zero for populations near steady state (see Chapter 23). This mathematical result conflicts with our intuitive appraisal at first (see Chapter 2), but the puzzle is resolved when we recognize that earlier reproduction often entails a greater rate of survival from birth until first reproduction, and that the latter will of course increase the growth rate. Biologically, it does seem generally to be the case that favorable environmental conditions (especially a favorable food supply) leads to noticeable decreases in α, so this change in age at first reproduction is taken to be indicative of population growth. It is worth pointing out, however, that even though this association is empirically valid, the actual increase in population growth rate relative to that in less favorable conditions is probably achieved more as a consequence of concurrent changes in the parameters l_α, m, or p than it is owing to the change in α in slowly reproducing organisms such as the large marine mammals. For rapidly growing or declining populations, the age at first reproduction is important in its own right, as discussed in Chapter 23.

Characterization of the life table in terms of the four parameters p, m, α, and l_α facilitates analysis of the dynamics of populations for which these parameters may be estimated. Even with uncertainty in a few of the parameters it is possible to explore the range of dynamics that are consistent with the plausible parameter space. For example, in Figure 1 we show an examination of rates of population increase and decrease compatible with ranges of adult survival rate and the ratio of adult to mature survival rates (this ratio, given α, of course determines l_α) that are thought to bracket the correct values for the spotted porpoise, *Stenella attenuata*. Similar graphs relating the growth rate to plausible ranges of values for m and α vividly illustrate the theoretically derived assertion that with this sort of life table the growth rate is more sensitive to small changes in p than in the other parameters (Goodman, 1980b). Furthermore, analyses like this, which show the difficulty of reconciling a high population growth rate with what is known about the reproductive biology of the porpoise, have been instrumental in management recommendations (Smith, 1979).

A second approach to the analytic approximation of the life table is to specify it, not in terms of explicit survival and fecundity rates, but in terms of the

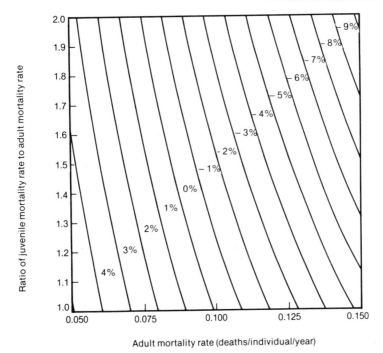

Figure 1   Population growth rate, expressed as percent increase per year, shown as a function of two variables, the adult mortality rate and the ratio of juvenile mortality rate to adult mortality rate. The calculation employs the geometric series approximation, with α set at 6, and m set at 0.1667 (see text).

parameters of a distribution which is fit to the observed net maternity function $\Phi_{(x)}$. The parameters used are R_0, the net replacement rate, T, the cohort generation time, and S, the dispersion in ages of a cohort's reproduction. These are calculated as

$$R_0 = \int_0^\infty \Phi_{(x)}\, dx \qquad (9)$$

$$T = \frac{1}{R_0} \int_0^\infty x\Phi_{(x)}\, dx \qquad (10)$$

$$S^2 = \frac{1}{R_0} \int_0^\infty x^2\Phi_{(x)}\, dx - T^2 \qquad (11)$$

These are immediately comprehensible as an area under the net maternity function (or specifically the expectation of reproduction over a lifetime), a simple mean in ages of reproduction, and a variance in ages of reproduction.

By assuming that these three parameters are parameters of a known mathematical distribution, that distribution may be substituted for the actual net maternity function. If the chosen distribution has much the same shape as the actual net maternity function, the substitution should not make any serious difference to the values of r calculated from solution of the integral in equation (4). If the functional form of the distribution permits analytic solution of the integral, a considerable simplification is achieved. Since distributions are available that increase gradually from zero and then decline, this approach can accommodate a gradual rise in fecundity with age, as contrasted with the abrupt onset of reproduction in the geometric series model.

Three distributions are commonly fit to human data: the normal (Lotka, 1939b), the gamma (Wicksell, 1931), and a special exponential function due to Hadwiger (1940). For each of these models, the integral of Lotka's equation is solved to yield a much simpler algebraic expression for r. It is the experience of demographers that all three of these distributions give extremely good fit to human life tables, and they provide calculated values of r that are correct to a fraction of a percent (Keyfitz, 1968).

The Hadwiger graduation is especially useful, for it yields convenient expressions for the complex as well as the real roots of Lotka's equation:

$$r_1 = \frac{\ln R_0}{T} + \frac{S^2}{2T}\left(\frac{\ln R_0}{T}\right)^2 \tag{12}$$

$$x_j = r_1 - \frac{2}{T}\left(\frac{(j-1)\pi S}{T}\right)^2 \tag{13}$$

and

$$y_j = \frac{i2\pi(j-1)}{T}\left(1 + \frac{S^2 \ln R_0}{T^2}\right) \tag{14}$$

where x_j is the real part of the complex root with the jth largest real part, and y_j is the associated imaginary part. Similar simplifications become possible for such quantities as reproductive value, or the scale factors Q_j used to represent the composition of an arbitrary age distribution. With the complex roots in hand, it finally becomes possible to make ready use of the spectral decomposition of the birth sequence, as developed in equation (3) or the approximation in equation (5).

With respect to the typical life tables of long-lived animals, we note that T and S^2 will be large by comparison with other life histories. The net replacement rate R_0, must, of course, vary about a value of 1 for any population that persists through time. Inspection of equations (11)-(13) shows, then, that the large value of T will have the consequences of diminishing r_1 and y, and it will increase the magnitude of x. Since $\ln R_0$ will be quite small, the effect of S^2 on r_1 and y will be slight. However, S will influence x strongly, and it will tend to counteract the effect of T: when T is much greater than S, x_2 will be close to r_1 in value; as S increases relative to T, x_2 will become progressively smaller relative to r_1. The real parts x of the complex roots clearly become smaller and smaller with increasing j, so these roots govern oscillations that decay faster and faster. The imaginary parts y give the frequencies of these respective transients, which clearly increase with j. For R_0 near 1, the wavelengths of the transients are essentially $T/(j - 1)$, so the first complex root contributes a transient with wavelength about equal to the generation time. The successively higher order roots contribute shorter period oscillations that drop out more quickly.

The amount by which r_1 exceeds x_2 indicates the rate at which the most refractory of the transients will decay relative to the dominant growth process, so this serves as a measure of the rate of convergence to stable age distribution. Thus increasing T will slow the rate of convergence to stable age distribution, whereas increasing S will increase the rate of convergence.

We see then that a typical long-lived mammal will tend to have unusually slow rates of increase and unusually low frequency transient oscillations. Deferred reproduction, which necessarily reduces S relative to T will result in slow convergence to stable age distribution; but for a given generation time, increasing the span of reproductive ages will increase the rate of convergence to stable age distribution.

If environmental conditions, or a harvest, affect survival rates, then the characteristic dynamics will be affected. An increase in mortality rates will diminish both T and S, leading to higher frequency fluctuations, and more rapid population growth or decline relative to a given value of R_0. Clearly, with a more detailed model for changes in the parameters R_0, T, and S, with respect to density, or with a more fully specified intervention, the consequences for the demography may be modeled in a fairly precise and very understandable fashion. One such application to a harvesting model is shown in Goodman (1980b).

To illustrate the capacity of the approximation calculated from the Hadwiger model to model transient dynamics, consider the fur seal life table discussed by Smith and Polacheck (Chapter 5). In Table 1 we reproduce schedules of fecundity and survivorship that are consistent with what is known about the life table of the population during its growth phase earlier in this century. The growth rate calculated from solution for the dominant eigenvalue of the full Leslie matrix of equation (13) with this life table is 0.0782, about an 8% annual rate of increase.

Instead of using the full life table, we may characterize the life history in terms of the parameters R_0, T, and S^2 which, from equations (9)-(11), take on the values 2.139, 10.179, and 13.005, respectively. Using the Hadwiger model, we substitute these parameter values in equation (12) to obtain 0.0784 as the approximate value for r, almost identical to the value computed from the full Leslie matrix.

Table 1   Fur seal life table growth phase[a]

x	p_x	m_x	x	ϕ_x
1	0.8786	0.0000	1.0000	0.0000
2	0.8786	0.0000	0.8786	0.0000
3	0.8837	0.0050	0.7721	0.0038
4	0.8888	0.0151	0.6823	0.0103
5	0.9039	0.2631	0.6064	0.1595
6	0.9191	0.3693	0.5482	0.2024
7	0.9342	0.4250	0.5038	0.2142
8	0.9443	0.4604	0.4707	0.2167
9	0.9494	0.4756	0.4445	0.2114
10	0.9443	0.4705	0.4220	0.1986
11	0.9292	0.4655	0.3985	0.1855
12	0.9039	0.4554	0.3703	0.1686
13	0.8786	0.4402	0.3347	0.1473
14	0.8484	0.4250	0.2941	0.1250
15	0.8029	0.4048	0.2495	0.1010
16	0.7524	0.3794	0.2003	0.0760
17	0.6918	0.3542	0.1507	0.0534
18	0.6262	0.3187	0.1043	0.0332
19	0.5454	0.2833	0.0653	0.0185
20	0.4494	0.2479	0.0356	0.0088
21	0.3282	0.2024	0.0160	0.0032
22	0.1009	0.1467	0.0052	0.0007
23	0.0000	0.0657	0.0005	0.0000

[a]Adapted from Smith and Polacheck, Chapter 5.

The next pair of roots, obtained from equations (13) and (14) with $j = 2$, add substantially and correctly to an understanding of the population processes. In Figure 2 we show a projection, over 30 years, of the birthrate (total censused births per year) to a hypothetical population initiated with 100 newborn fur seals. The continuous stepped line reproduces an actual projection of the population using the full Leslie matrix as per equation (1). The initial birthrate is, of course, 100; it drops to zero as these juvenile seals develop through their prereproductive period. Then the birthrate shows a first "baby boom" centered around eight years after initiation; and after one barely discernible "echo" of the "boom" the birthrate settles into exponential growth as the population vector converges on stable age distribution.

The early history of this population is obviously very different from simple exponential growth, so calculation or estimation of the growth rate r_1 above will not give an adequate representation of the population's behavior. In an actual population, the oscillation owing to perturbation of the age structure may have serious management consequences (cf. Fowler and Smith, 1973, in a discussion of elephant population dynamics), for the transient highs may result in habitat damage and transient lows increase vulnerability to inbreeding and chance local extinction. The solid dots in Figure 2 represent an approximation to the birth sequence, obtained by truncating equation (5) to just the first two terms and calculating the roots r_1 and r_2 and the scale factors Q_1 and Q_2 from the approximate formulae in equations (12)-(14). The damped oscillation introduced by the second root pair clearly reproduces the first birthrate "bust" and the ensuing "boom," and then settles into an exponential trajectory that closely matches that of the full matrix projection.

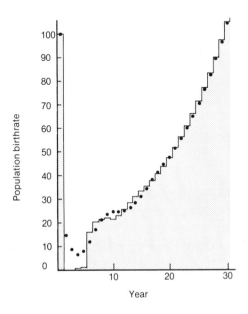

Figure 2   Simulation of population growth rate, as reflected in the total birthrate, in a model fur seal population initiated with 100 newborn. The life table confers an 8% annual rate of increase. The stepped line represents a calculation from the full matrix projection. The dots are calculated from a simple approximation explained in the text.

The point of this illustration is not to argue that a population projection is best carried out with an approximate formula rather than with the Leslie matrix. Surely, when the actual Leslie matrix is in hand, the matrix projection is at least as easily computed as the approximation. Rather, the point is that the approximate formulae for the first two roots in equations (12)-(14) faithfully capture the essence of the population dynamics, including a representation of the intensity, period, and rate of decay of the transient oscillatory tendencies. Therefore, in the absence of sufficient data to justify a full Leslie matrix model, we may confidently resort to the approximations when we can specify the life table only very generally in terms of the parameters R_0, T, and S; and we may use simple algebraic manipulations of the approximate formulas to gain insight into the relevance of each parameter value to the resultant dynamics.

Time-Varying Vital Rates

The models based on Lotka's equation or on the eigenanalysis of the Leslie matrix assume that the life table is constant in time. This assumption is routinely violated in almost any interesting system, so we are obliged to consider the problem of modeling population growth where the vital rates depend, to an important extent, on extrinsic environmental factors.

Where age specific vital rates are not constant, the population's age distribution will not converge to a unique stable form. One major general result known for growth processes under a time varying Leslie matrix is Lopez's (1961) theorem of weak ergodicity. Lopez proved that, given some minimal restrictions on the sequence of Leslie matrices, the populations's age distribution will be dominated by its recent history, and not by the long past history of its vital rates or age distribution. As a consequence, two populations experiencing the same sequence of changes in vital rates will converge in age distribution regardless of how different their age distributions were.

Stable population theory allows ready calculation of the ultimate age distribution on the basis of just the Leslie matrix, when that matrix is constant, but there is as yet no equivalent theory predicting the values for the age distribution where the vital rates are changing. All we know is that a kind of convergence will take place. In an abstract way, this is important knowledge, for it assures us that the behavior of such a population is not chaotic, or dependent (except for scale) on initial conditions. Thus there are reasons to believe that particular realizations of such a growth process, in numerical simulation, will reveal generalizable patterns.

For example, we may wish to examine the extent to which properties, such as the stable age distribution, calculated under the assumption of constant vital rates may serve as useful characterizations of a population even though vital rates are not constant but instead vary randomly about some mean. Consider a life table as a compilation of *average* survival rates and fecundity rates in a population where the actual vital rates show considerable variation over time. We can model population growth under such conditions by applying to the population vector a matrix that is the sum of this mean Leslie matrix and a matrix of temporally random variations from the mean for each element. The mean remains the same, but the matrix of deviations changes with each time unit. Sykes (1969b) has worked out the matrix manipulations for this sort of population projection. The random terms will cause variations in the rate of population growth and in the population's age distribution.

To illustrate computer simulation of such a process we substituted the equilibrium life table for the fur seal population, mentioned earlier, as the mean Leslie matrix. Sufficient variation was introduced via the random terms affecting fecundity that the population's per capita birthrate exhibited a coefficient of variation of 18% over a projection of 1000 time units. Despite this variation in realized vital rates, the age distribution was in many ways remarkably like the stable age distribution calculated from the mean Leslie matrix. The mean age distribution in the population, obtained by simply averaging the frequencies of each age class over the 1000 iterations, was virtually indistinguishable from the theoretical stable age distribution, though of course in any given time unit the age distribution usually departed noticeably from the stable form.

To gain some insight into how this average age distribution was maintained, the observed age distributions in each time unit were broken down in terms of the eigenvectors of the mean Leslie matrix, just as if we were calculating the scale factors for an initial population, beginning anew each time unit. The method of computing the values of the scale factors Q is described by Leslie (1945). To the extent that the scale factor for the first eigenvector remained almost exactly one, the population remained nominally in stable age distribution. The perturbations in age structure induced by variation in the Leslie matrix appeared almost exclusively in the form of shapes associated with the higher order roots, but these are the roots with the smallest real part (Table 2). In other words, the departures in shape of the age distribution from the stable vector corresponded to just those shapes that decay very rapidly under the influence of the mean Leslie matrix.

Table 2   Decomposition, in Terms of Eigenvectors of the Mean Matrix, of a Sequence of Population Vectors Projected by a Random Leslie Matrix.[a]

Root No.	Mean Modulus of Q	Real Part of Root	Half-life
1	1.0005	0	∞
2,3	0.056	−0.191	3.64
4,5	0.164	−0.255	2.71
6,7	0.356	−0.329	2.11
8,9	0.560	−0.391	1.77
10,11	0.727	−0.410	1.69
12,13	1.058	−0.432	1.60
14,15	1.958	−0.447	1.55
16,17	2.368	−0.454	1.53
18,19	2.046	−0.472	1.47
20	2.478	−0.512	1.35
21,22	3.455	−0.525	1.32
23	0.061	−2.925	0.23

[a]The mean Leslie matrix is a model life table for a fur seal population at equilibrium. The perturbations were introduced by sampling a normal distribution with a mean of zero and a standard deviation of 0.25; multiplying the fecundities in the mean matrix by one plus the sampled number and multiplying the survival rates by one plus one-tenth the sampled number to form that time unit's projection matrix. The mean modulus of Q_j indicates the mean departure from zero, in any direction on the complex plane, of the quantity indicating the intensity of the contribution of the shape associated with the jth eigenvector to the actual age distributions.

Under these dynamics the perturbations in age structure did not cumulate or expand, but appeared simply as random and short-lived excursions. Thus, for purposes of many sorts of calculation, one could have assumed that the population had the age distribution corresponding to the stable age distribution of the mean Leslie matrix, even though the actual age distribution was subject to random perturbations mediated through variations in the life table over time. This justifies a number of applications of stable population theory in using analysis of age distributions to calculate vital rates in a fluctuating environment (Goodman, 1979b).

On reflection, we may deduce why it should be that the life table typical of a long-lived mammal would so effectively damp out the perturbations in age structure induced by random variation that primarily affects the birth rate. An annual departure of the birthrate will itself introduce a departure in abundance of the first age class. Prior disturbances of the birthrate will be reflected in departures of the abundance of age classes that are now older, but since we are treating the initial disturbances as uncorrelated, we will in essence be superimposing short waveform deviations on the general underlying shape of the age distribution.

The equations for the eigenvectors into which we decompose the age distribution are given simply by the survivorship schedule discounted by the associated root. The survivorship schedule is a monotonic decreasing function which for a long-lived animal will be relatively shallow in decline over most of the life span. This will contribute a long wavelength form. The shorter components of the waveform must be due to the complex roots which introduce sine waves of a wavelength proportional to their imaginary parts. Recalling our discussion of approximate formulas for these roots, equations (13) and (14), we see that the shorter wavelengths will indeed be accounted for by the higher order roots, and of course these are the roots with the smaller real parts, so their contribution to the birth sequence will tend to damp out the fastest.

The short time span of perturbations caused by random disturbances in the birthrate, relative to fundamental wavelengths given by the generation time, will be most pronounced for organisms with long generation times; and the rate of decay of contributions due to higher order roots is greatest for life tables where the spread in ages of reproduction is large. Thus this tendency to remain in nominally stable age distribution, despite random perturbations in fecundity (or in the survival rates of the very young) will be characteristic of the population dynamics of many large mammals. We will discuss later how this tendency may be of adaptive significance. At this point we can surmise that large departures of the age distribution from stable form are not likely to occur except under disturbances so severe that they are rare events in the organism's evolutionary history.

OPTIMAL LIFE HISTORIES

The Trade-Off Between Survival and Reproduction

In evaluating a life table, we may wonder why a species has evolved that particular schedule of probabilities of birth and death. That is, we may think of the life table itself as an adaptation, the costs and benefits of which we should like to evaluate and compare with those of alternative life history patterns.

Imagine that there is a "possible" set of schedules of fecundity and survival rates. From an evolutionary perspective, this will be the set of biologically possible life history options available to the species. In a management context, this will be the set of life histories that will result under the range of possible policy options. In either case, the problem of interest will often be that of maximizing the consequent population's Malthusian parameter r_1. This is an awkward problem because, except for approximations, we do not have an explicit solution for r_1.

Recently, methods have been developed to decompose this problem into a sequence of simple maximization problems that may be considered one age class at a time (Goodman, 1974; Schaffer, 1974a; Taylor et al., 1974). We may illustrate an application in the matter of the optimal balance between reproduction and survival.

Events in age class j affect fitness in two ways. Reproduction contributes m_j newborn; and survival till the next age contributes the expected future reproduction of an individual one age class older. This latter is simply the reproductive value of an individual that is one age class older, V_{j+1} (Fisher, 1930). Reproductive value itself is scaled in units of newborn ($V_1=1$), so m_j and V_{j+1} are in commensurate units. The dimensionless quantity p_j gives the probability that the individual will survive to make its expected contribution as an older individual. Thus, the total contribution of events in age class j will be the sum

$$m_j + p_j V_{j+1} = m_j + p_j e^{r_1 j} \sum_{x=j+1}^{\omega} e^{-r_1 x} m_x \prod_{k=j+1}^{x-1} p_k \qquad (15)$$

and the optimal combination of fertility and survival rates for age class j will be achieved by maximizing this sum.

While this result is an immediate conceptual aid, it is not directly implementable, for the term V_{j+1} contains the unknown we are ultimately trying to maximize, namely r_1, and it also contains the fertility and survival rates of all older age classes. Neglecting, for the moment, the problem of finding the correct value of r_1, we see some advantage in the fact that only age classes older than j affect the optimization of vital rates for age class j. This suggests a sequential optimization of vital rates, one age class at a time, starting with the oldest. Then each maximization will have to deal only with variables pertaining to that one age class, as the terms involving the older age classes will already have been set.

We may deal with the unknown r_1 by generalizing the sum maximized in equation (15) to yield a function of a free variable s:

$$g_j(s) = m_j + p_j \sum_{x=j+1}^{\omega} e^{s(j-x)} m_x \prod_{k=j+1}^{x-1} p_k \qquad (16)$$

The recursive nature of the age class by age class maximization is immediately revealed by the relation

$$g_j(s) = m_j + p_j e^{-s} g_{j+1}(s) \tag{17}$$

For age class 1, we see that the function g takes on a form almost identical with the characteristic equation (6), so that if s were, in fact, equal to the Malthusian parameter r_1, we would have the identity

$$g_1(s) = e^s \tag{18}$$

Thus, the complete program of solution will involve finding the life table that maximizes the schedule g for some arbitrary value of s, and then checking the identity (18). If equation (18) is not satisfied, a new value is chosen for s, noting that g is a well-behaved decreasing function of s, so successive correction to improve the fit to equation (18) is straightforward. Once the identity is satisfied, the optimal life table has been found (Taylor et al., 1974).

The Adaptive Significance of Deferred Reproduction

It is common among long-lived animals that reproduction be deferred. Most often this is expressed as a puzzlingly late age of reproductive maturation and low reproductive rates. In some extremely long-lived organisms this is also expressed in a ready tendency to abandon reproduction during unfavorable times. This apparent deemphasis of reproduction is, at first, difficult to reconcile with our preceding suggestion that evolution should maximize population growth. On reflection, however, the optimization models described above allow us to identify the circumstances in which the life history that maximizes the Malthusian parameter actually involves a rather low annual reproductive rate in some age classes balanced by a high contribution to future fitness from other components of the life table.

One simple model for delayed maturity, and occasional abandonment of reproduction follows directly from the maximization of the sum of m_j and $p_j V_{j+1}$. In a typical long-lived mammal, which in any case does not have a large litter size, the fecundity, which represents litter size less mortality before the first census, will usually have a range of zero to perhaps two female offspring per female. By contrast, the reproductive value of an individual that is in, or approaching, its reproductive years is a discounted sum of reproduction over much of the reproductive span, with the first term discounted only by one multiple of the inverse of the finite rate of increase (e^r) and with successive terms discounted by higher powers of the rate of increase and by the probability of reaching that age. In a usual large mammal, the Malthusian parameter is invariably small, so discounting by e^{-r} will not be precipitous. Furthermore, adult survival rates are typically high in these animals, so discounting for adult mortality will not be very steep. Thus the reproductive value of a long-lived animal in, or approaching, its reproductive years will likely be several times larger than the fecundity achieved in the time span corresponding to one age class.

Since long-lived organisms have, by definition, the option of maintaining a high survival rate — that is, a p_j near one — the major contribution to the sum in equation (15) will be the term $p_j V_{j+1}$ and not m_j; and owing to the dominance of V_{j+1} relative to m_j, a fractional decrease in p will exact a cost which is of greater magnitude than the benefit which would accrue from increasing m_j by the same fraction. In fact, for some long-lived vertebrates the calculated sensitivity of r_1 to the adult survival rate is about 20 times that of the sensitivity to the annual fecundity (Goodman, 1974).

Now, for an organism operating at the limit of its time and energy resources, an increase in reproduction at some age must entail a decrease in survival, and vice versa. Therefore, we come to the conclusion that it will be disadvantageous for a long-lived vertebrate to incur this cost of reproduction when the return in fitness is lower than usual, nor will it benefit the organism to attempt to maintain the usual level of reproductive output when the cost to survival is higher than usual. This easily accounts for abandonment of reproduction during unfavorable times.

This same explanation can apply to account, in part, for the delayed maturity of many long-lived mammals. These animals are sufficiently advanced in their behavior that they may learn foraging and fighting and courting skills over a period of years. In those with complex social structures, advancing in the group hierarchy may similarly require several years. So, it is quite understandable that the potential benefits of an attempt at reproduction will be outweighed by the costs at the age of earliest physiologically possible maturity, but that after the passage of a few additional years, the potential return will have increased and the costs will have diminished to the point that a non-negligible investment in reproduction is optimal. The intervening years will be the period of delayed maturity in an optimal life table.

The preceding analysis, like the model on which it is based (see above section), confines all trade-offs within age class boundaries. Yet there is at least one important trait, namely growth, the effects of which carry over from one time period, and hence from one age class, to the next. Some of the energy allocated to reproduction or avoidance of death in a given time period might equally well be diverted to growth that will affect the individual's *capacity* to capture resources in the future, regardless of how they are later budgeted. Thus it is not really sufficient to describe an individual according to age alone, nor is it really sufficient in describing a possible set of future fertility and survival probabilities for a given individual just to specify its present age. In both instances some index of state, such as size, is also required. This extension is easily incorporated into the model (Taylor et al., 1974).

Now the sum $m_j + p_j V_{j+1}$ will be affected by the allocation between reproduction and survival in the terms m_j and p_j as before. The growth component will appear in V_{j+1}. If the individual grows in size between age j and age

$j + 1$, it may, in so doing, increase its resource gathering potential, for example. This should increase its capacity for future reproduction, thus increasing V_{j+1} beyond the value achieved simply through chronological aging from age j to age $j+1$, without the concomitant growth.

Let us assume that the increase in reproductive value is linearly related to growth. (This assumption is easier to defend if growth is slow.) Let f be the fraction of potential growth that gets allocated, instead, to reproduction. The realized growth, then, will be proportional to $(1 - f)$, so, neglecting changes in p_j, the term $p_j V_{j+1}$ may be represented as a linear function of f with a negative slope:

$$p_j V_{j+1} = k - cf \qquad (19)$$

where k is some constant.

Let $b(a)$ be the function relating fecundity at this age to the actual amount of resources committed to reproduction, where a is some measure of that resource commitment. In general, b will be sigmoid, increasing faster than linearly for low values of its argument and increasing slower than linearly for high values. The concave up portion will usually be due to the fact that there is an optimal size for the young (fractional offspring being of no biological worth); so the first increments of allocation to reproduction will reflect the approach to the optimal size of the first young. The time and size constraints on the adult will result in diminishing returns in the allocation to reproduction above some level, thus the function b must saturate, and become convex up for high values. In an intermediate range, increments in reproductive allocation are readily convertible to additional young or more robust young in a way that will result in a linear relationship between the investment and fitness.

The measure of the amount of the reproductive investment a is a product of f, the allocation fraction, and G, the total resource budget. When G is small, this will keep $b(fG)$ in the concave up region, for f cannot exceed one. Thus there are two possible reasons that the conversion function b is likely to be concave up for a large, long-lived animal: the necessary care and the size (and hence the high minimal cost) of individual offspring, and the fact that the resource budget G may be quite limited, especially for a young individual.

Replacing m_j with the function $b(fG)$, and replacing $p_j V_{j+1}$ with $(k - cf)$, we see that the optimal life table maximizes the sum $b(fG) - cf$, which is to say that it maximizes the difference between cf, an increasing linear function of f, and b, an increasing function of f that will probably be concave up in the region of interest. One can see immediately, from simply graphing two such functions, that the difference between them will be maximized at one extreme or the other. That is, the optimal f is zero or one. So, in the optimal life table the organism will not grow and reproduce in the same time interval. Biologically this means that the organism will not attempt reproduction until it has reached what will be its final size. For a large slow-growing animal, this growth period may be of considerable duration.

In this manner we have found two models that can account for deferred reproduction. One has to do with the balance between reproduction and survival, and the other has to do with the balance between reproduction and growth. In either case, the circumstances of large, long-lived animals, especially those that produce relatively large young, are likely to result in deferred reproduction being an adaptive feature of the optimal life table.

A third possible adaptive consequence of deferred reproduction coupled to a long reproductive span applies in a fluctuating environment. Here our first difficulty is to identify an appropriate measure of population growth. Stochastic Leslie matrix models very readily will generate estimates of the annual expectation of population growth (Sykes, 1969b), but it turns out that this kind of average is not nearly so informative as we would like. Population growth is a multiplicative process. Accordingly, the geometric mean, rather than the arithmetic mean, of a sequence of factors of increase provides the proper index to overall growth during the span of that sequence.

Lewontin and Cohen (1969) considered a sequence of simple multiplicative factors accounting for the growth of a population. By Taylor expansion they showed that the geometric mean of the sequence could be approximated as the arithmetic mean minus a term that was proportional to the variance in the sequence. This suggested that there are two considerations in maximizing population growth in a fluctuating environment: maximizing the arithmetic mean factor of increase and minimizing the variance. Where these two conflict, there must be some balance between them.

The model investigated by Lewontin and Cohen did not incorporate an explicit life table; per capita birth and death rates were treated as independent of age composition. Now for a population in stable age distribution, the per capita rates are constant, and the dominant eigenvalue or Malthusian parameter is the correct measure of the factor of increase applying in any given interval. However, in a fluctuating environment, the population will not be exactly in stable age distribution, and so the current eigenvalue or Malthusian parameter is not truly equivalent to the realized rate of increase (as the higher order roots and the age distribution itself must be taken into account). This problem has not been satisfactorily resolved, except for particular cases (Goodman, 1981), but the argument developed earlier (that for many long-lived mammals experiencing random variation in reproduction or immature survivorship the population may nevertheless remain quite close to nominal stable age distribution) implies that we may continue to use the dominant root as an interim metaphor for the actual factor of increase, and proceed with the heuristic hypothesis that optimizing a life table in a fluctuating environment will involve some degree of minimizing the variance in the dominant roots associated with the stochastic Leslie matrices, even at some cost to the average value.

One means of decreasing the variance is through risk spreading. This is achieved whenever it can be arranged that the individuals at risk are subjected to the full range of circumstances and thus experience, overall, an average of these circumstances, rather than letting the temporal fluctuations of the environment impose the full weight of their variability on the individuals at any one time. Intuitively, this would seem to imply an advantage in a longer life span as the realized net maternity function of an individual will then average over a longer history of environmental fluctuations, and thus will smooth out the effects of these fluctuations in the same way that a running average will smooth a sequence of random values.

Murphy's (1967) analysis of a fishery model illustrates the stabilizing effect of reproducing over a longer life span. The mathematical theory of risk spreading in life histories, treated in Goodman (1981), confirms the intuitive appraisal, developed above, of the advantages of repeated reproductive episodes as an adaptation to environmental fluctuations. This is not simply a matter of environmental fluctuations always favoring life tables with deferred reproduction and longer reproductive span. The reverse may hold in known cases, depending on whether the environmental fluctations more strongly affect the vital rates of young or of adults (Hairston et al., 1967; Schaffer, 1974b; Goodman, 1979a). However, for the typical long-lived animal, reproduction and the survivorship of young are the components of the life table most sensitive to environmental change. So, as a general rule, we would expect that, in addition to the adaptative consequences discussed earlier in this section, the long reproductive span and lower annual reproductive output of many long-lived animals can be an adaptive response to the problems posed by environmental fluctuation.

SUMMARY AND FURTHER IMPLICATIONS

Large mammals have a particular sort of life table. Survival rates are high, fecundity rates are low, and reproductive maturity is usually late. These features create problems for us, in our attempt to develop adequate models for the population growth of large mammal populations and in our attempt to learn the ways in which features of the life table and the environment affect the population dynamics. In the first section of this paper we reviewed some techniques for solution of these problems, and employed the techniques to explore some consequences of the typical life table of a large mammal.

The immediate consequences are low rates of increase and decrease, low frequency oscillatory transients under major perturbation, stability under moderate fluctuations in the effective fecundity, and greater sensitivity of the growth rate to changes in adult mortality than to other changes in the life table components. These sorts of behavior can be advantageous in some respect: they will improve the population's prospects for weathering short unfavorable periods of modest severity, and they will minimize the variance in the population growth rate. The same traits will be a nuisance to population management, as they will delay the attainment of equilibrium population composition after major intervention, slow the recovery from population collapse, spawn generally exasperating time lags in population responses, and in some cases lead to quite low resilience to harvest of adults.

Our analysis of life table optimization suggests that this balance of demographic traits is not accidental. Rather, it represents an evolutionary solution to a suite of ecological problems, given the biologically possible options that are available to the organism. We explored in particular the possible adaptive significance of delayed maturity and long reproductive life spans. It seems reasonable to suppose that other features are likewise adaptive. This perspective advises caution when contemplating interventions that will affect the realized survival and fecundity rates, for the changes are liable to push the life table into a very different dynamic regime.

An unselective harvest, or a specific adult harvest, for example, will lower the effective adult survival rates. If there is density compensation, we might expect some concurrent increase in fecundity and immature survival rates. Raising the effective fecundity and lowering the effective survival rate will actually change some of the character of the population dynamics away from that typical for long-lived animals. That is, the rates of change of population size, for the same net replacement rate will be quicker, the population oscillations will be of higher frequency, the population will be more sensitive to environmental fluctuations, and there may be a slowing of the rate of convergence to stable age distribution. The gravity of the outcome will depend on the magnitude of the change and on the severity of the ecological problems to which the original life table was an adaptive solution.

The presentation in this chapter has emphasized abstract manipulations and qualitative conclusions in order to illustrate the power of these demographic techniques in the extraction of generalizations even when the data coverage is sparse. These same models can be applied in a precise concrete way, in particular cases, whenever sufficient data are available.

LITERATURE CITED

Fisher, R. A. 1930. The genetical theory of natural selection. Oxford University Press, London.

Fowler, C. W., and T. Smith. 1973. Characterizing stable populations: An application to the African elephant population. J. Wildl. Manage. 37:513-523.

Goodman, D. 1974. Natural selection and a cost-ceiling on reproductive effort. Am. Nat. 108:247-268.

Goodman, D. 1979a. Regulating reproductive effort in a changing environment. Am. Nat. 113:735-748.

Goodman, D. 1979b. On the interpretation of age distributions. IUCN (International Union for the Conservation of Nature) and WWF (World Wildlife Fund) Workshop on Biology and Management of Northwest Atlantic Harp Seals, Working Paper 7.

Goodman, D. 1980a. Demographic intervention for closely managed populations. *In:* M. Soule and B. Wilcox (eds.). Conservation Biology. Sinauer, Stamford, Connecticut.

Goodman, D. 1980b. The maximum yield problem distortion in the yield curve due to age structure. Theor. Pop. Biol. 18:160-174.

Goodman, D. 1981. Risk spreading as an adaptive strategy in iteroparous life histories. Theor. Pop. Biol. (In press).

Goodman, L. A. 1967. On the age-sex composition of the population that would result from given fertility and mortality conditions. Demography 4:423-441.

Gourley, R. S., and C. E. Lawrence. 1977. Stable population analysis in periodic environments. Theor. Pop. Biol. 11:49-59.

Hadwiger, H. 1940. Eine analytische Reproduktionsfunktion fur biologische Gestheiten. Skandinavisk Aktuarietidskrift 23:101-113.

Hairston, N. G., D. W. Tinkle, and H. M. Wilbur. 1970. Natural selection and the parameters of population growth. J. Wildl. Manage. 34:681-690.

Keyfitz, N. 1968. Introduction to the Mathematics of Population. Addison-Wesley, Reading, Massachusetts.

Leslie, P. H. 1945. On the use of matrices in certain population mathematics. Biometrika 33:183-212.

Leslie, P. H. 1948. Some further notes on the use of matrices in population mathematics. Biometrika 35:213-245.

Lewontin, R. C., and D. Cohen. 1969. On population growth in a randomly varying environment. Proc. Natl. Acad. Sci. U.S. 62:1056-1060.

Lopez, A. 1961. Problems in Stable Population Theory. Office of Population Research, Princeton, New Jersey.

Lotka, A. J. 1939a. A contribution to the theory of self-renewing aggregates, with special reference to industrial replacement. Ann. Math. Stat. 10:1-25.

Lotka, A. J. 1939b. Theorie Analytique des Associations Biologiques. Part II. Analyse Demographique avec Application Partibuliere a L'Espece Humaine. Herman, Paris.

Murphy, G. I. 1968. Pattern in life history and the environment. Am. Nat. 102:390-404.

Schaffer, W. M. 1974a. Selection for optimal life histories: The effects of age structure. Ecology 55:291-303.

Schaffer, W. M. 1974b. Optimal reproductive effort in fluctuating environments. Am. Nat. 108:783-790.

Smith, T. D. (ed.) 1979. Report of the Status of Porpoise Stocts Workshop, August 27-31, 1979. Southwest Fish. Ctr. Admin. Rep. LJ-79-41, Nat. Mar. Fish. Serv., La Jolla, California.

Sykes, Z. M. 1969a. On discrete stable population theory. Biometrics 25:285-293.

Sykes, Z. M. 1969b. Some stochastic versions of the matrix model for population dynamics. J. Am. Stat. Assoc. 64:111-130.

Taylor, H. M., R. S. Gourley, C. E. Lawrence, and R. S. Kaplan. 1974. Natural selection of life history attributes: An analytical approach. Theor. Pop. Biol. 5:104-122.

Walters, C. J., and R. Hilborn. 1978. Ecological optimization and adaptive management. Annu. Rev. Ecol. Syst. 9:157-188.

Wicksell, S. D. 1931. Nuptiality, fertility, and reproductivity. Skandinavisk Aktuarietidskrift 14:125-157.

Yellin, J., and P. A. Samuelson. 1977. Comparison of linear and nonlinear models for human population dynamics. Theor. Pop. Bio. 11:105-126.

TWENTY-THREE

Comparative Population Dynamics
in Large Mammals

CHARLES W. FOWLER ⎯⎯⎯⎯⎯⎯⎯⎯⎯⎯⎯⎯⎯⎯⎯⎯

INTRODUCTION

Progress in many fields of science has been stimulated by comparative studies. Population biology has profited from the current interest in life history phenomena (e.g., see Pianka, 1970; Stearns, 1976) stimulated by Cole's (1954) work. Most of these efforts have been directed at determining the evolutionary processes that have produced various life history strategies. Relatively less effort has been spent in determining the factors of importance in producing patterns in the density-dependent dynamics of populations.

Beverton and Holt (1959), Holt (1962), Blueweiss et al. (1979), Ohsumi (1979), and Bunnell and Tait (Chapter 4) have used comparative concepts to determine the extent to which vital rates (e.g., birth, maturation, mortality) are correlated with each other as well as such features as body size. This work serves the very useful role of providing better contexts within which to evaluate independent estimates (or even produce rough guesses) of values for the vital rates of typical populations for various species. This work, however, is characterized by a tendency to think more in terms of approximate mean rates associated with specific species rather than in terms of the patterns in the variability these rates may exhibit in response to the variability in their environment. Specifically, such approaches have not been extended to making comparisons for finding patterns in the ways vital rates vary in response to changes in population density.

As the physical and biological environments change in reaction to altered density it is to be expected that dynamic properties will change according to specific patterns as shown in Chapters 2 and 3 and in Eberhardt (1977). The same is true for changes in the composition of populations. Additionally, the process of evolution may have resulted in specific types of dynamics being correlated with specific life history strategies (Gilpin et al., 1976; Fowler, 1981). Similar patterns are expected on the basis of ecological or trophic status (May, 1973; Fowler, 1981). The knowledge of such patterns is of both theoretical and practical value. Comparative work helps demonstrate these patterns and helps understand the process and mechanisms behind such patterns.

In the following sections examples of comparisons of relevance to large mammal populations are presented. An overview of some concepts of importance to comparative population dynamics contains examples of features of population dynamics which may be compared along with examples of characteristics of populations over which comparisons of such dynamics may be made. This is followed by examples of progress made in making such comparisons. It is concluded (in parallel with Chapter 2) that research in comparative population dynamics of large mammals is necessary, possible, and productive.

THE PROCESS OF COMPARING POPULATION DYNAMICS

Comparisons involve two basic sets of elements. First there is the set of characteristics being compared. Second there are the features used to categorize items over which the first set of characteristics are compared. The first plays a role similar to that of a dependent variable, the second functions much like an independent variable. Comparisons may be made across continuous dimensions (productivity as a function of body size) as well as discrete differences (mechanism for expression of density dependence as a function of habitat type, e.g., marine and terrestrial). In this section some of the elements of population dynamics that may be compared are discussed. This is followed by a discussion of some examples of features of populations over which useful comparisons may be made.

The Compared Elements

The logistic model, as well as Graham's (1935) and Schaefer's (1954) stock recruitment model, assume that there is a linear relationship between specific vital rates (or combinations thereof) and population size. Such a relationship (e.g., the solid line in Figure 1) may serve best, however, only as a point of reference for comparison. Those species or populations for which such relationships curve upward (negative second derivative), as shown by the broken line in Figure 1., may also have other categorical differences from those species with such curves that curve downward. In other words, the second derivative of these curves may be determined by (or at least correlated with) other characteristics of the species involved.

The solid line in Figure 1 is an example of only one possible relationship between birthrate and density (or survival and density) and serves well as a point of reference for comparison. With the slope reversed it could represent the relationship between age at maturation, calving interval or mortality and density (Figure 2). Such relationships may be specific to certain portions of the population as in the case of the survival of juveniles.

Relationships such as exemplified in Figure 1 and 2 (and discussed above) translate to specific growth or productivity rates calculated as $(N_{t+1} - N_t)/N_t$, where the subscripts represent time and N the population size. This expression represents the production or growth of the population as expected on a per capita basis. When such growth is expressed as a linear function of the popula-

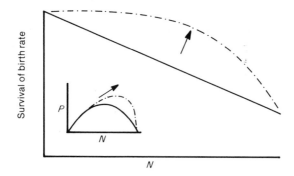

Figure 1   Relationships between survival or birth rate as they hypothetically vary with population size N. The inset shows how productivity (P, as defined in the text) is altered in its relationship to the population size as dependent upon the degree of nonlinearity in the basic relationship. See text for details.

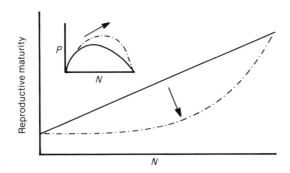

Figure 2   The same as Figure 1 where the density dependence involves age at first reproduction.

tion density (Graham, 1935; Schaefer, 1959) we have a model for population dynamics which serves as a reference point for comparison. It is to be expected that across taxonomic groups, trophic levels or social systems, we will find orderly patterns in the shape of such relationships.

Relationships of the types shown in Figures 1 and 2 may be translated to productivity or rate of growth of the population as a function of population size, shown in the insets of these same figures. The logistic and Graham-Schaefer models imply that peak productivity occurs at population levels equivalent to one-half the level of the equilibrium or unexploited population. Rather than give rise to any rule of thumb for characterizing populations in general, however, this point should serve as a point of reference.

Within the context of population dynamics, then, there are many characteristics of populations which may be compared. Examples of these include the nature of the relationship in which density may act to produce changes in birthrates, survival among adults, survival among juveniles, sexual or reproductive maturation, survival of females, survival of males, calving intervals, and change in the variance of any of the above (see Boyce, 1977, Chapter 8). These examples relate to the basic birth and death processes which have been shown to be density dependent in large mammals (Fowler et al. 1980a). There are many other basic elements of population dynamics (such as those due to migration); which should also be considered for comparative work. Some are discussed in Chapters 2 and 20.

The Elements Over Which Comparisons Are Made

Having identified dynamic properties of populations which may be compared the next step is to look for patterns. Patterns will usually be found to be related to other characteristics (the independent variable) such as the structure of a population by age or sex. One way of categorizing populations for comparison involves the mechanisms through which density dependence is expressed.

Different mechanisms may be expected to produce different types of dynamic relationships of the types shown in Figures 1 and 2. At the level of resolution of our considerations here these may be viewed as the second-order effects of Caughley (Chapter 18). Starvation, predatory regulation, resource levels (as regulating factors), territoriality, shortage of denning sites, parasites, and social stress are recognized mechanisms for population regulation. Each of these mechanism can be expected to produce its own pattern in the dynamics of the vital rates, and hence productivity, at the level of the individual as well as at the level of the population. There may be a tendency for predators such as mountain lions to be regulated more through territoriality than an herbivore, such as elk, and hence a tendency to exhibit different population dynamics. Further research is needed to determine how the dynamics of such widely different groups produce patterns in population dynamics, especially with regard to the shape of the relationships exemplified in Figures 1 and 2.

Other ways of categorizing populations for comparison involve the sector of the population most involved in the expression of density dependence (through any specific mechanism). For example, density-dependent juvenile survival might involve competition within the population as a whole when adults as well as juveniles depend on the same limiting resources. It may be a function of adult density more than anything else (see Chapter 9). On the other hand, the predominant causes of juvenile mortality may involve other juveniles as has been postulated for fur seals (Chapman, 1961). Segregation by sex may mean that density dependence is most closely related to one or the other sex as is assumed in some whale models (Chapter 13). Thus, patterns of similarity in dynamic properties being compared may be expected with respect to the portion of the population most influential in producing competitive density-dependent responses.

In translating relationships of the type shown in Figures 1 and 2 to that of the productivity curves shown in the insets, the structure of the population and the related life history play an important role as may be seen in a careful study of the work of Cole (1954). There are consistent patterns of curvilinearity. These patterns will be examined in greater detail in the next section. Related to this are the ways specific harvest regimes influence such dynamics.

It is obvious that there are a great number of potential bases for characterizing populations such that their dynamics might be compared. These include mechanisms through which density dependence is expressed such as competition, cannibalism, territoriality, diseases, predation, starvation, and aggression. Other examples include the portion of the population responsible for eliciting a density-dependent response such as sex, age, reproductive condition, and social status. The life-history-type, degree and type of social structure, taxonomic category, habitat type (e.g., marine versus terrestrial), and trophic status of various species will all very likely be of importance in producing patterns in the dynamics exhibited by their populations.

Some of the bases for comparison mentioned in this section may be used either as the characteristic being compared (dependent variable) or as the characteristic over which comparisons are being conducted (independent variable). For example, correlations may exist between the mechanism for density dependence and the type of social structure. Killing of conspecifics may be directed toward producing changes in juvenile survival as opposed to that of older age classes (see Chapter 6 and 9, for example). Cannibalistic and noncannibalistic species may show quite distinct productivity curves.

In the following sections examples of comparisons are presented. Some involve theory only; others involve both theory and empirical information. The first two deal with comparisons that affect the shape of productivity curves (insets of Figures 1 and 2). The third involves the shape of the density dependence curves (main part of Figures 1 and 2). As examples, these comparative studies are presented to emphasize the potential, the need and the utility of comparative population dynamics in large mammals.

MODE OF REGULATION AND THE NATURE OF PRODUCTIVITY CURVES

Birthrate

As an example of the types of generalizations that may emerge from a study of comparative population dynamics, we first consider a linear relationship between birthrates and population size. The consequence of such a relationship can be examined in the context of a model. The model used will be a variable projection matrix (Smith, 1973; Fowler and Smith, 1973) as shown in Figure 3. In this example the top row consists, in part, of fixed elements that distribute reproductive rates according to age class. These are then multiplied by a factor derived from a linear relationship such as depicted by the solid line in Figure 1. Specific productivity for the various populations at equilibrium may be found from the equation

$$\frac{A(N_t) - N_t}{N_t} = \lambda(N_t) - 1$$

where $\lambda(N_t)$ is the density-dependent eigenvalue of the matrix A as a function of N_t, the population size. Note that at equilibrium it is assumed that the removal of this production brings the population back to its original stable age distribution (eigenvector).

$$\begin{bmatrix} f_{11}(N_t) & f_{12}(N_t) & \cdots & f_{1m}(N_t) \\ f_{21}(N_t) & 0 & \cdots & 0 \\ 0 & f_{32}(N_t) & \cdots & 0 \\ & & & \\ \cdot & \cdot & & \cdot \\ 0 & 0 & f_{m,m-1}(N_t) & 0 \end{bmatrix} \begin{bmatrix} n_1 \\ n_2 \\ \cdot \\ \cdot \\ \cdot \\ n_m \end{bmatrix}_t = \begin{bmatrix} n_1 \\ n_2 \\ \cdot \\ \cdot \\ \cdot \\ n_m \end{bmatrix}_{t+1}$$

$$A(N_t) \qquad\qquad N_t \qquad N_{t+1}$$

Figure 3   A density-dependent projection matrix A in which the elements of the matrix (f_{ij}) are functions of the population N_t. These functions use the vector N_t which in one simple form is collapsed to its sum.

If the birthrate is a linear function of population size (as shown in Figure 1) the relationship between $\lambda(N)$ and N is of the general shape shown in Figure 4 when translated by the application of any matrix representing a realistic life history. Note that the relationship is curvilinear. This curvilinearity derives from the fact that populations at lower densities will be characterized by a lower mean age than at higher densities because of the higher population growth rate (eigenvalue). This is due to the effect of the increased birthrate as discussed in Chapter 21. Thus, a smaller portion of the population is reproductive at lower densities. The reproductive portion of the population increases toward population levels near the carrying capacity. To an extent, this increase offsets the reduced birthrate. At higher densities the decrease in birthrate becomes the dominant factor in causing the population to exhibit a reduced ability to increase.

The construction of various matrices representing various life history strategies gives rise to the conclusion that the more reproduction is delayed (as a fraction of the total life span) the greater the degree of nonlinearity (see Figure 4). Translating this to population productivity produces an asymmetric produc-

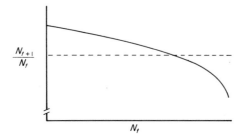

Figure 4 A plot of the eigenvalue of a density-dependent matrix for a typical large mammal in which the birthrate (or juvenile survival) is linearly dependent on the total population size N_t. The eigenvalue is equivalent to N_{t+1}/N_t if the age structure is maintained equivalent to the eigenvector through a harvest strategy that maintains equilibrium.

tivity curve. In this relationship the peak is shifted toward the carrying capacity. Again, this shift is dependent on the age at maturation, being greater for species that mature and reproduce not long before they die.

Tables 1 and 2 contain a list of species of large mammals for which there is evidence of density dependence (Fowler et al., 1980a; Fowler, 1981). As can be seen, birthrates have been shown to be density dependent in a number of cases. Because of the early age at maturation in large mammals (when expressed as a fraction of the life span) the effect of linear density dependence in bithrates is to move the population level at which maximal productivity occurs to a level of only 55 to 60% of the carrying capacity.

Juvenile Survival

Tables 1 and 2 also contain a list of large mammals for which evidence has been collected to indicate that juvenile survival is density dependent. If juvenile survival is incorporated in the top row of the matrix as being linearly density dependent, the results will be the same as just discussed for density-dependent birthrates. Similar results are obtained if juvenile survival is defined as the first element of the subdiagonal. Thus, linear density dependence in either the birthrate or juvenile survival implies an asymmetric productivity curve. In both cases the population level which produces the greatest increment in growth (at equilibrium) will be greater than one-half the mean unexploited population level.

Age at Maturation

A list of large mammals for which there is evidence that age at reproductive maturation changes in a density-dependent fashion is included in Tables 1 and

Table 1 A List of Various Species of Large Terrestrial Mammals for Which There Is Evidence of Density Dependence[a, b]

Species	Common Name	Birthrate	Age at First Reproduction	Juvenile Survival	Adult Survival
Phacochoerus aethiopicus	Wart hogs	*		*	
Odocoileus virginianus	White-tailed deer		*	+	
O. hemionus	Mule deer		+		
Cervus canadensis	Elk	*	*	*	*
C. elephus[c]	Red deer				
Bison bison	Bison	*			
Bos taurus	Longhorn cattle	*			
Alces alces	Moose	*	*	*	
Ovis canadensis	Bighorn sheep	*	+	+	
O. dalli	Dall sheep	*		+	
O. aries	Soay sheep	*		*	*
Capra aegagous	Ibex	+			
Syncerus caffer	African buffalo	+		*	*
Chonochaetes taurinus	Wildebeest		+	+	*
Hemitragus jemlachicus	Himalayan tahr		*	+	+
Loxodonta africana	African elephants	*	*	+	
Hippopotamus amphibius	Hippos		+		
Canis latrans	Coyotes	+		*	
C. lupus	Wolves			+	
Panthera leo	Lion	+		*	
Ursus arctos	Grizzly bears	*		*	
Homo sapiens	Humans				*

[a]See Fowler et al. (1980a) and Fowler (1980d, 1981) for details.
[b]A star indicates statistically significant evidence and a plus sign indicates anecdotal evidence.
[c]Red deer are represented by evidence for recruitment as a combination of the factors identified individually for the other species.

Table 2   A List of Various Species of Marine Mammals for Which There Is Evidence of Density Dependence[a, b]

Species	Common Name	Birthrate	Age at First Reproduction	Juvenile Survival
Dugong dugon	Dugong		+	
Halichoerus grypus	Gray seals			*
Pagophilus groenlandicus	Harp seals	+	+	*
Callorhinus ursinus	Northern fur seals		*	*
Arctocephalus gazella[c]	Antarctic fur seals		+	
Leptonychotes weddelli	Weddell seals		+	
Mirounga angustirostris	Northern elephant seal			+
M. leonina	Southern elephant seals		*	
Stenella attenuata	Spotted porpoise	+	+	
S. coeruleoalba	Striped dolphin	+		
Balaenoptera musculus	Blue whales	*	*	
B. physalus	Fin whales	*	*	
B. borealis	Sei whales	*	*	
Physeter catedon	Sperm whales	+		

[a]See Fowler et al. (1980a) and Fowler (1980a, 1981) for details.
[b]A star indicates statistically significant evidence and a plus sign indicates anecdotal evidence.
[c]The evidence for the Antarctic fur seals involve a comparison with the northern fur seal.

2. If the density-dependent relationship shown by the solid line in Figure 2 is used to determine the age at first reproduction in a density-dependent matrix, the specific productivity curve to be expected shows a positive second derivative (opposite to that in Figure 4). The total productivity of such a population, as it would depend on density, is shown in the inset of Figure 2. Note that the effect is the opposite of that produced by birth and survival as linear functions of density, less change occurs at equilibrium than for growing populations (see Chapter 22). For the range of changes that are reasonable for large mammals, changes in age at maturation alone produce small changes in the eigenvalue of a Leslie matrix. Thus, age at first reproduction may not be as important as other factors in acting as a regulating mechanism for such species (contrary to Law's conclusions, Chapter 2, see also Chapter 22). For those species characterized by especially early maturation this mode of regulation is more effective but does not exhibit the regulatory effect that equally realistic changes in birthrate or juvenile survival show.

Adult Survival

Another factor that acts in the regulation of animal populations is adult survival. Being relatively inflexible in populations of large mammals, and subject to error in measurement, adult survival has not been shown to change in many large mammal populations (Table 1). If defined as the elements of the subdiagonal of a projection matrix a linear relationship between adult survival and density is translated to a nearly symmetric productivity curve. Similarly, any nonlinearity in this relationship is reflected directly in the productivity relationships. If most density-dependent change in adult survival occurs at population levels close to the carrying capacity, so does most change in specific productivity. As a result, the most productive populations are those close to the carrying capacity.

Patterns

On the basis of the observations above, then, a case can be made for expecting patterns based on the mode of expression of density dependence. All else being equal, density-dependent birthrates and juvenile survival show similar tendencies in their expression at the level of productivity of the population. The way adult survival in large mammal populations gets translated to the shape of productivity curves is relatively unaffected by the age structure of populations, while age at maturation has an effect on equilibrium productivity opposite to that of birth and juvenile survival. Simultaneous linear changes in all these properties, as a function of density, produce a tendency to cause populations which exhibit peak productivity to occur closer to the carrying capacity than would be expected in a population without age structure. We have evidence for density dependence in juvenile survival, adult survival, age at maturation, and birthrates among adults for large mammals. Analytical results show that each of these factors (when linearly density dependent) produces its own characteristic affect on the dynamics of populations involved.

It is obvious that it is important to know to what degree each mode of regulation contributes to the regulation of specific populations. Further work to examine for correlations between each mode of regulation and other factors is needed.

THE INFLUENCE OF HARVEST STRATEGY ON THE DENSITY DEPENDENCE OF PRODUCTIVITY

To further exemplify the utility and need for comparative studies we examine, in this section, the way harvest strategies influence the productivity curves exemplified in the insets of Figures 1 and 2. As examples of possible harvest strategies consider the harvest of specific age classes, or of one or the other of the two sex classes. In this section an example of each will be used to demonstrate the comparative potentials based on specific types of mortality patterns to which populations may be subjected.

Harvest of Juveniles

First, as is the case with the harp seals of the northwest Atlantic, or as exemplified in model form in Chapter 20, an age-specific harvest may concentrate on juveniles. If mortality due to harvest is induced during the first year of life (specified as the first element of the subdiagonal of a projection matrix) and the matrix used for simulation is constructed such that the natural survival of juveniles is the only density-dependent regulatory factor (also the first element of the subdiagonal), the harvest may be expressed by the equation:

$$C = \left[\frac{(\ln S)\,(t_2 - t_1)}{\ln S'} + 1\right]^{-1} \left[1 - S'S^{\,(t_2 - t_1)}\right] S^{t_1}N_\circ$$

where C = harvest (in numbers) of juveniles

S = natural survival of juveniles (as a function of density) if acting alone and spread evenly over time 0 to 1

S' = survival of juveniles if harvesting were the only source of mortality (harvest would be: $N_\circ\,[1 - S']$). (when applied, S' is restricted to the time interval t_1 to t_2)

t_1 = time during first year of life at which harvest is started (beginning of season). $0 \leqslant t_1 \leqslant t_2 \leqslant 1$

t_2 = time during first year of life at which harvest is terminated (end of season)

$N_\circ$ = abundance of juveniles (i.e., number of pups born)

This equation may be isolated from the context of the matrix as a result of the condition that the eigenvalue of the matrix involved will always be 1 (condition for equilibrium) and the age structure derived from the eigenvector may be scaled either by population size or any single age class. It is assumed that at equilibrium for each population level the combined effect of harvest and density-dependent juvenile survival is constant. We are thus dealing with changes in the composition but not value of only one element (first subdiagonal). The structure of the population will remain the same at equilibrium and only the population size will change.

Figure 5 shows the productivity of juveniles to be expected from such a population, as a function of population size, if natural survival is a linear function of the total population size (which can be translated into the number of young born in this example). Note that the timing of the harvest of young during the first year of life is critical in determining both the level of sustainable harvest as well as at which population level this maximum is expected to be produced. Note also that for linear density dependent juvenile survival, the *minimum* population level at which maximum productivity occurs is that corresponding to approximately fifty percent of the equilibrium level (carrying capacity). Curvilinearity in the underlying relationship between survival and population size will affect the position of this point. Those with negative second derivatives (see next section) tend to cause this point to shift even further to the right (toward the equilibrium population levels).

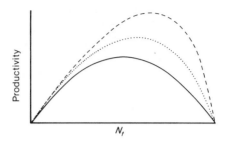

Figure 5   The nature of changes in the production of harvestable juveniles as dependent on the timing of the harvest when juvenile survival is linearly density dependent on the total population size N_t. The lower curve is for a harvest at the end of the first year, the upper curve represents a harvest following birth. The intermediate curve is for a harvest during the year following birth.

Sex-Specific Harvest

A harvest regime in which males are taken preferentially to females is not un-common in large mammals and provides a further example of the potential for comparative studies. Most ungulate management, and the harvest of fur seals and sperm whales, is based on the premise that a sex ratio in which the females outnumber the males is not detrimental. Whether or not such a strategy is harm-ful in any way is open to further research. Given that many species exhibit a social structure in which each adult male mates with several females, however, the dynamics of populations managed under such regimens may be examined. As an example, we may again consider a hypothetical population regulated naturally by juvenile survival (or birth rate) which is linearly density dependent on the total population size. If males older than a specified age are considered mature, as in the case of the females, and a particular and predetermined ratio of mature males to mature females (less than 1) is acceptable, a simple variable matrix model may be used to represent the females of such a population. A similar model, without the top row, can be used to represent the males.

Figure 6 shows the total harvest at equilibrium to be expected from such a population. In the lower lines of Figure 6 the total harvest is decomposed to show the corresponding harvest by sex. There are several points of interest in the rela-tionships shown. First, it is to be noted that the male harvest always exceeds the female harvest. The peak in male production occurs at a population level very near the carrying capacity relative to that of the females (or the total population if both sexes were harvested equally). Finally, it must be noted that a small female harvest is required to maintain a population at its maximally productive level.

The model that was used to produce Figure 6 (taken from Fowler et al., 1980b, after Smith, 1977) was a general matrix model as described above. In working with this model it is seen that some species (sperm whales for example)

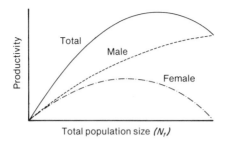

Figure 6   The productivity of a hypothetical sperm whale population from which males and females may be removed at different rates. The sex ratio of adults involves 10 females per male and density dependence is based on the model of Smith (1977).

may actually reach a peak production of males at population levels equivalent to those at the carrying capacity! This conclusion may appear counterintuitive. However, it may be understood by considering a natural unexploited population with an even sex ratio in which the total population contributes to the expression of density dependence involved in the juvenile mortality or birthrate. By reducing the population through harvesting only males, the composition of the population may be altered to one in which females predominate. This larger population of females will, of course, produce more offspring than would have been produced by a population of the same size but of even sex ratio. Having grown in numbers to levels equivalent to naturally determined equilibrium levels the "excess" population growth that results from this production of young must then be compensated for by removal of males (to bring the population back to its equilibrium size). Left to its own dynamics, a population with such a skewed sex ratio would at first grow through the addition of males (as discussed in Chapter 12). Then if males contribute to density dependence, the population will decline to its equilibrium level as the sex structure and density-dependent mechanisms reachieved stability.

Patterns

In both the case of the harvest of males and the harvest of juveniles there are situations in which productivity curves reach a maximum at levels close to the carrying capacity. The dynamics of these types of populations, *when compared* to otherwise similar populations harvested by other means, are expected to show different patterns in population dynamics. These differences are shown here on the basis of theoretical comparisons for which the assumptions are probably met in many natural situations. Further discussion and development of these comparisons, as well as some empirical information, may be found in Chapters 11 and 12, and Fowler et al (1980b).

The two examples of different harvest strategies discussed above show the potential for the comparison of the population dynamics of species subjected to different harvest strategies. It also shows the value of research conducted in the context of systems subjected to experimental manipulation. As examples they are meant to demonstrate the importance of considering harvest strategies as bases for comparing population dynamics. These examples are by no means exhaustive; other such strategies should receive similar examination.

THE EFFECT OF LIFE HISTORY STRATEGY ON LINEARITY

As mentioned earlier, linear relationships such as those shown in Figures 1 and 2, may be used for bases of comparison and serve as points of reference. The relationships observed in naturally occurring populations will, in most cases, be nonlinear. Patterns in such deviations are of importance in comparative work as well as of utility in management and research. Until recently little information existed in sufficient quantity to show any pattern in such deviations for large mammals. In this section it is argued that recent studies in comparative population dynamics involving large mammals (in the context of species of other life history strategies) have produced useful results. The examples of this section serve to emphasize again the importance of comparative studies.

Evolutionary considerations have been brought to bear to determine the kinds of density-dependent relationships expected on the basis of comparing across life history strategies (Gilpin et al., 1976; Fowler, 1981). These studies conclude that K-selected species, or species regulated predominately by resource levels, are expected to show most density-dependent changes at levels close to the carrying capacity. These dynamics would be of the type shown by the dashed lines in Figures 1 and 2.

Other theoretical considerations involve trophic interaction (see Chapter 18). Considering the dynamic relationships between a population and its resources, Fowler (1980b) describes how populations regulated by resource levels will tend to show dynamics in which most density-dependent change occurs at high levels, parallel to that expected for K-selected species as outlined above.

These theoretical arguments bring us to the conclusion that among large mammals most density-dependent change is expected to occur at population levels quite close to the carrying capacity. To examine this hypothesis published data as referred to in Tables 1 and 2 may be examined as was done in Fowler (1981). Those cases that contained sufficiently large sample size for tests resulting in statistically significant relationships with measured values of density were used to test for nonlinearity. These included data representing density dependence in birthrates, survival of both juveniles and adults, as well as age at maturation. There are over 26 cases that show such significant density dependence. Of these only four show any sign that most density dependent change is experienced at low population levels (below 50% of the carrying capacity). Of five cases with a sample size of 12 or more, all support the hypothesis that most density-dependent change occurs at levels above 50% of the carrying capacity. These results are statistically significant and indicate that large mammals show most density-dependent change at levels close to the carrying capacity (as shown in the broken lines in Figures 1 and 2).

What are the consequences of such nonlinearity? As shown in the insets of Figures 1 and 2, such nonlinearity will result in a shift of the peak productivity toward those population levels near the carrying capacity. If the relationships in Figure 1 exhibit negative second derivatives, a similar curvature is characteristic of the relationship between specific production and population size. In other words, the curvature of Figure 4 becomes exaggerated. This translates to a productivity curve that reaches a maximum at population levels close to the carrying capacity. Population growth of large mammals tends to be exponential (in time) up to levels quite close to the carrying capacity (see also Chapters 10, 14, 20).

Thus, both evolutionary and systems level arguments lead us to the conclusion that managing large mammal populations at levels below or near 50% of the carrying capacity results in productivity that is lower than can be sustained by such populations when maintained at levels close to the carrying capacity. Empirical information, as outlined above (discussed in Fowler, 1981), as well as similar conclusions reached by McCullough (1979; Chapter 9), substantiate these conclusions.

Comparative studies of the population dynamics of large mammals has thus produced information in sufficient quantity to characterize large mammals as a group. The practical importance of the dynamics characteristic of this group is only now beginning to be appreciated.

DISCUSSION

In the preceding sections several examples of comparative studies have been presented as they relate to the population dynamics of large mammals. These involve (1) the differences between the effects of different vital rates when they show density-dependent changes, (2) the differences between populations subjected to different harvest strategies, and (3) the differences between K-selected (resource-limited) and r-selected species (limited by other factors), and the common pattern characteristic of large mammals as K-selected species. These examples were presented in a context produced by a history of work based on models. The complexity of the interactions among various factors, the synergistic effects of such interactions, and the lack of empirical data make such approaches necessary.

Simple single-species models provide many of the basic elements of potential utility in making comparisons. However, the nature and use of the equations involved in these models have produced a tendency to think in terms of the specific restricted type of dynamics represented by each model rather than in terms of variable properties of use in comparison. For example, the use of the logistic model (see Hutchinson, 1978, for discussion of the history and influence of this model in population biology) has produced the tendency to conclude that maximal growth or productivity is to be observed at that population level equivalent to one-half the level observed at the carrying capacity (mean unexploited level). Various stock-recruitment models are structurally limited to produce the conclusion that maximum productivity occurs at population levels equal to or below one-half of the carrying capacity (Graham, 1935; Eberhardt, 1977; Ricker, 1958; Schaefer, 1954; Beverton and Holt, 1957). The generalized growth model (Pella and Tomlinson, 1969; Chapman, 1960; Richards, 1955) is not as inflexible, but has not been widely used.

One property common to all of these models is a population level (variable among populations) at which it is expected that, on the average, maximum productivity will occur, an important insight (see Chapter 6) into population dynamics. The size of that population, relative to an unexploited population, is determined by many factors as discussed above, This characteristic of population dynamics has served as the primary focus for comparison in this chapter.

We can draw several conclusions. The models mentioned above, as developed in the context of work on species other than large mammals, have restricted applicability. Carelessly applying them to large mammals prevents the discovery of differences, both among large mammals and between large mammals and other groups. Taking a more open-minded comparative approach is necessary. With the accumulation of more information and through the collection and study of published information we can extract ourselves from the characterization of individual cases to produce more realistic views by way of comparative studies.

As seen in the previous sections, the various harvest strategies, life history strategies, and modes of density-dependent regulation pertinent to large mammal populations produces a general tendency for maximal growth and production to occur at levels very close to the carrying capacity within such populations. This generalization (of which there naturally may be exceptions) was developed on the basis of comparisons both theoretical and empirical in nature. The examples leading to this generalization serve to show that comparisons are neither impossible nor impractical.

Logistically, the collection of the necessary data through field work is the greatest problem faced in the progress of comparative studies of large mammals. Our progressively better understanding of the population dynamics of large mammals will help guide this type of work more efficiently. Owing to the long-lived nature of such species, however, field studies remain a slow and painful process for collecting the necessary data (see Chapter 2). As alternatives to collecting new data there are two related possibilities. First, the general literature contains a great deal of information that can be usefully compiled for comparison. The use of this information is exemplified in Chapter 4, in the comparative work present in the last section, and in Fowler (1981) and Fowler et al. (1980c). Second, theoretical models can be used to help push forward our understanding as exemplified in many other chapters in this book. Combined efforts along both lines will prove especially fruitful.

As individual investigators become involved with a particular problem it should be their responsibility to look at that problem in a general comparative context as based on information in the literature. The relative importance of age at maturation as a regulatory mechanism, for example, (see Chapters 2, 22), should be compared across various categories or spectra of types of large mammals. How important is this process in various habitats, for various taxa, and within various trophic levels? How does its importance vary across such bases for comparison? What are the differences in dynamics between populations of social animals when compared to relatively asocial groups? Do various social structures express different types of dynamics? These types of questions need to be addressed on the basis of existing data before costly field studies are begun.

Because of the complex synergistic properties of structured populations, and the interactions of populations with their environment, models must be used to help guide our thinking in comparative work. Models are specific and restricted, however, and subject to the assumptions they contain. Only with an appreciation of these qualities of models can they be the useful tools that they are. As such they provide the means for amalgamating knowledge of dynamics of the components that contribute to the whole.

By carefully combining information gleaned from published studies and cautiously using this information in the context of models, it is possible to gain a much better concept of the nature of the dynamics being examined. This type of activity can help focus and define the type of field work that will be most useful. It can help avoid costly mistakes and wasted time if conducted before field studies are started. It involves a multidisciplinary approach and may require several individuals with varied backgrounds but, in the long run, remains an advisable approach.

Given the success observed in other fields, and given the limited progress observed in the field of population dynamics of large mammals, further comparative work is to be encouraged in this field as suggested in Chapter 2. Through such work we can develop a holistic perspective or framework of use in conducting work involving specific cases. Such perspectives can help remove much of the guesswork all too commonly characteristic of current management and research.

LITERATURE CITED

Beverton, R. J. H., and S. J. Holt. 1957. On the dynamics of exploited fish populations. Fish. Invest., Lond., Sep. 2, 19:533 pp.

Beverton, R. J. H., and S. J. Holt. 1959. A review of the lifespans and mortality rates of fish in nature, and their relation to growth and other physiological characteristics. CIBA Found. Colloq. on ageing. V. The life span of animals. Churchill, London. Pp. 142-177.

Blueweiss, L., H. Fox, V. Kadsma, D. Nakashima, R. Peters, and S. Sams. 1979. Relationships between body size and some life history parameters. Oecologia 37:257-272.

Boyce, M. S. 1977. Population growth with stochastic fluctuations in the life table. Theor. Pop. Biol. 12:366-373.

Chapman, D. G. 1960. Statistical problems in dynamics of exploited fisheries populations. Proc. Berkeley Math. Symp. Stat. Prob. 4:153-168.

Chapman, D. G. 1961. Population dynamics of the Alaska fur seal herd. Trans. N. Am. Wildl. Conf. 26:356-369.

Cole, L. C. 1954. The population consequences of life history phenomena. Q. Rev. Biol. 29:103-137.

Eberhardt, L. L. 1977. Relationship between two stock-recruitment curves. J. Fish. Res. Board Can. 34:425-428.

Eberhardt, L. L. 1977. "Optimal" management policies for marine mammals. Wildl. Soc. Bull. 5:163-169.

Fowler, C. W. 1980a. Non-linearity in population dynamics with special reference to large mammals. Appendix C in: Fowler et al. (1980c).

Fowler, C. W. 1980b. Exploited populations of predator and prey: Implications of a model. Appendix F in: Fowler et al. (1980c).

Fowler, C. W. 1981. Density dependence as related to life history strategy. Ecology. (In press).

Fowler, C. W., and R. J. Ryel. 1980. Life history aspects of population dynamics. Appendix E in: Fowler et al. (1980c).

Fowler, C. W., W. T. Bunderson, R. J. Ryel, and B. B. Steele. 1980a. A preliminary review of density dependent reproduction and survival in large mammals. Appendix B in: Fowler et al. (1980c).

Fowler, C. W., R. J. Ryel, and B. B. Steele. 1980b. Animal population dynamics as influenced by sex ratio. Appendix D in: Fowler et al. (1980c).

Fowler, C. W., W. T. Bunderson, M. B. Cherry, R. J. Ryel, and B. B. Steele. 1980c. Comparative population dynamics of large mammals: A search for management criteria. Report to the U.S. Marine Mammal Commission. Contract #MM7AC013. NTIS #PB80-178627. National Technical Information Service.

Fowler, C. W., and T. D. Smith. 1973. Characterizing stable populations; an application in the African elephant population. J. Wild. Manage. 37:513-523.

Gilpin, M. E., T. J. Case, and F. J. Ayala. 1976. θ-selection. Math. Biosci. 32:131-135.

Graham, M. 1935. Modern theory of exploiting a fishery, an application to North Sea trawling. J. Cons. Perm. Inter. Explor. Mer 10:264-274.

Holt, S. J. 1962. The application of comparative population studies to fishery biology—An exploration. Pp. 51-71 in: E. D. Lecren, and M. W. Holdgate (eds.). The Exploration of Natural Animal Populations. John Wiley and Sons. New York.

Hutchinson, G. E. 1978. An Introduction to Population Ecology. Yale University Press, New Haven, Connecticut.

May, R. M. 1973. Stability and Complexity in Model Ecosystems. Princeton University Press, Princeton, New Jersey.

McCullough, D. 1979. The George Reserve Deer Herd: Population Ecology of a K-Selected Species. University of Michigan Press, Ann Arbor.

Ohsumi, S. 1979. Interspecies relationships among some biological parameters in cetaceans and estimation of the natural mortality coefficent of the southern hemisphere minke whale. Rep. Int. Whaling Comm. 29:397-406.

Pella, J. J., and P. K. Tomlinson. 1969. A generalized stock production model. Inter-Am. Trop. Tuna Comm. Bull. 13:420-456.

Pianka, E. R. 1970. On r- and K-selection. Am. Nat. 104:592-597.

Richards, F. J. 1955. A flexible growth function for empirical use. J. Exp. Bot. 10:290-300.

Ricker, W. E. 1958. Handbook of Computations for Biological Statistics of Fish Populations. Information Canada, Ottawa.

Schaefer, M. B. 1954. Some aspects of the dynamics of population important to the management of the commercial marine fisheries. Bull. Inter-Am. Trop. Tuna Comm. 1:25-56.

Smith, T. D. 1973. Variable population projection matrices: Theory and application to the evaluation of harvesting strategy. Ph.D. Dissertation, University of Washington, Seattle.

Smith, T. D. 1977. A matrix model of sperm whale populations. Int. Whaling Comm. Rep. 27:337-342.

Stearns, S. C. 1976. Life-history tactics: A review of ideas. Q. Rev. Biol. 51:3-47.